Global Climate Change
Impacts in the United States

A State of Knowledge Report from the
U.S. Global Change Research Program

Recommended Citation:

Global Climate Change Impacts in the United States, Thomas R. Karl, Jerry M. Melillo, and Thomas C. Peterson, (eds.). Cambridge University Press, 2009.

The bars at the bottom of the front cover show the global annual average temperature from 1900-2008, see page 17.

Federal Advisory Committee Authors

Co-Chairs and Editors-in-Chief

Thomas R. Karl,
NOAA National Climatic Data Center

Jerry M. Melillo,
Marine Biological Laboratory

Thomas C. Peterson,
NOAA National Climatic Data Center

Author Team

David M. Anderson,
NOAA World Data Center for
Paleoclimatology

Donald F. Boesch,
University of Maryland Center for
Environmental Science

Virginia R. Burkett,
U.S. Geological Survey

Lynne M. Carter,
Adaptation Network, Louisiana
State University

Stewart J. Cohen,
Environment Canada and University of
British Columbia

Nancy B. Grimm,
Arizona State University

Jerry L. Hatfield,
U.S. Department of Agriculture

Katharine Hayhoe,
Texas Tech University

Anthony C. Janetos,
Pacific Northwest National Laboratory

Jack A. Kaye,
National Aeronautics and Space
Administration

Jay H. Lawrimore,
NOAA National Climatic Data Center

James J. McCarthy,
Harvard University

A. David McGuire,
U.S. Geological Survey/University of
Alaska Fairbanks

Edward L. Miles,
University of Washington

Evan Mills,
Lawrence Berkeley National Laboratory

Jonathan T. Overpeck,
University of Arizona

Jonathan A. Patz,
University of Wisconsin at Madison

Roger S. Pulwarty,
NOAA Climate Program Office and Earth
System Research Laboratory

Benjamin D. Santer,
Lawrence Livermore National Laboratory

Michael J. Savonis,
U.S. Department of Transportation

H. Gerry Schwartz, Jr.,
Consultant/Transportation

Eileen L. Shea,
NOAA National Climatic Data Center/
Integrated Data and Environmental
Applications Center

John M.R. Stone,
Carleton University

Bradley H. Udall,
University of Colorado/NOAA Earth
System Research Laboratory

John E. Walsh,
University of Alaska Fairbanks

Michael F. Wehner,
Lawrence Berkeley National Laboratory

Thomas J. Wilbanks,
Oak Ridge National Laboratory

Donald J. Wuebbles,
University of Illinois

Senior Science Writer and Lead Graphic Designer

Senior Science Writer
Susan J. Hassol,
Climate Communication, LLC

Lead Graphic Designer
Sara W. Veasey,
NOAA National Climatic Data Center

Key Support Personnel

Jessica Blunden, Editorial Assistant,
STG, Inc.

Marta Darby, Copy Editor, STG, Inc.

David Dokken, CCSP Technical
Advisor, USGCRP

Byron Gleason, Data Analysis/Visualization,
NOAA National Climatic Data Center

Glenn M. Hyatt, Graphics Support,
NOAA National Climatic Data Center

Clare Keating, Editorial Support,
Texas Tech University

Staci Lewis, Technical Advisor,
NOAA

Jolene McGill, Logistical Support,
NOAA National Climatic Data Center

Deborah J. Misch, Graphics Support,
STG, Inc.

William Murray, Technical Support,
STG, Inc.

Susan Osborne, Copy Editor,
STG, Inc.

Tim Owen, Logistical Support,
NOAA National Climatic Data Center

Ray Payne, Printing Support
NOAA Nationa Climatic Data Center

Deborah Riddle, Graphics Support,
NOAA National Climatic Data Center

Susanne Skok, Copy Editor,
STG, Inc.

Mara Sprain, Editorial Support,
STG, Inc.

Michael Squires, Cartographic Support,
NOAA National Climatic Data Center

Jeff VanDorn, Technical and Graphics Support,
ATMOS Research

David Wuertz, Data Analysis/Visualization,
NOAA National Climatic Data Center

Christian Zamarra, Graphics Support,
STG, Inc.

Federal Executive Team

Acting Director, U.S. Global Change Research Program: ....Jack A. Kaye

Director, U.S. Global Change Research Program Office: ....Peter A. Schultz

Lead Agency Principal Representative to CCSP (through January 2009),
National Oceanic and Atmospheric Administration: ....Mary M. Glackin

Lead Agency Principal Representative to CCSP,
Product Lead, National Oceanic and Atmospheric Administration: ....Thomas R. Karl

Lead Agency Principal Representative to CCSP,
Group Chair Synthesis and Assessment Products, Environmental Protection Agency: ....Michael W. Slimak

Synthesis and Assessment Product Coordinator,
U.S. Global Change Research Program Office: ....Fabien J.G. Laurier

Communications Advisor/Coordinator/Editor,
U.S. Global Change Research Program Office: ....Anne M. Waple

Special Advisor, National Oceanic and Atmospheric Administration: ....Chad A. McNutt

Federal Advisory Committee Designated Federal Official,
National Oceanic and Atmospheric Administration: ....Christopher D. Miller

---

Reviewers

Blue Ribbon Reviewers

Robert W. Corell,
   Global Change Program, H. John Heinz III
   Center for Science, Economics and the
   Environment

Robert A. Duce,
   Department of Atmospheric Sciences,
   Texas A&M University

Kristie L. Ebi,
   Independent consultant, ESS, LLC
   Alexandria, VA

Christopher B. Field,
   Carnegie Institution

William H. Hooke,
   Atmospheric Policy Program, American
   Meteorological Society

Michael C. MacCracken,
   Climate Institute

Linda O. Mearns,
   Environmental and Societal Impacts
   Group, National Center for Atmospheric
   Research

Gerald A. Meehl,
   Climate and Global Dynamics Division,
   National Center for Atmospheric Research

John Reilly,
   Sloan School of Management,
   Massachusetts Institute of Technology

Susan Solomon,
   NOAA, Earth System Research
   Laboratory

Steven C. Wofsy,
   Harvard University

Communication Reviewers

Robert Henson,
   University Corporation for Atmospheric Research

Jack W. Williams,
   American Meteorological Society

June 2009

Members of Congress:

On behalf of the National Science and Technology Council, the U.S. Global Change Research Program is pleased to transmit to the President and the Congress this state of knowledge report: "*Global Climate Change Impacts in the United States.*" This report summarizes the science of climate change and the impacts of climate change on the United States, now and in the future.

As our nation strives to develop effective policies to respond to climate change, it is critical to have the latest and best scientific information to inform decision making. More than a year in the making, this report provides that information. It is the first report in almost a decade to provide an extensive evaluation of climate change impacts on the United States at the regional level.

An expert team of scientists operating under the authority of the Federal Advisory Committee Act, assisted by communication specialists, wrote the document. The report was reviewed and revised based on comments from experts and the public in accordance with the Information Quality Act guidelines issued by the Department of Commerce and the National Oceanic and Atmospheric Administration.

We highly commend the authors and support personnel of both this report and the underlying Synthesis and Assessment Products for the outstanding quality of their work in providing sound and thorough science-based information for policy formulation and climate change research priority setting. We intend to use the essential information contained in this report as we make policies and decisions about the future, and we recommend others do the same.

Sincerely,

Dr. John Holdren
Director,
Office of Science and Technology Policy

Dr. Jane Lubchenco
Administrator,
National Oceanic and Atmospheric Administration

CONTENTS

CONTENTS

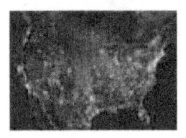

Regional Climate Change Impacts

An Agenda for Climate Impacts Science

Concluding Thoughts

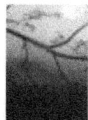

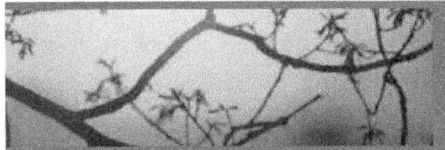

About this Report

What is this report?

This report summarizes the science of climate change and the impacts of climate change on the United States, now and in the future. It is largely based on results of the U.S. Global Change Research Program (USGCRP),[a] and integrates those results with related research from around the world. This report discusses climate-related impacts for various societal and environmental sectors and regions across the nation. It is an authoritative scientific report written in plain language, with the goal of better informing public and private decision making at all levels.

Who called for it, who wrote it, and who approved it?

The USGCRP called for this report. An expert team of scientists operating under the authority of the Federal Advisory Committee Act, assisted by communication specialists, wrote the document. The report was extensively reviewed and revised based on comments from experts and the public. The report was approved by its lead USGCRP Agency, the National Oceanic and Atmospheric Administration, the other USGCRP agencies, and the Committee on the Environment and Natural Resources on behalf of the National Science and Technology Council.[b] This report meets all Federal requirements associated with the Information Quality Act, including those pertaining to public comment and transparency.

What are its sources?

The report draws from a large body of scientific information. The foundation of this report is a set of 21 Synthesis and Assessment Products (SAPs), which were designed to address key policy-relevant issues in climate science (see page 161); several of these were also summarized in the *Scientific Assessment of the Effects of Climate Change on the United States* published in 2008. In addition, other peer-reviewed scientific assessments were used, including those of the Intergovernmental Panel on Climate Change, the U.S. National Assessment of the Consequences of Climate Variability and Change, the Arctic Climate Impact Assessment, the National Research Council's Transportation Research Board report on the Potential Impacts of Climate Change on U.S. Transportation, and a variety of regional climate impact assessments. These assessments were augmented with government statistics as necessary (such as population census and energy usage) as well as publicly available observations and peer-reviewed research published through the end of 2008. This new work was carefully selected by the author team with advice from expert reviewers to update key aspects of climate change science relevant to this report. The icons on the bottom of this page represent some of the major sources drawn upon for this synthesis report.

On the first page of each major section, the sources primarily drawn upon for that section are shown using these icons. Endnotes, indicated by superscript numbers and compiled at the end of the book, are used for specific references throughout the report.

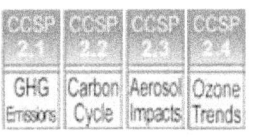

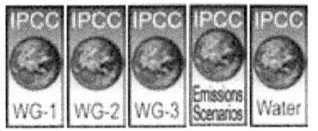

See page 161 for descriptions of these sources.

[a.] The U.S. Global Change Research Program (USGCRP), which was established in 1990 by the Global Change Research Act, encompasses the Climate Change Science Program (CCSP).

[b.] A description of the National Science and Technology Council (NSTC) can be found at www.ostp.gov/cs/nstc.

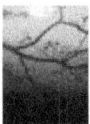
Does this report deal with options for responding to climate change?

While the primary focus of this report is on the impacts of climate change in the United States, it also deals with some of the actions society is already taking or can take to respond to the climate challenge. Responses to climate change fall into two broad categories. The first involves "mitigation" measures to reduce climate change by, for example, reducing emissions of heat-trapping gases and particles, or increasing removal of heat-trapping gases from the atmosphere. The second involves "adaptation" measures to improve our ability to cope with or avoid harmful impacts and take advantage of beneficial ones, now and in the future. Both of these are necessary elements of an effective response strategy. These two types of responses are linked in that more effective mitigation measures reduce the amount of climate change, and therefore the need for adaptation.

This report underscores the importance of mitigation by comparing impacts resulting from higher versus lower emissions scenarios. The report shows that choices made about emissions in the next few decades will have far-reaching consequences for climate change impacts. Over the long term, lower emissions will lessen both the magnitude of climate change impacts and the rate at which they appear.

While the report underscores the importance of mitigation as an essential part of the nation's climate change strategy, it does not evaluate mitigation technologies or undertake an analysis of the effectiveness of various approaches. These issues are the subject of ongoing studies by the U.S. Government's Climate Change Technology Program and several federal agencies including the Department of Energy, Environmental Protection Agency, National Oceanic and Atmospheric Administration, Department of Transportation, and Department of Agriculture. The range of mitigation responses being studied includes more efficient production and use of energy, increased use of non-carbon-emitting energy sources, and carbon capture and storage.

Adaptation options also have the potential to moderate harmful impacts of current and future climate variability and change. While this report does address adaptation, it does not do so comprehensively.

Rather, in the context of impacts, this report identifies examples of actions currently being pursued in various sectors and regions to address climate change, as well as other environmental problems that could be exacerbated by climate change such as urban air pollution and heat waves. In most cases, there is currently insufficient peer-reviewed information to evaluate the practicality, effectiveness, costs, or benefits of these measures, highlighting a need for research in this area. Thus, the discussion of various public and private adaptation examples should not be viewed as an endorsement of any particular option, but rather as illustrative examples of approaches being tried.

How is the likelihood of various outcomes expressed given that the future is not certain?

When it is considered necessary to express a range of possible outcomes and identify the likelihood of particular impacts, this report takes a plain-language approach to expressing the expert judgment of the author team based on the best available evidence. For example, an outcome termed "likely" has at least a two-thirds chance of occurring; an outcome termed "very likely," at least a 90 percent chance.[1] In using these terms, the Federal Advisory Committee has taken into consideration a wide range of information, including the strength and consistency of the observed evidence, the range and consistency of model projections, the reliability of particular models as tested by various methods, and most importantly, the body of work addressed in earlier synthesis and assessment reports. Key sources of information used to develop these characterizations of uncertainty are referenced in endnotes.

How does this report address incomplete scientific understanding?

This assessment identifies areas in which scientific uncertainty limits our ability to estimate future climate change and its impacts. The section on *An Agenda for Climate Impacts Science* at the end of this report highlights some of these areas.

Executive Summary

Observations show that warming of the climate is unequivocal. The global warming observed over the past 50 years is due primarily to human-induced emissions of heat-trapping gases. These emissions come mainly from the burning of fossil fuels (coal, oil, and gas), with important contributions from the clearing of forests, agricultural practices, and other activities.

Warming over this century is projected to be considerably greater than over the last century. The global average temperature since 1900 has risen by about 1.5°F. By 2100, it is projected to rise another 2 to 11.5°F. The U.S. average temperature has risen by a comparable amount and is very likely to rise more than the global average over this century, with some variation from place to place. Several factors will determine future temperature increases. Increases at the lower end of this range are more likely if global heat-trapping gas emissions are cut substantially. If emissions continue to rise at or near current rates, temperature increases are more likely to be near the upper end of the range. Volcanic eruptions or other natural variations could temporarily counteract some of the human-induced warming, slowing the rise in global temperature, but these effects would only last a few years.

Reducing emissions of carbon dioxide would lessen warming over this century and beyond. Sizable early cuts in emissions would significantly reduce the pace and the overall amount of climate change. Earlier cuts in emissions would have a greater effect in reducing climate change than comparable reductions made later. In addition, reducing emissions of some shorter-lived heat-trapping gases, such as methane, and some types of particles, such as soot, would begin to reduce warming within weeks to decades.

Climate-related changes have already been observed globally and in the United States. These include increases in air and water temperatures, reduced frost days, increased frequency and intensity of heavy downpours, a rise in sea level, and reduced snow cover, glaciers, permafrost, and sea ice. A longer ice-free period on lakes and rivers, lengthening of the growing season, and increased water vapor in the atmosphere have also been observed. Over the past 30 years, temperatures have risen faster in winter than in any other season, with average winter temperatures in the Midwest and northern Great Plains increasing more than 7°F. Some of the changes have been faster than previous assessments had suggested.

These climate-related changes are expected to continue while new ones develop. Likely future changes for the United States and surrounding coastal waters include more intense hurricanes with related increases in wind, rain, and storm surges (but not necessarily an increase in the number of these storms that make landfall), as well as drier conditions in the Southwest and Caribbean. These changes will affect human health, water supply, agriculture, coastal areas, and many other aspects of society and the natural environment.

This report synthesizes information from a wide variety of scientific assessments (see page 7) and recently published research to summarize what is known about the observed and projected consequences of climate change on the United States. It combines analysis of impacts on various sectors

such as energy, water, and transportation at the national level with an assessment of key impacts on specific regions of the United States. For example, sea-level rise will increase risks of erosion, storm surge damage, and flooding for coastal communities, especially in the Southeast and parts of Alaska. Reduced snowpack and earlier snow melt will alter the timing and amount of water supplies, posing significant challenges for water resource management in the West.

Society and ecosystems can adjust to some climatic changes, but this takes time. The projected rapid rate and large amount of climate change over this century will challenge the ability of society and natural systems to adapt. For example, it is difficult and expensive to alter or replace infrastructure designed to last for decades (such as buildings, bridges, roads, airports, reservoirs, and ports) in response to continuous and/or abrupt climate change.

Impacts are expected to become increasingly severe for more people and places as the amount of warming increases. Rapid rates of warming would lead to particularly large impacts on natural ecosystems and the benefits they provide to humanity. Some of the impacts of climate change will be irreversible, such as species extinctions and coastal land lost to rising seas.

Unanticipated impacts of increasing carbon dioxide and climate change have already occurred and more are possible in the future. For example, it has recently been observed that the increase in atmospheric carbon dioxide concentration is causing an increase in ocean acidity. This reduces the ability of corals and other sea life to build shells and skeletons out of calcium carbonate. Additional impacts in the future might stem from unforeseen changes in the climate system, such as major alterations in oceans, ice, or storms; and unexpected consequences of ecological changes, such as massive dislocations of species or pest outbreaks. Unexpected social or economic changes, including major shifts in wealth, technology, or societal priorities would also affect our ability to respond to climate change. Both anticipated and unanticipated impacts become more challenging with increased warming.

Projections of future climate change come from careful analyses of outputs from global climate models run on the world's most advanced computers. The model simulations analyzed in this report used plausible scenarios of human activity that generally lead to further increases in heat-trapping emissions. None of the scenarios used in this report assumes adoption of policies explicitly designed to address climate change. However, the level of emissions varies among scenarios because of differences in assumptions about population, economic activity, choice of energy technologies, and other factors. Scenarios cover a range of emissions of heat-trapping gases, and the associated climate projections illustrate that lower emissions result in less climate change and thus reduced impacts over this century and beyond. Under all scenarios considered in this report, however, relatively large and sustained changes in many aspects of climate are projected by the middle of this century, with even larger changes by the end of this century, especially under higher emissions scenarios.

In projecting future conditions, there is always some level of uncertainty. For example, there is a high degree of confidence in projections that future temperature increases will be greatest in the Arctic and in the middle of continents. For precipitation, there is high confidence in projections of continued increases in the Arctic and sub-Arctic (including Alaska) and decreases in the regions just outside the tropics, but the precise location of the transition between these is less certain. At local to regional scales and on time frames up to a few years, natural climate variations can be relatively large and can temporarily mask the progressive nature of global climate change. However, the science of making skillful projections at these scales has progressed considerably, allowing useful information to be drawn from regional climate studies such as those highlighted in this report.

This report focuses on observed and projected climate change and its impacts on the United States. However, a discussion of these issues would be incomplete without mentioning some of the actions society can take to respond to the climate challenge. The two major categories are "mitigation" and "adaptation." Mitigation refers to options for limiting climate change by, for example, reducing

heat-trapping emissions such as carbon dioxide, methane, nitrous oxide, and halocarbons, or removing some of the heat-trapping gases from the atmosphere. Adaptation refers to changes made to better respond to present or future climatic and other environmental conditions, thereby reducing harm or taking advantage of opportunity. Effective mitigation measures reduce the need for adaptation. Mitigation and adaptation are both essential parts of a comprehensive climate change response strategy.

Carbon dioxide emissions are a primary focus of mitigation strategies. These include improving energy efficiency, using energy sources that do not produce carbon dioxide or produce less of it, capturing and storing carbon dioxide from fossil fuel use, and so on. Choices made about emissions reductions now and over the next few decades will have far-reaching consequences for climate-change impacts. The importance of mitigation is clear in comparisons of impacts resulting from higher versus lower emissions scenarios considered in this report. Over the long term, lower emissions will lessen both the magnitude of climate-change impacts and the rate at which they appear. Smaller climate changes that come more slowly make the adaptation challenge more tractable.

However, no matter how aggressively heat-trapping emissions are reduced, some amount of climate change and resulting impacts will continue due to the effects of gases that have already been released. This is true for several reasons. First, some of these gases are very long-lived and the levels of atmospheric heat-trapping gases will remain elevated for hundreds of years or more. Second, the Earth's vast oceans have absorbed much of the heat added to the climate system due to the increase in heat-trapping gases, and will retain that heat for many decades. In addition, the factors that determine emissions, such as energy-supply systems, cannot be changed overnight. Consequently, there is also a need for adaptation.

Adaptation can include a wide range of activities. Examples include a farmer switching to growing a different crop variety better suited to warmer or drier conditions; a company relocating key business centers away from coastal areas vulnerable to sea-level rise and hurricanes; and a community

altering its zoning and building codes to place fewer structures in harm's way and making buildings less vulnerable to damage from floods, fires, and other extreme events. Some adaptation options that are currently being pursued in various regions and sectors to deal with climate change and/or other environmental issues are identified in this report. However, it is clear that there are limits to how much adaptation can achieve.

Humans have adapted to changing climatic conditions in the past, but in the future, adaptations will be particularly challenging because society won't be adapting to a new steady state but rather to a rapidly moving target. Climate will be continually changing, moving at a relatively rapid rate, outside the range to which society has adapted in the past. The precise amounts and timing of these changes will not be known with certainty.

In an increasingly interdependent world, U.S. vulnerability to climate change is linked to the fates of other nations. For example, conflicts or mass migrations of people resulting from food scarcity and other resource limits, health impacts, or environmental stresses in other parts of the world could threaten U.S. national security. It is thus difficult to fully evaluate the impacts of climate change on the United States without considering the consequences of climate change elsewhere. However, such analysis is beyond the scope of this report.

Finally, this report identifies a number of areas in which inadequate information or understanding hampers our ability to estimate future climate change and its impacts. For example, our knowledge of changes in tornadoes, hail, and ice storms is quite limited, making it difficult to know if and how such events have changed as climate has warmed, and how they might change in the future. Research on ecological responses to climate change is also limited, as is our understanding of social responses. The section titled *An Agenda for Climate Impacts Science* at the end of this report offers some thoughts on the most important ways to improve our knowledge. Results from such efforts would inform future assessments that continue building our understanding of humanity's impacts on climate, and climate's impacts on us.

Key Findings

1. Global warming is unequivocal and primarily human-induced.
Global temperature has increased over the past 50 years. This observed increase is due primarily to human-induced emissions of heat-trapping gases. (p. 13)

2. Climate changes are underway in the United States and are projected to grow.
Climate-related changes are already observed in the United States and its coastal waters. These include increases in heavy downpours, rising temperature and sea level, rapidly retreating glaciers, thawing permafrost, lengthening growing seasons, lengthening ice-free seasons in the ocean and on lakes and rivers, earlier snowmelt, and alterations in river flows. These changes are projected to grow. (p. 27)

3. Widespread climate-related impacts are occurring now and are expected to increase.
Climate changes are already affecting water, energy, transportation, agriculture, ecosystems, and health. These impacts are different from region to region and will grow under projected climate change. (p. 41-106, 107-152)

4. Climate change will stress water resources.
Water is an issue in every region, but the nature of the potential impacts varies. Drought, related to reduced precipitation, increased evaporation, and increased water loss from plants, is an important issue in many regions, especially in the West. Floods and water quality problems are likely to be amplified by climate change in most regions. Declines in mountain snowpack are important in the West and Alaska where snowpack provides vital natural water storage. (p. 41, 129, 135, 139)

5. Crop and livestock production will be increasingly challenged.
Many crops show positive responses to elevated carbon dioxide and low levels of warming, but higher levels of warming often negatively affect growth and yields. Increased pests, water stress, diseases, and weather extremes will pose adaptation challenges for crop and livestock production. (p. 71)

6. Coastal areas are at increasing risk from sea-level rise and storm surge.
Sea-level rise and storm surge place many U.S. coastal areas at increasing risk of erosion and flooding, especially along the Atlantic and Gulf Coasts, Pacific Islands, and parts of Alaska. Energy and transportation infrastructure and other property in coastal areas are very likely to be adversely affected. (p. 111, 139, 145, 149)

7. Risks to human health will increase.
Harmful health impacts of climate change are related to increasing heat stress, waterborne diseases, poor air quality, extreme weather events, and diseases transmitted by insects and rodents. Reduced cold stress provides some benefits. Robust public health infrastructure can reduce the potential for negative impacts. (p. 89)

8. Climate change will interact with many social and environmental stresses.
Climate change will combine with pollution, population growth, overuse of resources, urbanization, and other social, economic, and environmental stresses to create larger impacts than from any of these factors alone. (p. 99)

9. Thresholds will be crossed, leading to large changes in climate and ecosystems.
There are a variety of thresholds in the climate system and ecosystems. These thresholds determine, for example, the presence of sea ice and permafrost, and the survival of species, from fish to insect pests, with implications for society. With further climate change, the crossing of additional thresholds is expected. (p. 76, 82, 115, 137, 142)

10. Future climate change and its impacts depend on choices made today.
The amount and rate of future climate change depend primarily on current and future human-caused emissions of heat-trapping gases and airborne particles. Responses involve reducing emissions to limit future warming, and adapting to the changes that are unavoidable. (p. 25, 29)

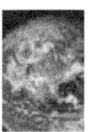

Global Climate Change

Key Messages:

- Human activities have led to large increases in heat-trapping gases over the past century.
- Global average temperature and sea level have increased, and precipitation patterns have changed.
- The global warming of the past 50 years is due primarily to human-induced increases in heat-trapping gases. Human "fingerprints" also have been identified in many other aspects of the climate system, including changes in ocean heat content, precipitation, atmospheric moisture, and Arctic sea ice.
- Global temperatures are projected to continue to rise over this century; by how much and for how long depends on a number of factors, including the amount of heat-trapping gas emissions and how sensitive the climate is to those emissions.

Key Sources

| CCSP 1.1 Temperature Trends | CCSP 1.3 Reanalysis | CCSP 2.1 GHG Emissions | CCSP 2.2 Carbon Cycle | CCSP 2.3 Aerosol Impacts | CCSP 2.4 Ozone Trends | CCSP 3.1 Climate Models | CCSP 3.2 Climate Projections | CCSP 3.3 Extremes | CCSP 3.4 Abrupt Climate Change | CCSP 4.1 Sea-Level Rise | CCSP 4.3 Impacts | IPCC WG-1 | IPCC Emissions Scenarios |

This introduction to global climate change explains very briefly what has been happening to the world's climate and why, and what is projected to happen in the future. While this report focuses on climate change impacts in the United States, understanding these changes and their impacts requires an understanding of the global climate system.

Many changes have been observed in global climate over the past century. The nature and causes of these changes have been comprehensively chronicled in a variety of recent reports, such as those by the Intergovernmental Panel on Climate Change (IPCC) and the U.S. Climate Change Science Program (CCSP). This section does not intend to duplicate these comprehensive efforts, but rather to provide a brief synthesis, and to integrate more recent work with the assessments of the IPCC, CCSP, and others.

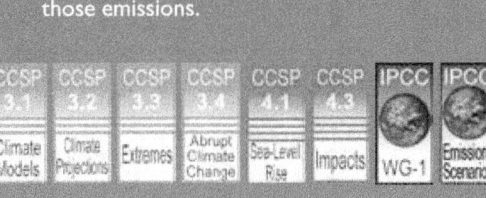

800,000 Year Record of Carbon Dioxide Concentration

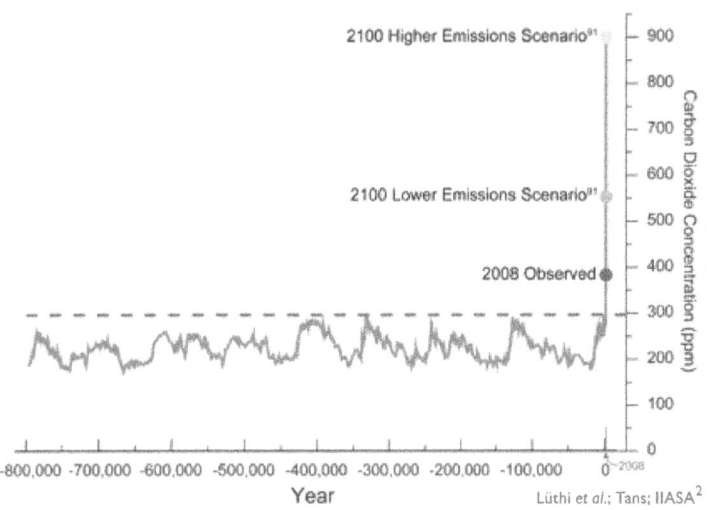

Lüthi *et al.*; Tans; IIASA[2]

Analysis of air bubbles trapped in an Antarctic ice core extending back 800,000 years documents the Earth's changing carbon dioxide concentration. Over this long period, natural factors have caused the atmospheric carbon dioxide concentration to vary within a range of about 170 to 300 parts per million (ppm). Temperature-related data make clear that these variations have played a central role in determining the global climate. As a result of human activities, the present carbon dioxide concentration of about 385 ppm is about 30 percent above its highest level over at least the last 800,000 years. In the absence of strong control measures, emissions projected for this century would result in the carbon dioxide concentration increasing to a level that is roughly 2 to 3 times the highest level occurring over the glacial-interglacial era that spans the last 800,000 or more years.

Human activities have led to large increases in heat-trapping gases over the past century.

The Earth's climate depends on the functioning of a natural "greenhouse effect." This effect is the result of heat-trapping gases (also known as greenhouse gases) like water vapor, carbon dioxide, ozone, methane, and nitrous oxide, which absorb heat radiated from the Earth's surface and lower atmosphere and then radiate much of the energy back toward the surface. Without this natural greenhouse effect, the average surface temperature of the Earth would be about 60°F colder. However, human activities have been releasing additional heat-trapping gases, intensifying the natural greenhouse effect, thereby changing the Earth's climate.

Climate is influenced by a variety of factors, both human-induced and natural. The increase in the carbon dioxide concentration has been the principal factor causing warming over the past 50 years. Its concentration has been building up in the Earth's atmosphere since the beginning of the industrial era in the mid-1700s, primarily due to the burning of fossil fuels (coal, oil, and natural gas) and the clearing of forests. Human activities have also increased the emissions of other greenhouse gases, such as methane, nitrous oxide, and halocarbons.[3]

These emissions are thickening the blanket of heat-trapping gases in Earth's atmosphere, causing surface temperatures to rise.

Heat-trapping gases

Carbon dioxide concentration has increased due to the use of fossil fuels in electricity generation, transportation, and industrial and household uses. It is also produced as a by-product during the manufacturing of cement. Deforestation provides a source of carbon dioxide and reduces its uptake by trees and other plants. Globally, over the past several decades, about 80 percent of human-induced carbon dioxide emissions came from the burning of fossil fuels, while about 20 percent resulted from deforestation and associated agricultural practices. The concentration of carbon dioxide in the atmosphere has increased by roughly 35 percent since the start of the industrial revolution.[3]

Methane concentration has increased mainly as a result of agriculture; raising livestock (which produce methane in their digestive tracts); mining, transportation, and use of certain fossil fuels; sewage; and decomposing garbage in landfills. About 70 percent of the emissions of atmospheric methane are now related to human activities.[4]

Nitrous oxide concentration is increasing as a result of fertilizer use and fossil fuel burning.

Halocarbon emissions come from the release of certain manufactured chemicals to the atmosphere. Examples include chlorofluorocarbons (CFCs), which were used extensively in refrigeration and for other industrial processes before their presence in the atmosphere was found to cause stratospheric ozone depletion. The abundance of these gases in the atmosphere is now decreasing as a result of international regulations designed to protect the ozone layer. Continued decreases in ozone-depleting halocarbon emissions are expected to reduce their relative influence on climate change in the future.[3,5] Many halocarbon replacements, however, are potent greenhouse gases, and their concentrations are increasing.[6]

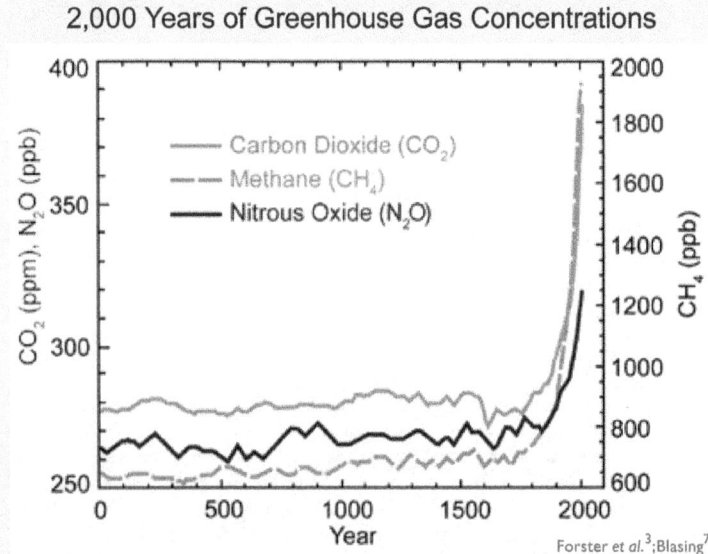

2,000 Years of Greenhouse Gas Concentrations

— Carbon Dioxide (CO$_2$)
— Methane (CH$_4$)
— Nitrous Oxide (N$_2$O)

Forster et al.[3];Blasing[7]

Increases in concentrations of these gases since 1750 are due to human activities in the industrial era. Concentration units are parts per million (ppm) or parts per billion (ppb), indicating the number of molecules of the greenhouse gas per million or billion molecules of air.

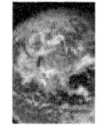

Ozone is a greenhouse gas, and is continually produced and destroyed in the atmosphere by chemical reactions. In the troposphere, the lowest 5 to 10 miles of the atmosphere near the surface, human activities have increased the ozone concentration through the release of gases such as carbon monoxide, hydrocarbons, and nitrogen oxides. These gases undergo chemical reactions to produce ozone in the presence of sunlight. In addition to trapping heat, excess ozone in the troposphere causes respiratory illnesses and other human health problems.

In the stratosphere, the layer above the troposphere, ozone exists naturally and protects life on Earth from exposure to excessive ultraviolet radiation from the Sun. As mentioned previously, halocarbons released by human activities destroy ozone in the stratosphere and have caused the ozone hole over Antarctica.[8] Changes in the stratospheric ozone layer have contributed to changes in wind patterns and regional climates in Antarctica.[9]

Water vapor is the most important and abundant greenhouse gas in the atmosphere. Human activities produce only a very small increase in water vapor through irrigation and combustion processes.[3] However, the surface warming caused by human-produced increases in other greenhouse gases leads to an increase in atmospheric water vapor, since a warmer climate increases evaporation and allows the atmosphere to hold more moisture. This creates an amplifying "feedback loop," leading to more warming.

Other human influences
In addition to the global-scale climate effects of heat-trapping gases, human activities also produce additional local and regional effects. Some of these activities partially offset the warming caused by greenhouse gases, while others increase the warming. One such influence on climate is caused by tiny particles called "aerosols" (not to be confused with aerosol spray cans). For example, the burning of coal produces emissions of sulfur-containing compounds. These compounds form "sulfate aerosol" particles, which reflect some of the incoming sunlight away from the Earth, causing a cooling influence at the surface. Sulfate aerosols also tend to make clouds more efficient at reflecting sunlight, causing an additional indirect cooling effect.

Another type of aerosol, often referred to as soot or black carbon, absorbs incoming sunlight and traps heat in the atmosphere. Thus, depending on their type, aerosols can either mask or increase the warming caused by increased levels of greenhouse gases.[13] On a globally averaged basis, the sum of these aerosol effects offsets some of the warming caused by heat-trapping gases.[10]

The effects of various greenhouse gases and aerosol particles on Earth's climate depend in part on how long these gases and particles remain in the atmosphere. After emission, the atmospheric concentration of carbon dioxide remains elevated for thousands of years, and that of methane for decades, while the elevated concentrations of aerosols only persist for days to weeks.[11,12] The climate effects of reductions in emissions of carbon dioxide and other long-lived gases do not become apparent for at least several decades. In contrast, reductions in emissions of short-lived compounds can have a rapid, but complex effect since the geographic patterns of their climatic influence and the resulting surface temperature responses are quite different. One modeling study found that while the greatest emissions of short-lived pollutants in summertime by late this century are projected to come from Asia, the strongest climate response is projected to be over the central United States.[13]

Human activities have also changed the land surface in ways that alter how much heat is reflected or absorbed by the surface. Such changes include the cutting and burning of forests, the replacement of other areas of natural vegetation with agriculture and cities, and large-scale irrigation. These transformations of the land surface can cause local (and even regional) warming or cooling. Globally, the net effect of these changes has probably been a slight cooling of the Earth's surface over the past 100 years.[14,15]

Natural influences
Two important natural factors also influence climate: the Sun and volcanic eruptions. Over the past three decades, human influences on climate have become increasingly obvious, and global temperatures have risen sharply. During the same period, the Sun's energy output (as measured by satellites since 1979) has followed its historical 11-year cycle

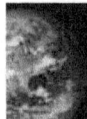

of small ups and downs, but with no net increase (see figure page 20).[16] The two major volcanic eruptions of the past 30 years have had short-term cooling effects on climate, lasting 2 to 3 years.[17] Thus, these natural factors cannot explain the warming of recent decades; in fact, their net effect on climate has probably been a slight cooling influence over this period. Slow changes in Earth's orbit around the Sun and its tilt toward or away from the Sun are also a purely natural influence on climate, but are only important on timescales from thousands to many tens of thousands of years.

The climate changes that have occurred over the last century are not solely caused by the human and natural factors described above. In addition to these influences, there are also fluctuations in climate that occur even in the absence of changes in human activities, the Sun, or volcanoes. One example is the El Niño phenomenon, which has important influences on many aspects of regional and global climate. Many other modes of variability have been identified by climate scientists and their effects on climate occur at the same time as the effects of human activities, the Sun, and volcanoes.

Carbon release and uptake

Once carbon dioxide is emitted to the atmosphere, some of it is absorbed by the oceans and taken up by vegetation, although this storage may be temporary. About 45 percent of the carbon dioxide emitted by human activities in the last 50 years is now stored in the oceans and vegetation. The rest has remained in the air, increasing the atmospheric concentration.[2,3,18] It is thus important to understand not only how much carbon dioxide is emitted, but also how much is taken up, over what time scales, and how these sources and "sinks" of carbon dioxide might change as climate continues to warm. For example, it is known from long records of Earth's climate history that under warmer conditions, carbon tends to be released, for instance, from thawing permafrost, initiating a feedback loop in which more carbon release leads to more warming which leads to further release, and so on.[19,20]

Global emissions of carbon dioxide have been accelerating. The growth rate increased from 1.3 percent per year in the 1990s to 3.3 percent per year between 2000 and 2006.[21] The increasing emissions of carbon dioxide are the primary cause of the increased concentration of carbon dioxide observed in the atmosphere. There is also evidence that a smaller fraction of the annual human-induced emissions is now being taken up than in the past, leading to a greater fraction remaining in the atmosphere and an accelerating rate of increase in the carbon dioxide concentration.[21]

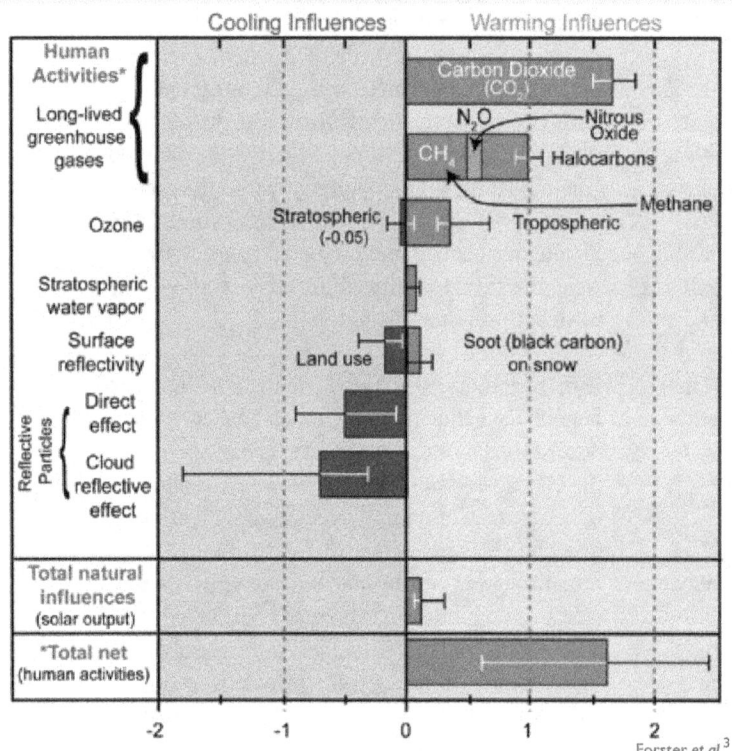

The figure above shows the amount of warming influence (red bars) or cooling influence (blue bars) that different factors have had on Earth's climate over the industrial age (from about 1750 to the present). Results are in watts per square meter. The longer the bar, the greater the influence on climate. The top part of the box includes all the major human-induced factors, while the second part of the box includes the Sun, the only major natural factor with a long-term effect on climate. The cooling effect of individual volcanoes is also natural, but is relatively short-lived (2 to 3 years), thus their influence is not included in this figure. The bottom part of the box shows that the total net effect (warming influences minus cooling influences) of human activities is a strong warming influence. The thin lines on each bar provide an estimate of the range of uncertainty.

Ocean acidification

As the ocean absorbs carbon dioxide from the atmosphere, seawater is becoming less alkaline (its pH is decreasing) through a process generally referred to as ocean acidification. The pH of seawater has decreased significantly since 1750,[22,23] and is projected to drop much more dramatically by the end of the century if carbon dioxide concentrations continue to increase.[24] Such ocean acidification is essentially irreversible over a time scale of centuries. As discussed in the *Ecosystems* sector and *Coasts* region, ocean acidification affects the process of calcification by which living things create shells and skeletons, with substantial negative consequences for coral reefs, mollusks, and some plankton species important to ocean food chains.[25]

Global average temperature and sea level have increased, and precipitation patterns have changed.

Temperatures are rising

Global average surface air temperature has increased substantially since 1970.[26] The estimated change in the average temperature of Earth's surface is based on measurements from thousands of weather stations, ships, and buoys around the world, as well as from satellites. These measurements are independently compiled, analyzed, and processed by different research groups. There are a number of important steps in the data processing. These include identifying and adjusting for the effects of changes in the instruments used to measure temperature, the measurement times and locations, the local environment around the measuring site, and such factors as satellite orbital drift. For instance, the growth of cities can cause localized "urban heat island" effects.

A number of research groups around the world have produced estimates of global-scale changes in surface temperature. The warming trend that is apparent in all of these temperature records is confirmed by other independent observations, such as the melting of Arctic sea ice, the retreat of mountain glaciers on every continent,[27] reductions in the extent of snow cover, earlier blooming of plants in spring, and increased melting of the Greenland and Antarctic ice sheets.[28,29] Because snow and ice

Global Temperature and Carbon Dioxide

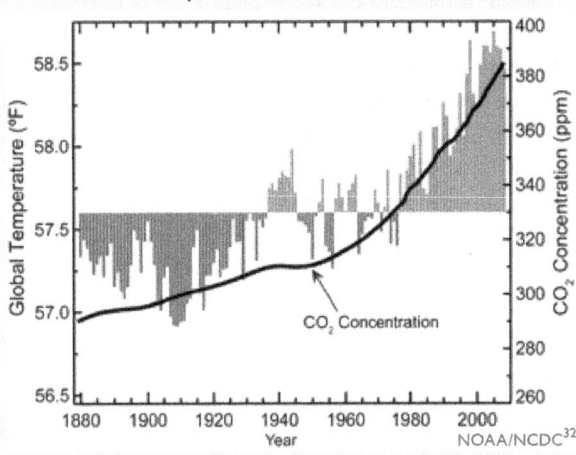

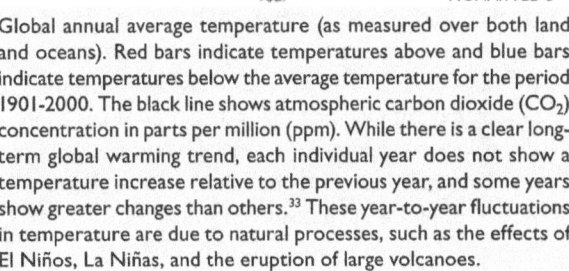

Global annual average temperature (as measured over both land and oceans). Red bars indicate temperatures above and blue bars indicate temperatures below the average temperature for the period 1901-2000. The black line shows atmospheric carbon dioxide (CO_2) concentration in parts per million (ppm). While there is a clear long-term global warming trend, each individual year does not show a temperature increase relative to the previous year, and some years show greater changes than others.[33] These year-to-year fluctuations in temperature are due to natural processes, such as the effects of El Niños, La Niñas, and the eruption of large volcanoes.

reflect the Sun's heat, this melting causes more heat to be absorbed, which causes more melting, resulting in another feedback loop.[20]

Additionally, temperature measurements above the surface have been made by weather balloons since the late 1940s, and from satellites since 1979. These measurements show warming of the troposphere, consistent with the surface warming.[30,31] They also reveal cooling in the stratosphere.[30] This pattern of tropospheric warming and stratospheric cooling agrees with our understanding of how atmospheric temperature would be expected to change in response to increasing greenhouse gas concentrations and the observed depletion of stratospheric ozone.[14]

Precipitation patterns are changing

Precipitation is not distributed evenly over the globe. Its average distribution is governed primarily by atmospheric circulation patterns, the availability of moisture, and surface terrain effects. The first two of these factors are influenced by temperature. Thus, human-caused changes in temperature are expected to alter precipitation patterns.

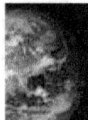

Observations show that such shifts are occurring. Changes have been observed in the amount, intensity, frequency, and type of precipitation. Pronounced increases in precipitation over the past 100 years have been observed in eastern North America, southern South America, and northern Europe. Decreases have been seen in the Mediterranean, most of Africa, and southern Asia. Changes in the geographical distribution of droughts and flooding have been complex. In some regions, there have been increases in the occurrences of both droughts and floods.[28] As the world warms, northern regions and mountainous areas are experiencing more precipitation falling as rain rather than snow.[34] Widespread increases in heavy precipitation events have occurred, even in places where total rain amounts have decreased. These changes are associated with the fact that warmer air holds more water vapor evaporating from the world's oceans and land surface.[31] This increase in atmospheric water vapor has been observed from satellites, and is primarily due to human influences.[35,36]

Sea level is rising

After at least 2,000 years of little change, sea level rose by roughly 8 inches over the past century. Satellite data available over the past 15 years show sea level rising at a rate roughly double the rate observed over the past century.[37]

There are two principal ways in which global warming causes sea level to rise. First, ocean water expands as it warms, and therefore takes up more space. Warming has been observed in each of the world's major ocean basins, and has been directly linked to human influences.[38,39]

Second, warming leads to the melting of glaciers and ice sheets, which raises sea level by adding water to the oceans. Glaciers have been retreating worldwide for at least the last century, and the rate of retreat has increased in the past decade.[29,40] Only a few glaciers are actually advancing (in locations that were

well below freezing, and where increased precipitation has outpaced melting). The total volume of glaciers on Earth is declining sharply. The progressive disappearance of glaciers has implications not only for the rise in global sea level, but also for water supplies in certain densely populated regions of Asia and South America.

The Earth has major ice sheets on Greenland and Antarctica. These ice sheets are currently losing ice volume by increased melting and calving of icebergs, contributing to sea-level rise. The Greenland Ice Sheet has also been experiencing record amounts of surface melting, and a large increase in the rate of mass loss in the past decade.[41] If the entire Greenland Ice Sheet melted, it would raise sea level by about 20 feet. The Antarctic Ice Sheet consists of two portions, the West Antarctic Ice Sheet and the East Antarctic Ice Sheet. The West Antarctic Ice Sheet, the more vulnerable to melting of the two, contains enough water to raise global sea levels by about 16 to 20 feet.[29] If the East Antarctic Ice Sheet melted entirely, it would raise global sea level by about 200 feet. Complete melting of these ice sheets over this century or the next is thought to be virtually impossible, although past climate records provide precedent for very significant decreases in ice volume, and therefore increases in sea level.[42,43]

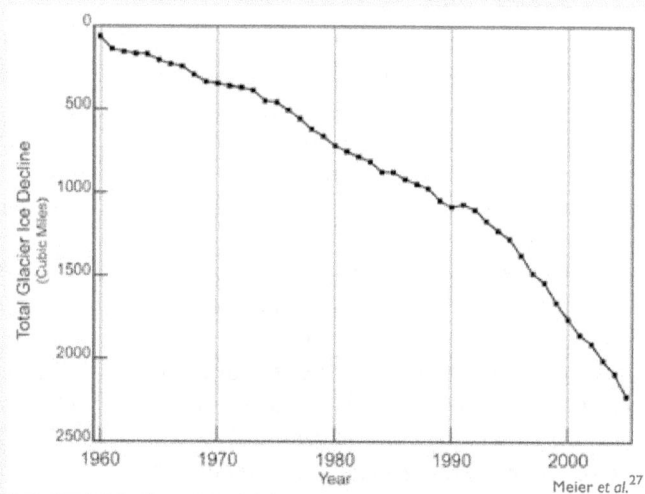

As temperatures have risen, glaciers around the world have shrunk. The graph shows the cumulative decline in glacier ice worldwide.

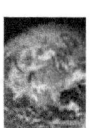

The global warming of the past 50 years is due primarily to human-induced increases in heat-trapping gases. Human "fingerprints" also have been identified in many other aspects of the climate system, including changes in ocean heat content, precipitation, atmospheric moisture, and Arctic sea ice.

In 1996, the IPCC Second Assessment Report[44] cautiously concluded that "the balance of evidence suggests a discernible human influence on global climate." Since then, a number of national and international assessments have come to much stronger conclusions about the reality of human effects on climate. Recent scientific assessments find that most of the warming of the Earth's surface over the past 50 years has been caused by human activities.[45,46]

This conclusion rests on multiple lines of evidence. Like the warming "signal" that has gradually emerged from the "noise" of natural climate variability, the scientific evidence for a human influence on global climate has accumulated over the past several decades, from many hundreds of studies. No single study is a "smoking gun." Nor has any single study or combination of studies undermined the large body of evidence supporting the conclusion that human activity is the primary driver of recent warming.

The first line of evidence is our basic physical understanding of how greenhouse gases trap heat, how the climate system responds to increases in greenhouse gases, and how other human and natural factors influence climate. The second line of evidence is from indirect estimates of climate changes over the last 1,000 to 2,000 years. These records are obtained from living things and their remains (like tree rings and corals) and from physical quantities (like the ratio between lighter and heavier isotopes of oxygen in ice cores) which change in measurable ways as climate changes. The lesson from these data is that global surface temperatures over the last several decades are clearly unusual, in that they were higher than at any time during at least the past 400 years.[47] For the Northern Hemisphere, the recent temperature rise is clearly unusual in at least the last 1,000 years.[47,48]

The third line of evidence is based on the broad, qualitative consistency between observed changes in climate and the computer model simulations of how climate would be expected to change in response to human activities. For example, when climate models are run with historical increases in greenhouse gases, they show gradual warming of the Earth and ocean surface, increases in ocean heat content and the temperature of the lower atmosphere, a rise in global sea level, retreat of sea ice and snow cover, cooling of the stratosphere, an increase in the amount of atmospheric water vapor, and changes in large-scale precipitation and pressure patterns. These and other aspects of modeled climate change are in agreement with observations.[14,49]

Finally, there is extensive statistical evidence from so-called "fingerprint" studies. Each factor that affects climate produces a unique pattern of climate response, much as each person has a unique fingerprint. Fingerprint studies exploit these unique signatures, and allow detailed comparisons of modeled and observed climate change patterns.[44] Scientists rely on such studies to attribute observed changes in climate to a particular cause or set of causes. In the real world, the climate changes that have occurred since the start of the Industrial Revolution are due to a complex mixture of human and natural causes. The importance of each individual influence in this mixture changes over time. Of course, there are not multiple Earths, which would allow an experimenter to change one factor at a time on each Earth, thus helping to isolate different fingerprints. Therefore, climate models are used to study how individual factors affect climate. For example, a single factor (like greenhouse gases) or a set of factors can be varied, and the response of the modeled climate system to these individual or combined changes can thus be studied.[50]

For example, when climate model simulations of the last century include all of the major influences on climate, both human-induced and natural, they can reproduce many important features of observed climate change patterns. When human influences are removed from the model experiments, results suggest that the surface of the Earth would actually have cooled slightly over the last 50 years. The clear message from fingerprint studies is that the

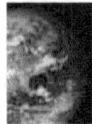

Separating Human and Natural Influences on Climate

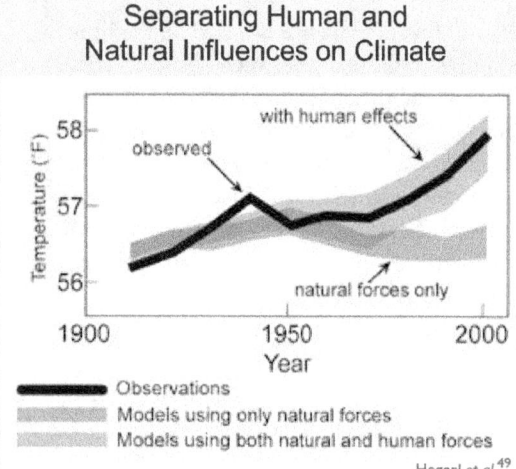

Hegerl et al.[49]

The blue band shows how global average temperatures would have changed due to natural forces only, as simulated by climate models. The red band shows model projections of the effects of human and natural forces combined. The black line shows actual observed global average temperatures. As the blue band indicates, without human influences, temperature over the past century would actually have first warmed and then cooled slightly over recent decades.[58]

observed warming over the last half-century cannot be explained by natural factors, and is instead caused primarily by human factors.[14,50]

Another fingerprint of human effects on climate has been identified by looking at a slice through the layers of the atmosphere, and studying the pattern of temperature changes from the surface up through the stratosphere. In all climate models, increases in carbon dioxide cause warming at the surface and in the troposphere, but lead to cooling of the stratosphere. For straightforward physical reasons, models also calculate that the human-caused depletion of stratospheric ozone has had a strong cooling effect in the stratosphere. There is a good match between the model fingerprint in response to combined carbon dioxide and ozone changes and the observed pattern of tropospheric warming and stratospheric cooling (see figure on next page).[14]

In contrast, if most of the observed temperature change had been due to an increase in solar output rather than an increase in greenhouse gases, Earth's atmosphere would have warmed throughout its full vertical extent, including the stratosphere.[9] The observed pattern of atmospheric temperature changes, with its pronounced cooling in the stratosphere, is therefore inconsistent with the hypothesis that changes in the Sun can explain the warming of recent decades. Moreover, direct satellite measurements of solar output show slight decreases during the recent period of warming.

The earliest fingerprint work[51] focused on changes in surface and atmospheric temperature. Scientists then applied fingerprint methods to a whole range of climate variables,[50,52] identifying human-caused climate signals in the heat content of the oceans,[38,39] the height of the tropopause[53] (the boundary between the troposphere and stratosphere, which has shifted upward by hundreds of feet in recent decades), the geographical patterns of precipitation,[54] drought,[55] surface pressure,[56] and the runoff from major river basins.[57]

Studies published after the appearance of the IPCC Fourth Assessment Report in 2007 have also found human fingerprints in the increased levels of atmospheric moisture[35,36] (both close to the surface and over the full extent of the atmosphere), in the

Measurements of Surface Temperature and Sun's Energy

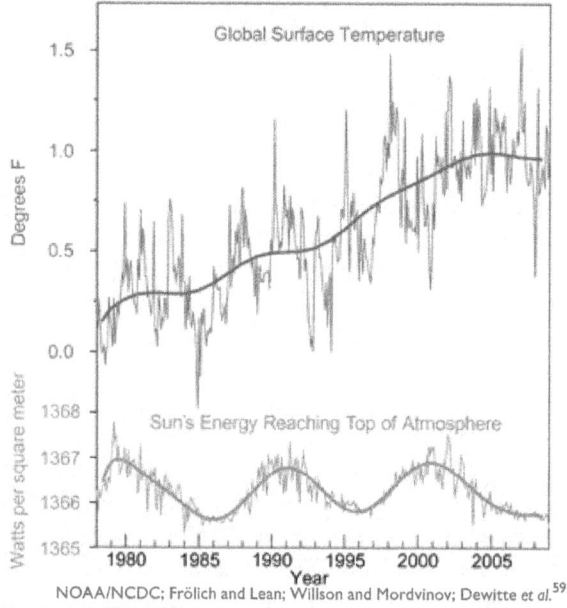

NOAA/NCDC; Frölich and Lean; Willson and Mordvinov; Dewitte et al.[59]

The Sun's energy received at the top of Earth's atmosphere has been measured by satellites since 1978. It has followed its natural 11-year cycle of small ups and downs, but with no net increase (bottom). Over the same period, global temperature has risen markedly (top).[60]

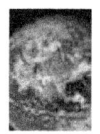

decline of Arctic sea ice extent,[61] and in the patterns of changes in Arctic and Antarctic surface temperatures.[62]

The message from this entire body of work is that the climate system is telling a consistent story of increasingly dominant human influence – the changes in temperature, ice extent, moisture, and circulation patterns fit together in a physically consistent way, like pieces in a complex puzzle.

Increasingly, this type of fingerprint work is shifting its emphasis. As noted, clear and compelling scientific evidence supports the case for a pronounced human influence on global climate. Much of the recent attention is now on climate changes at continental and regional scales,[64,65] and on variables that can have large impacts on societies. For example, scientists have established causal links between human activities and the changes in snowpack, maximum and minimum temperature, and the seasonal timing of runoff over mountainous regions of the western United States.[34] Human activity is likely to have made a substantial contribution to ocean surface temperature changes in hurricane formation regions.[66-68] Researchers are also looking beyond the physical climate system, and are beginning to tie changes in the distribution and seasonal behavior of plant and animal species to human-caused changes in temperature and precipitation.[69,70]

For over a decade, one aspect of the climate change story seemed to show a significant difference between models and observations.[14]

In the tropics, all models predicted that with a rise in greenhouse gases, the troposphere would be expected to warm more rapidly than the surface. Observations from weather balloons, satellites, and surface thermometers seemed to show the opposite behavior (more rapid warming of the surface than the troposphere). This issue was a stumbling block in our understanding of the causes of climate change. It is now largely resolved.[71] Research showed that there were large uncertainties in the satellite and weather balloon data. When uncertainties in models and observations are properly accounted for, newer observational data sets (with better treatment of known problems) are in agreement with climate model results.[31,72-75]

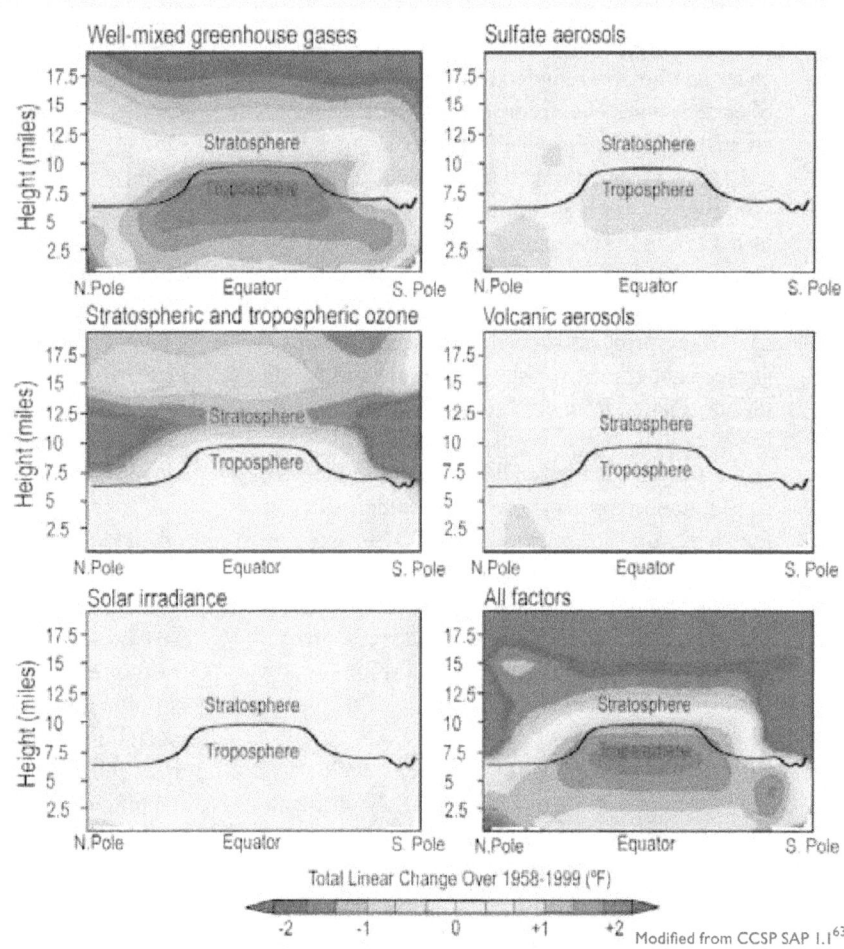

Patterns of Temperature Change Produced by Various Atmospheric Factors, 1958-1999

Climate simulations of the vertical profile of temperature change due to various factors, and the effect due to all factors taken together. The panels above represent a cross-section of the atmosphere from the north pole to the south pole, and from the surface up into the stratosphere. The black lines show the location of the tropopause, the boundary between the lower atmosphere (troposphere) and the stratosphere.

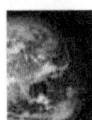

This does not mean, however, that all remaining differences between models and observations have been resolved. The observed changes in some climate variables, such as Arctic sea ice,[61,76] some aspects of precipitation,[54,77] and patterns of surface pressure,[56] appear to be proceeding much more rapidly than models have projected. The reasons for these differences are not well understood. Nevertheless, the bottom-line conclusion from climate fingerprinting is that most of the observed changes studied to date are consistent with each other, and are also consistent with our scientific understanding of how the climate system would be expected to respond to the increase in heat-trapping gases resulting from human activities.[14,49]

Scientists are sometimes asked whether extreme weather events can be linked to human activities.[24] Scientific research has concluded that human influences on climate are indeed changing the likelihood of certain types of extreme events. For example, an analysis of the European summer heat wave of 2003 found that the risk of such a heat wave is now roughly four times greater than it would have been in the absence of human-induced climate change.[68,78]

Like fingerprint work, such analyses of human-caused changes in the risks of extreme events rely on information from climate models, and on our understanding of the physics of the climate system. All of the models used in this work have imperfections in their representation of the complexities of the "real world" climate system.[79,80] These are due to both limits in our understanding of the climate system, and in our ability to represent its complex behavior with available computer resources. Despite this, models are extremely useful, for a number of reasons.

First, despite remaining imperfections, the current generation of climate models accurately portrays many important aspects of today's weather patterns and climate.[79,80] Models are constantly being improved, and are routinely tested against many observations of Earth's climate system. Second, the fingerprint work shows that models capture not only our present-day climate, but also key features of the observed climate changes over the past century.[47] Third, many of the large-scale observed cli-

mate changes (such as the warming of the surface and troposphere, and the increase in the amount of moisture in the atmosphere) are driven by very basic physics, which is well-represented in models.[35] Fourth, climate models can be used to predict changes in climate that can be verified in the real world. Examples include the short-term global cooling subsequent to the eruption of Mount Pinatubo and the stratospheric cooling with increasing carbon dioxide. Finally, models are the only tools that exist for trying to understand the climate changes likely to be experienced over the course of this century. No period in Earth's geological history provides an exact analogue for the climate conditions that will unfold in the coming decades.[20]

Global temperatures are projected to continue to rise over this century; by how much and for how long depends on a number of factors, including the amount of heat-trapping gas emissions and how sensitive the climate is to those emissions.

Some continued warming of the planet is projected over the next few decades due to past emissions. Choices made now will influence the amount of future warming. Lower levels of heat-trapping emissions will yield less future warming, while higher levels will result in more warming, and more severe impacts on society and the natural world.

Emissions scenarios
The IPCC developed a set of scenarios in a Special Report on Emissions Scenarios (SRES).[81] These have been extensively used to explore the potential for future climate change. None of these scenarios, not even the one called "lower", includes implementation of policies to limit climate change or to stabilize atmospheric concentrations of heat-trapping gases. Rather, differences among these scenarios are due to different assumptions about changes in population, rate of adoption of new technologies, economic growth, and other factors.

The IPCC emission scenarios also do not encompass the full range of possible futures: emissions can change less than those scenarios imply, or they can change more. Recent carbon dioxide emissions are, in fact, above the highest emissions scenario developed by the IPCC[82] (see figure below). Whether this will continue is uncertain.

There are also lower possible emissions paths than those put forth by the IPCC. The Framework Convention on Climate Change, to which the United States and 191 other countries are signatories, calls for stabilizing concentrations of greenhouse gases in the atmosphere at a level that would avoid dangerous human interference with the climate system. What exactly constitutes such interference is subject to interpretation.

A variety of research studies suggest that a further 2°F increase (relative to the 1980-1999 period) would lead to severe, widespread, and irreversible impacts.[83-85] To have a good chance (but not a guarantee) of avoiding temperatures above those levels,

it has been estimated that atmospheric concentration of carbon dioxide would need to stabilize in the long term at around today's levels.[86-89]

Reducing emissions of carbon dioxide would reduce warming over this century and beyond. Implementing sizable and sustained reductions in carbon dioxide emissions as soon as possible would significantly reduce the pace and the overall amount of climate change, and would be more effective than reductions of the same size initiated later. Reducing emissions of some shorter-lived greenhouse gases, such as methane, and some types of particles, such as soot, would begin to reduce the warming influence within weeks to decades.[13]

The graphs below show emissions scenarios and resulting carbon dioxide concentrations for three IPCC scenarios[90,91] and one stabilization scenario.[25]

Scenarios of Future Carbon Dioxide Global Emissions and Concentrations

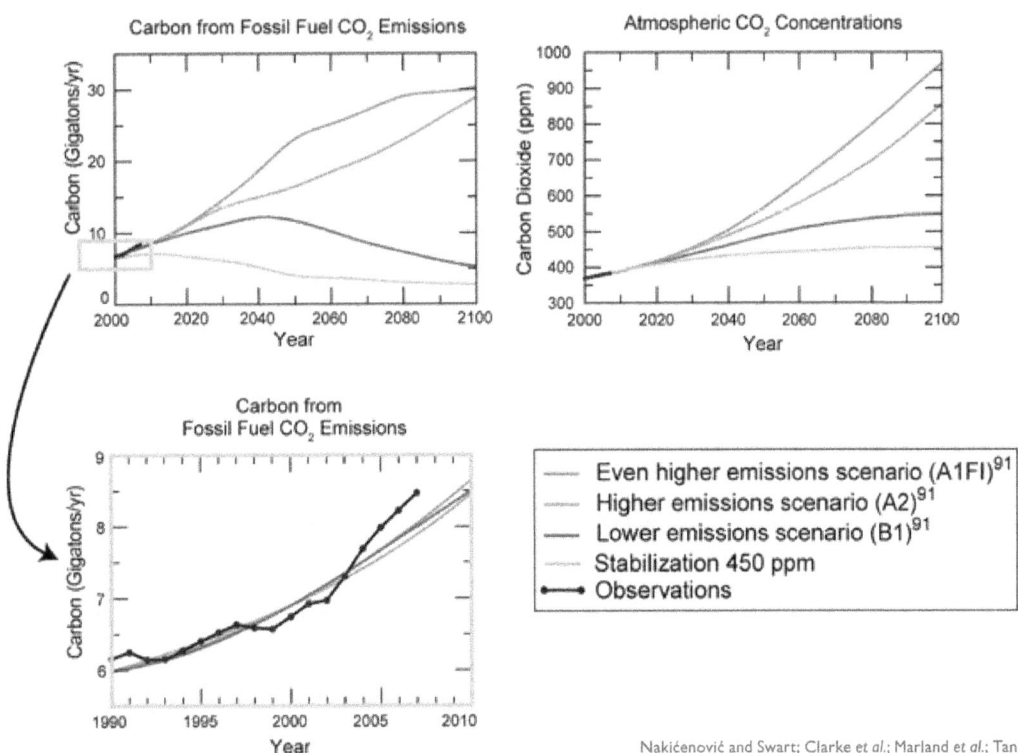

Nakićenović and Swart; Clarke et al.; Marland et al.; Tans[92]

The graphs show recent and projected global emissions of carbon dioxide in gigatons of carbon, on the left, and atmospheric concentrations on the right under five emissions scenarios. The top three in the key are IPCC scenarios that assume no explicit climate policies (these are used in model projections that appear throughout this report). The bottom line is a "stabilization scenario," designed to stabilize atmospheric carbon dioxide concentration at 450 parts per million. The inset expanded below these charts shows emissions for 1990-2010 under the three IPCC scenarios along with actual emissions to 2007 (in black).

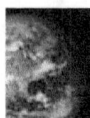

The stabilization scenario is aimed at stabilizing the atmospheric carbon dioxide concentration at roughly 450 parts per million (ppm); this is 70 ppm above the 2008 concentration of 385 ppm. Resulting temperature changes depend on atmospheric concentrations of greenhouse gases and particles and the climate's sensitivity to those concentrations.[87] Of those shown on the previous page, only the 450 ppm stabilization target has the potential to keep the global temperature rise at or below about 3.5°F from pre-industrial levels and 2°F above the current average temperature, a level beyond which many concerns have been raised about dangerous human interference with the climate system.[88,89] Scenarios that stabilize carbon dioxide below 450 ppm (not shown in the figure) offer an increased chance of avoiding dangerous climate change.[88,89]

Carbon dioxide is not the only greenhouse gas of concern. Concentrations of other heat-trapping gases like methane and nitrous oxide and particles like soot will also have to be stabilized at low enough levels to prevent global temperatures from rising higher than the level mentioned above. When these other gases are added, including the offsetting cooling effects of sulfate aerosol particles, analyses suggest that stabilizing concentrations around 400 parts per million of "equivalent carbon dioxide" would yield about an 80 percent chance of avoiding exceeding the 2°F above present temperature threshold. This would be true even if concentrations temporarily peaked as high as 475 parts per million and then stabilized at 400 parts per million roughly a century later.[72,88,89,93-95] Reductions in sulfate aerosol particles would necessitate lower equivalent carbon dioxide targets.

Rising global temperature

All climate models project that human-caused emissions of heat-trapping gases will cause further warming in the future. Based on scenarios that do not assume explicit climate policies to reduce greenhouse gas emissions, global average temperature is projected to rise by 2 to 11.5°F by the end of this century[90] (relative to the 1980-1999 time period). Whether the actual warming in 2100 will be closer to the low or the high end of this range depends primarily on two factors: first, the future level of emissions of heat-trapping gases, and second, how sensitive climate is to past and future

emissions. The range of possible outcomes has been explored using a range of different emissions scenarios, and a variety of climate models that encompass the known range of climate sensitivity.

Changing precipitation patterns

Projections of changes in precipitation largely follow recently observed patterns of change, with overall increases in the global average but substantial shifts in where and how precipitation falls.[90] Generally, higher latitudes are projected to receive more precipitation, while the dry belt that lies just outside the tropics expands further poleward,[96,97] and also receives less rain. Increases in tropical precipitation are projected during rainy seasons (such as monsoons), and especially over the tropical Pacific. Certain regions, including the U.S. West (especially the Southwest) and the Mediterranean, are expected to become drier. The widespread trend toward more heavy downpours is expected to continue, with precipitation becoming less frequent but more intense.[90] More precipitation is expected to fall as rain rather than snow.

Currently rare extreme events are becoming more common

In a warmer future climate, models project there will be an increased risk of more intense, more frequent, and longer-lasting heat waves.[90] The European heat wave of 2003 is an example of the type of extreme heat event that is likely to become much more common.[90] If greenhouse gas emissions continue to increase, by the 2040s more than half of European summers will be hotter than the summer of 2003, and by the end of this century, a summer as hot as that of 2003 will be considered unusually cool.[78]

Increased extremes of summer dryness and winter wetness are projected for much of the globe, meaning a generally greater risk of droughts and floods. This has already been observed,[55] and is projected to continue. In a warmer world, precipitation tends to be concentrated into heavier events, with longer dry periods in between.[90]

Models project a general tendency for more intense but fewer storms overall outside the tropics, with more extreme wind events and higher ocean waves in a number of regions in association with those

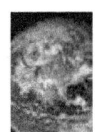

storms. Models also project a shift of storm tracks toward the poles in both hemispheres.[90]

Changes in hurricanes are difficult to project because there are countervailing forces. Higher ocean temperatures lead to stronger storms with higher wind speeds and more rainfall.[98] But changes in wind speed and direction with height are also projected to increase in some regions, and this tends to work against storm formation and growth.[99-101] It currently appears that stronger, more rain-producing tropical storms and hurricanes are generally

Global Average Temperature
1900 to 2100

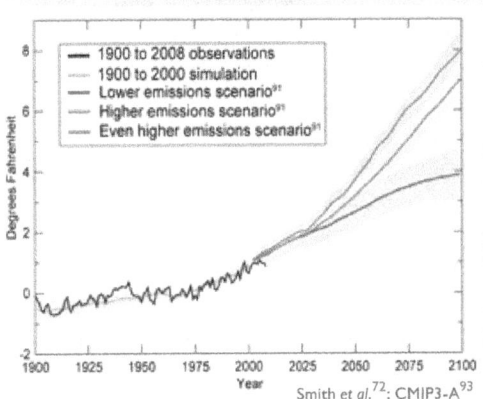

Smith et al.[72]; CMIP3-A[93]

Observed and projected changes in the global average temperature under three IPCC no-policy emissions scenarios. The shaded areas show the likely ranges while the lines show the central projections from a set of climate models. A wider range of model types shows outcomes from 2 to 11.5°F.[90] Changes are relative to the 1960-1979 average.

Global Increase in Heavy Precipitation
1900 to 2100

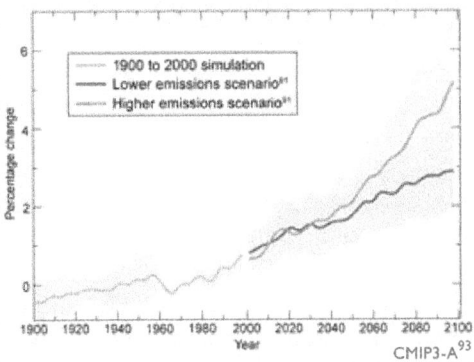

CMIP3-A[93]

Simulated and projected changes in the amount of precipitation falling in the heaviest 5 percent of daily events. The shaded areas show the likely ranges while the lines show the central projections from a set of climate models. Changes are relative to the 1960-1979 average.

more likely, though more research is required on these issues.[68] More discussion of Atlantic hurricanes, which most affect the United States, appears on page 34 in the *National Climate Change* section.

Sea level will continue to rise

Projecting future sea-level rise presents special challenges. Scientists have a well-developed understanding of the contributions of thermal expansion and melting glaciers to sea-level rise, so the models used to project sea-level rise include these processes. However, the contributions to past and future sea-level rise from ice sheets are less well understood. Recent observations of the polar ice sheets show that a number of complex processes control the movement of ice to the sea, and thus affect the contributions of ice sheets to sea-level rise.[29] Some of these processes are already producing substantial loss of ice mass. Because these processes are not well understood it is difficult to predict their future contributions to sea-level rise.[102]

Because of this uncertainty, the 2007 assessment by the IPCC could not quantify the contributions to sea-level rise due to changes in ice sheet dynamics, and thus projected a rise of the world's oceans from 8 inches to 2 feet by the end of this century.[90]

More recent research has attempted to quantify the potential contribution to sea-level rise from the accelerated flow of ice sheets to the sea[27,42] or to estimate future sea level based on its observed relationship to temperature.[103] The resulting estimates exceed those of the IPCC, and the average estimates under higher emissions scenarios are for sea-level rise between 3 and 4 feet by the end of this century. An important question that is often asked is, what is the upper bound of sea-level rise expected over this century? Few analyses have focused on this question. There is some evidence to suggest that it would be virtually impossible to have a rise of sea level higher than about 6.5 feet by the end of this century.[42]

The changes in sea level experienced at any particular location along the coast depend not only on the increase in the global average sea level, but also on changes in regional currents and winds, proximity to the mass of melting ice sheets, and on the vertical movements of the land due to geological

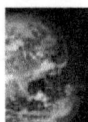

forces.[104] The consequences of sea-level rise at any particular location depend on the amount of sea-level rise relative to the adjoining land. Although some parts of the U.S. coast are undergoing uplift (rising), most shorelines are subsiding (sinking) to various degrees – from a few inches to over 2 feet per century.

Abrupt climate change

There is also the possibility of even larger changes in climate than current scenarios and models project. Not all changes in the climate are gradual. The long record of climate found in ice cores, tree rings, and other natural records show that Earth's climate patterns have undergone rapid shifts from one stable state to another within as short a period as a decade. The occurrence of abrupt changes in climate becomes increasingly likely as the human disturbance of the climate system grows.[90] Such changes can occur so rapidly that they would challenge the ability of human and natural systems to adapt.[105] Examples of such changes are abrupt shifts in drought frequency and duration. Ancient climate records suggest that in the United States, the Southwest may be at greatest risk for this kind of change, but that other regions including the Midwest and Great Plains have also had these kinds of abrupt shifts in the past and could experience them again in the future.

Rapid ice sheet collapse with related sea-level rise is another type of abrupt change that is not well understood or modeled and that poses a risk for the future. Recent observations show that melting on the surface of an ice sheet produces water that flows down through large cracks that create conduits through the ice to the base of the ice sheet where it lubricates ice previously frozen to the rock below.[29] Further, the interaction with warm ocean water, where ice meets the sea, can lead to sudden losses in ice mass and accompanying rapid global sea-level rise. Observations indicate that ice loss has increased dramatically over the last decade, though scientists are not yet confident that they can project how the ice sheets will respond in the future.

There are also concerns regarding the potential for abrupt release of methane from thawing of frozen soils, from the sea floor, and from wetlands in the tropics and the Arctic. While analyses suggest that an abrupt release of methane is very unlikely to occur within 100 years, it is very likely that warming will accelerate the pace of chronic methane emissions from these sources, potentially increasing the rate of global temperature rise.[106]

A third major area of concern regarding possible abrupt change involves the operation of the ocean currents that transport vast quantities of heat around the globe. One branch of the ocean circulation is in the North Atlantic. In this region, warm water flows northward from the tropics to the North Atlantic in the upper layer of the ocean, while cold water flows back from the North Atlantic to the tropics in the ocean's deep layers, creating a "conveyor belt" for heat. Changes in this circulation have profound impacts on the global climate system, from changes in African and Indian monsoon rainfall, to atmospheric circulation relevant to hurricanes, to changes in climate over North America and Western Europe.

Recent findings indicate that it is very likely that the strength of this North Atlantic circulation will decrease over the course of this century in response to increasing greenhouse gases. This is expected because warming increases the melting of glaciers and ice sheets and the resulting runoff of freshwater to the sea. This additional water is virtually salt-free, which makes it less dense than sea water. Increased precipitation also contributes fresh, less-dense water to the ocean. As a result, less surface water is dense enough to sink, thereby reducing the conveyor belt's transport of heat. The best estimate is that the strength of this circulation will decrease 25 to 30 percent in this century, leading to a reduction in heat transfer to the North Atlantic. It is considered very unlikely that this circulation would collapse entirely during the next 100 years or so, though it cannot be ruled out. While very unlikely, the potential consequences of such an abrupt event would be severe. Impacts would likely include sea-level rise around the North Atlantic of up to 2.5 feet (in addition to the rise expected from thermal expansion and melting glaciers and ice sheets), changes in atmospheric circulation conditions that influence hurricane activity, a southward shift of tropical rainfall belts with resulting agricultural impacts, and disruptions to marine ecosystems.[76]

National Climate Change

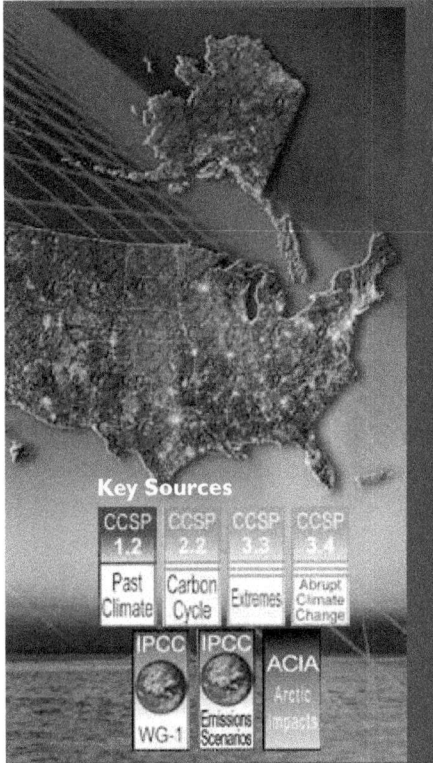

Key Sources

CCSP 1.2 — Past Climate
CCSP 2.2 — Carbon Cycle
CCSP 3.3 — Extremes
CCSP 3.4 — Abrupt Climate Change

IPCC WG-1
IPCC Emissions Scenarios
ACIA Arctic Impacts

Key Messages:

- U.S. average temperature has risen more than 2°F over the past 50 years and is projected to rise more in the future; how much more depends primarily on the amount of heat-trapping gases emitted globally and how sensitive the climate is to those emissions.
- Precipitation has increased an average of about 5 percent over the past 50 years. Projections of future precipitation generally indicate that northern areas will become wetter, and southern areas, particularly in the West, will become drier.
- The amount of rain falling in the heaviest downpours has increased approximately 20 percent on average in the past century, and this trend is very likely to continue, with the largest increases in the wettest places.
- Many types of extreme weather events, such as heat waves and regional droughts, have become more frequent and intense during the past 40 to 50 years.
- The destructive energy of Atlantic hurricanes has increased in recent decades. The intensity of these storms is likely to increase in this century.
- In the eastern Pacific, the strongest hurricanes have become stronger since the 1980s, even while the total number of storms has decreased.
- Sea level has risen along most of the U.S. coast over the last 50 years, and will rise more in the future.
- Cold-season storm tracks are shifting northward and the strongest storms are likely to become stronger and more frequent.
- Arctic sea ice is declining rapidly and this is very likely to continue.

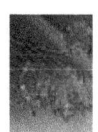

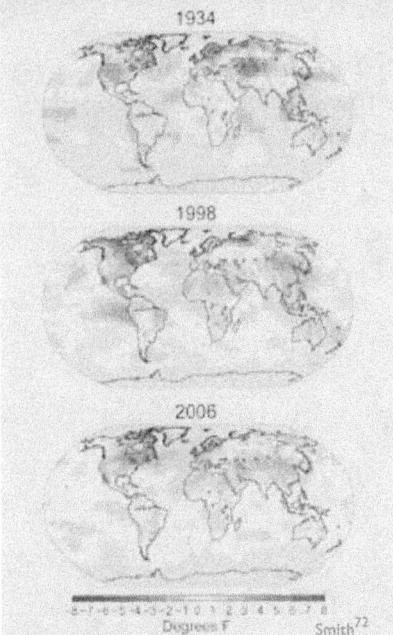

1934

1998

2006

Degrees F   Smith[72]

The maps show annual temperature difference from the 1961-1990 average for the 3 years that were the hottest on record in the United States: 1998, 1934 and 2006 (in rank order). Red areas were warmer than average, blue were cooler than average. The 1930s were very warm in much of the United States, but they were not unusually warm globally. On the other hand, the warmth of 1998 and 2006, as for most years in recent decades, has been global in extent.

Like the rest of the world, the United States has been warming significantly over the past 50 years in response to the build up of heat-trapping gases in the atmosphere. When looking at national climate, however, it is important to recognize that climate responds to local, regional, and global factors. Therefore, national climate varies more than the average global climate. While various parts of the world have had particularly hot or cold periods earlier in the historical record, these periods have not been global in scale, whereas the warming of recent decades has been global in scale – hence the term *global* warming. It is also important to recognize that at both the global and national scales, year-to-year fluctuations in natural weather and climate patterns can produce a period that does not follow the long-term trend. Thus, each year will not necessarily be warmer than every year before it, though the warming trend continues.

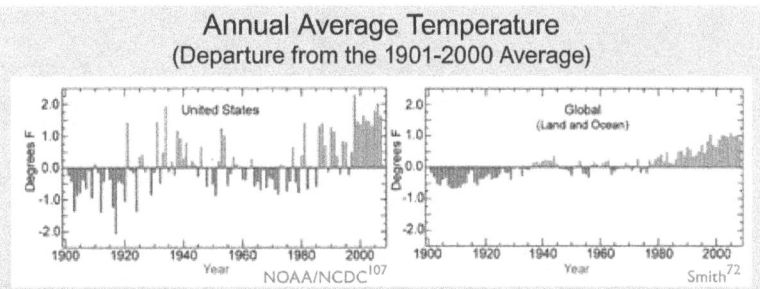

Annual Average Temperature
(Departure from the 1901-2000 Average)

United States — NOAA/NCDC[107]

Global (Land and Ocean) — Smith[72]

From 1901 to 2008, each year's temperature departure from the long-term average is one bar, with blue bars representing years cooler than the long-term average and red bars representing years warmer than that average. National temperatures vary much more than global temperatures, in part because of the moderating influence of the oceans on global temperatures.

27

U.S. average temperature has risen more than 2°F over the past 50 years and is projected to rise more in the future; how much more depends primarily on the amount of heat-trapping gases emitted globally and how sensitive the climate is to those emissions.

The series of maps and thermometers on these two pages shows the magnitude of the observed and projected changes in annual average temperature. The map for the period around 2000 shows that most areas of the United States have warmed 1 to 2°F compared to the 1960s and 1970s. Although not reflected in these maps of annual average temperature, this warming has generally resulted in longer warm seasons and shorter, less intense cold seasons.

The remaining maps show projected warming over the course of this century under a lower emissions scenario and a higher emissions scenario[91] (see *Global Climate Change* section, page 23). Tempera-

tures will continue to rise throughout the century under both emissions scenarios,[91] although higher emissions result in more warming by the middle of the century and significantly more by the end of the century.

Temperature increases in the next couple of decades will be primarily determined by past emissions of heat-trapping gases. As a result, there is little difference in projected temperature between the higher and lower emissions scenarios[91] in the near-term (around 2020), so only a single map is shown for this timeframe. Increases after the next couple of decades will be primarily determined by future emissions.[90] This is clearly evident in greater projected warming in the higher emissions scenario[91] by the middle (around 2050) and end of this century (around 2090).

On a seasonal basis, most of the United States is projected to experience greater warming in summer than in winter, while Alaska experiences far more warming in winter than summer.[108]

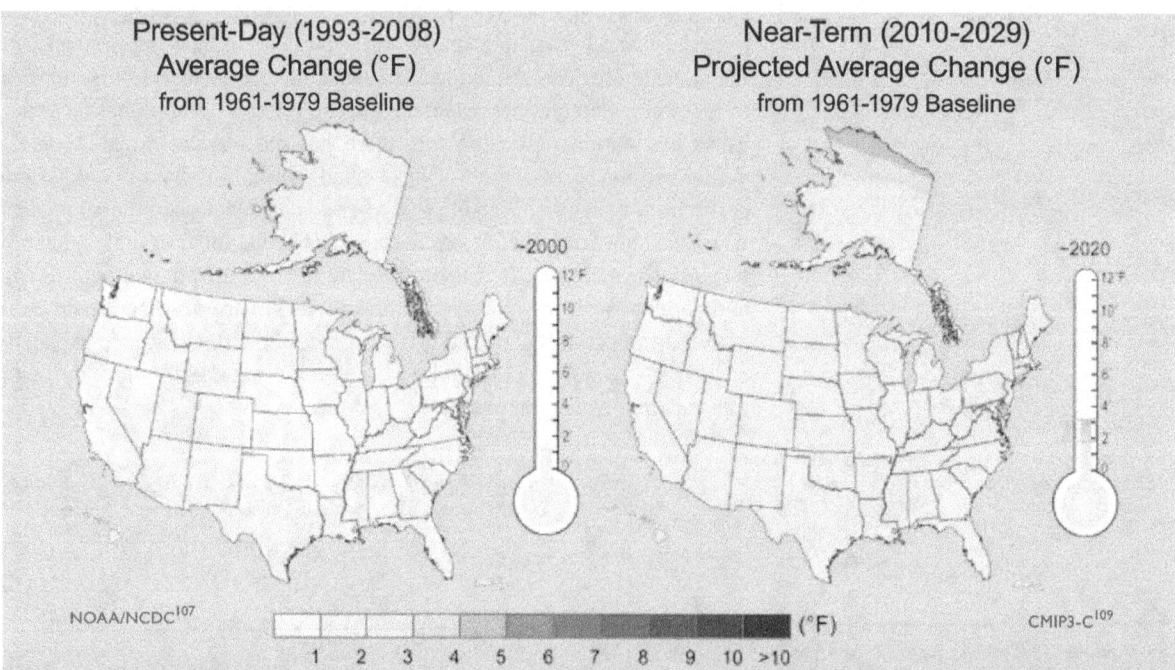

Present-Day (1993-2008)
Average Change (°F)
from 1961-1979 Baseline

Near-Term (2010-2029)
Projected Average Change (°F)
from 1961-1979 Baseline

NOAA/NCDC[107]

(°F)

CMIP3-C[109]

1 2 3 4 5 6 7 8 9 10 >10

The maps and thermometers on this page and the next page show temperature differences (either measured or projected) from conditions as they existed during the period from 1961-1979. Comparisons to this period are made because the influence on climate from increasing greenhouse gas emissions has been greatest during the past five decades. The present-day map is based on the average observed temperatures from 1993-2008 minus the average from 1961-1979. Projected temperatures are based on results from 16 climate models for the periods 2010-2029, 2040-2059, and 2080-2099. The brackets on the thermometers represent the likely range of model projections, though lower or higher outcomes are possible. The mid-century and end-of-century maps show projections for both the higher and lower emission scenarios.[91] The projection for the near-term is the average of the higher and lower emission scenarios[91] because there is little difference in that timeframe.

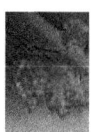

The average warming for the country as a whole is shown on the thermometers adjacent to each map. By the end of the century, the average U.S. temperature is projected to increase by approximately 7 to 11°F under the higher emissions scenario[91] and by approximately 4 to 6.5°F under the lower emissions scenario.[91] These ranges are due to differences among climate model results for the same emissions scenarios. Emissions scenarios even lower than the lower scenario shown here, such as the 450 ppm stabilization scenario described on pages 23-24, would yield lower temperature increases than those shown below.[25]

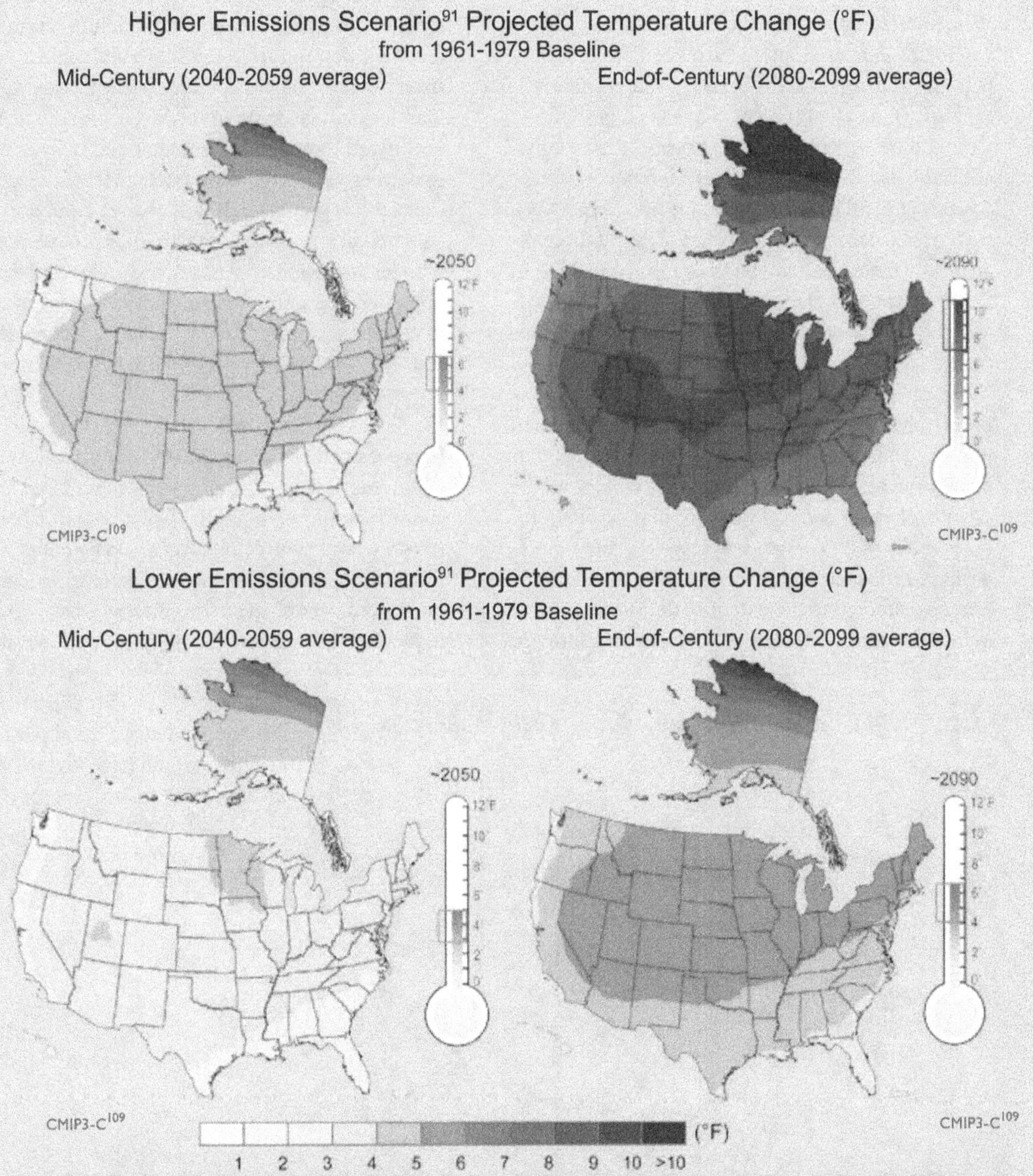

Higher Emissions Scenario[91] Projected Temperature Change (°F)
from 1961-1979 Baseline

Mid-Century (2040-2059 average)   End-of-Century (2080-2099 average)

~2050

~2090

CMIP3-C[109]

CMIP3-C[109]

Lower Emissions Scenario[91] Projected Temperature Change (°F)
from 1961-1979 Baseline

Mid-Century (2040-2059 average)   End-of-Century (2080-2099 average)

~2050

~2090

CMIP3-C[109]

CMIP3-C[109]

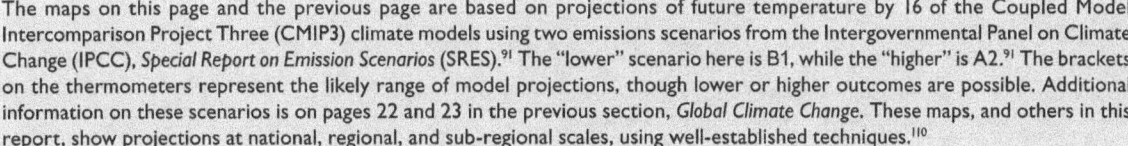

1 2 3 4 5 6 7 8 9 10 >10   (°F)

The maps on this page and the previous page are based on projections of future temperature by 16 of the Coupled Model Intercomparison Project Three (CMIP3) climate models using two emissions scenarios from the Intergovernmental Panel on Climate Change (IPCC), *Special Report on Emission Scenarios* (SRES).[91] The "lower" scenario here is B1, while the "higher" is A2.[91] The brackets on the thermometers represent the likely range of model projections, though lower or higher outcomes are possible. Additional information on these scenarios is on pages 22 and 23 in the previous section, *Global Climate Change*. These maps, and others in this report, show projections at national, regional, and sub-regional scales, using well-established techniques.[110]

Precipitation has increased an average of about 5 percent over the past 50 years. Projections of future precipitation generally indicate that northern areas will become wetter, and southern areas, particularly in the West, will become drier.

While precipitation over the United States as a whole has increased, there have been important regional and seasonal differences. Increasing trends throughout much of the year have been predominant in the Northeast and large parts of the Plains and Midwest. Decreases occurred in much of the Southeast in all but the fall season and in the Northwest in all seasons except spring. Precipitation also generally decreased during the summer and fall in the Southwest, while winter and spring, which are the wettest seasons in states such as California and Nevada, have had increases in precipitation.[111]

Future changes in total precipitation due to human-induced warming are more difficult to project than changes in temperature. In some seasons, some areas will experience an increase in precipitation, other areas will experience a decrease, and others will see little discernible change. The difficulty arises in predicting the extent of those areas and the amount of change. Model projections of future pre-

cipitation generally indicate that northern areas will become wetter, and southern areas, particularly in the West, will become drier.[97,108]

Confidence in projected changes is higher for winter and spring than for summer and fall. In winter and spring, northern areas are expected to receive significantly more precipitation than they do now, because the interaction of warm and moist air coming from the south with colder air from the north is projected to occur farther north than it did on average in the last century. The more northward incursions of warmer and moister air masses are expected to be particularly noticeable in northern regions that will change from very cold and dry atmospheric conditions to warmer but moister conditions.[68] Alaska, the Great Plains, the upper Midwest, and the Northeast are beginning to experience such changes for at least part of the year, with the likelihood of these changes increasing over time.

In some northern areas, warmer conditions will result in more precipitation falling as rain and less as snow. In addition, potential water resource benefits from increasing precipitation could be countered by the competing influences of increasing evaporation and runoff. In southern areas, significant reductions in precipitation are projected in winter and spring as the subtropical dry belt expands.[108] This is particularly pronounced in the Southwest, where it would have serious ramifications for water resources.

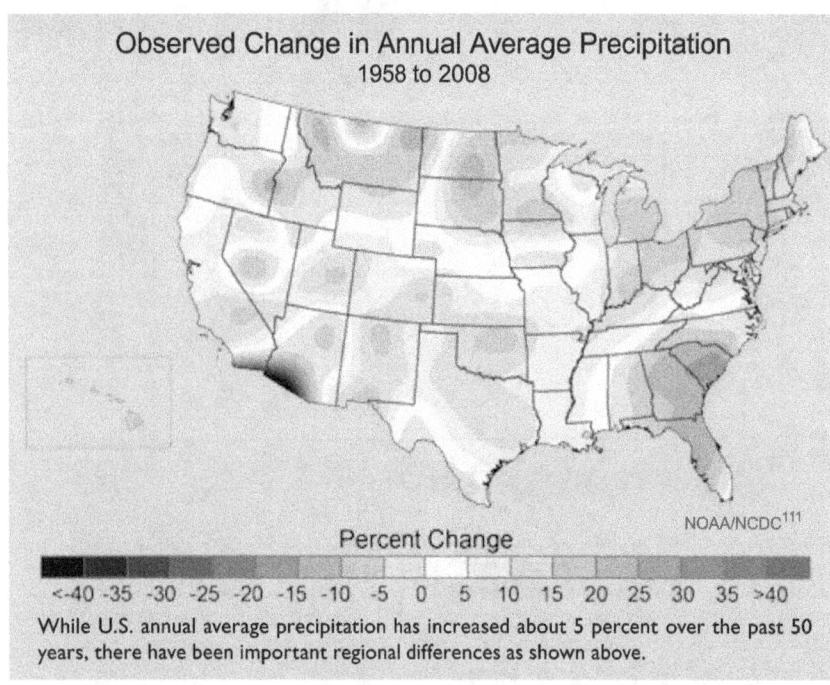

Observed Change in Annual Average Precipitation
1958 to 2008

Percent Change

| <-40 | -35 | -30 | -25 | -20 | -15 | -10 | -5 | 0 | 5 | 10 | 15 | 20 | 25 | 30 | 35 | >40 |

NOAA/NCDC[111]

While U.S. annual average precipitation has increased about 5 percent over the past 50 years, there have been important regional differences as shown above.

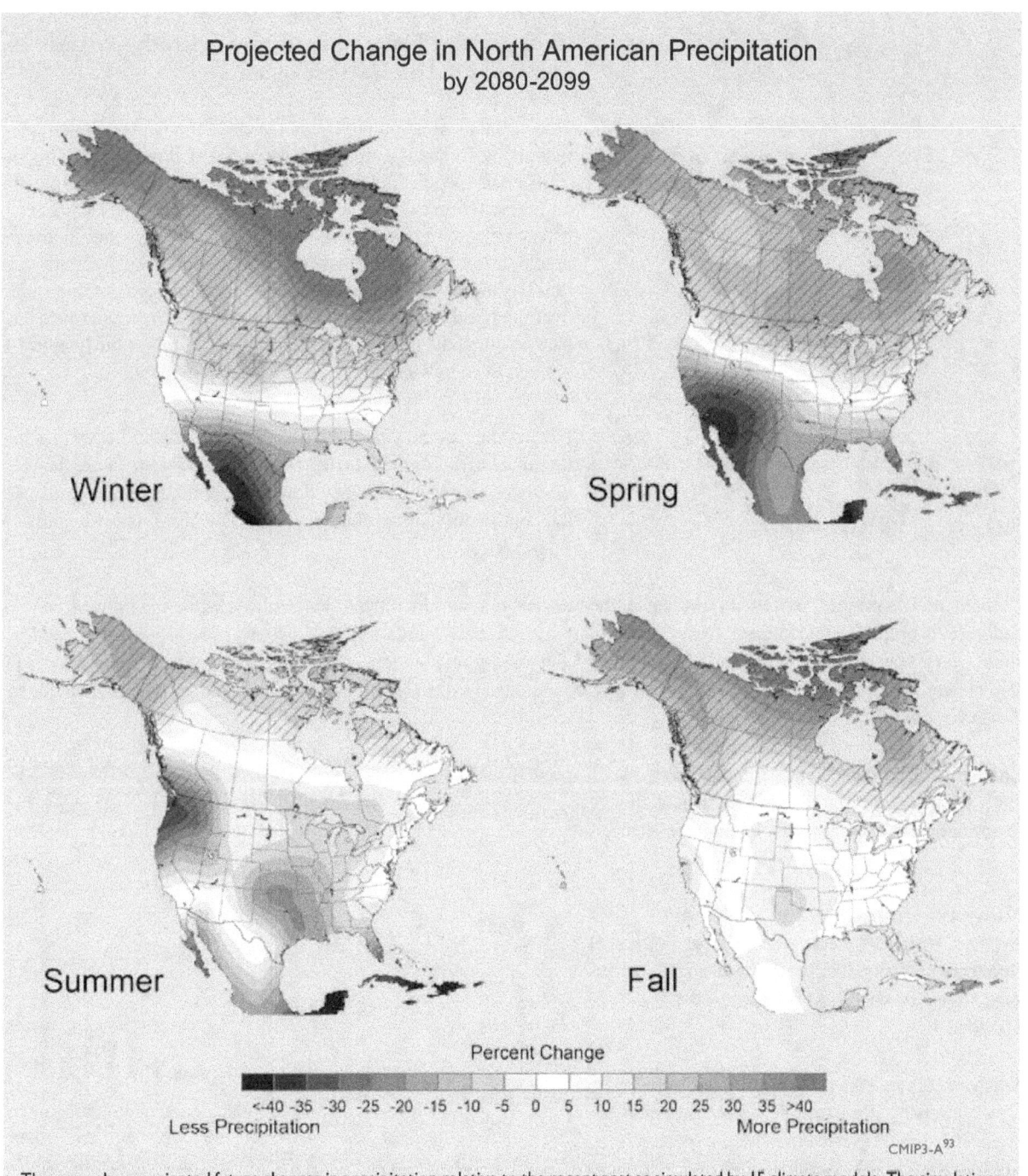

Projected Change in North American Precipitation
by 2080-2099

Winter

Spring

Summer

Fall

Percent Change

<-40 -35 -30 -25 -20 -15 -10 -5   0   5   10   15   20   25   30   35 >40

Less Precipitation

More Precipitation

CMIP3-A[93]

The maps show projected future changes in precipitation relative to the recent past as simulated by 15 climate models. The simulations are for late this century, under a higher emissions scenario.[91] For example, in the spring, climate models agree that northern areas are likely to get wetter, and southern areas drier. There is less confidence in exactly where the transition between wetter and drier areas will occur. Confidence in the projected changes is highest in the hatched areas.

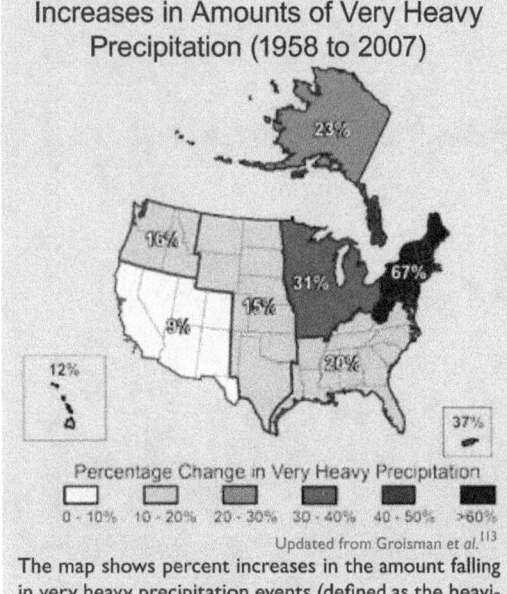

Increases in Amounts of Very Heavy Precipitation (1958 to 2007)

Percentage Change in Very Heavy Precipitation

0 - 10%   10 - 20%   20 - 30%   30 - 40%   40 - 50%   >60%

Updated from Groisman et al.[113]

The map shows percent increases in the amount falling in very heavy precipitation events (defined as the heaviest 1 percent of all daily events) from 1958 to 2007 for each region. There are clear trends toward more very heavy precipitation for the nation as a whole, and particularly in the Northeast and Midwest.

The amount of rain falling in the heaviest downpours has increased approximately 20 percent on average in the past century, and this trend is very likely to continue, with the largest increases in the wettest places.

One of the clearest precipitation trends in the United States is the increasing frequency and intensity of heavy downpours. This increase was responsible for most of the observed increase in overall precipitation during the last 50 years. In fact, there has been little change or a decrease in the frequency of light and moderate precipitation during the past 30 years, while heavy precipitation has increased. In addition, while total average precipitation over the nation as a whole increased by about 7 percent over the past century, the amount of precipitation falling in the heaviest 1 percent of rain events increased nearly 20 percent.[112]

During the past 50 years, the greatest increases in heavy precipitation occurred in the Northeast and the Midwest. There have also been increases in heavy downpours in the other regions of the continental United States, as well as Alaska, Hawaii, and Puerto Rico.[112]

Climate models project continued increases in the heaviest downpours during this century, while the lightest precipitation is projected to decrease. Heavy downpours that are now 1-in-20-year occurrences are projected to occur about every 4 to 15 years by the end of this century, depending on location, and the intensity of heavy downpours is also expected to increase. The 1-in-20-year heavy downpour is expected to be between 10 and 25 percent heavier by the end of the century than it is now.[112]

Changes in these kinds of extreme weather and climate events are among the most serious challenges to our nation in coping with a changing climate.

Many types of extreme weather events, such as heat waves and regional droughts, have become more frequent and intense during the past 40 to 50 years.

Many extremes and their associated impacts are now changing. For example, in recent decades most of North America has been experiencing more unusually hot days and nights, fewer unusually cold days and nights, and fewer frost days. Droughts are becoming more severe in some regions. The power and frequency of Atlantic hurricanes have increased substantially in recent decades. The number of North American mainland landfalling hurricanes does

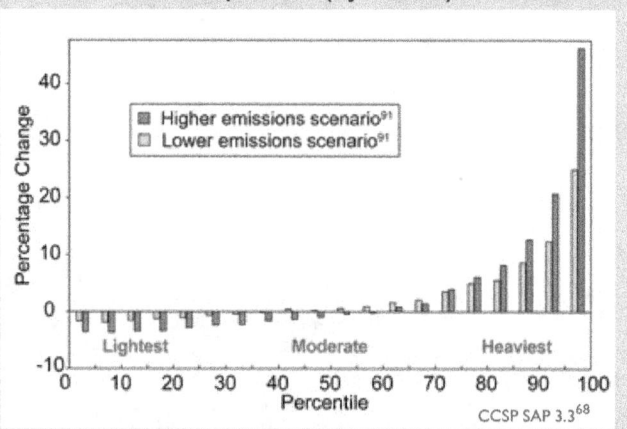

Projected Changes in Light, Moderate, and Heavy Precipitation (by 2090s)

Higher emissions scenario[91]
Lower emissions scenario[91]

Lightest   Moderate   Heaviest

CCSP SAP 3.3[68]

The figure shows projected changes from the 1990s average to the 2090s average in the amount of precipitation falling in light, moderate, and heavy events in North America. Projected changes are displayed in 5 percent increments from the lightest drizzles to the heaviest downpours. As shown here, the lightest precipitation is projected to decrease, while the heaviest will increase, continuing the observed trend. The higher emission scenario[91] yields larger changes. Projections are based on the models used in the IPCC 2007 Fourth Assessment Report.

not appear to have increased over the past century. Outside the tropics, cold-season storm tracks are shifting northward and the strongest storms are becoming even stronger. These trends in storms outside the tropics are projected to continue throughout this century.[68,112,114]

Drought

Like precipitation, trends in drought have strong regional variations. In much of the Southeast and large parts of the West, the frequency of drought has increased coincident with rising temperatures over the past 50 years. In other regions, such as the Midwest and Great Plains, there has been a reduction in drought frequency.

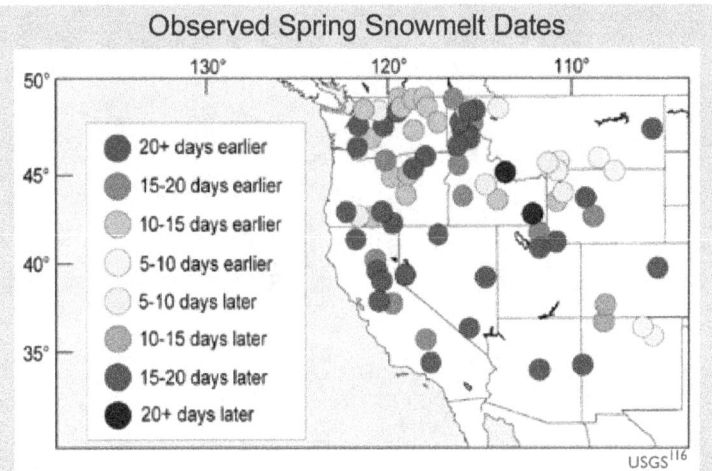

Observed Spring Snowmelt Dates

- 20+ days earlier
- 15-20 days earlier
- 10-15 days earlier
- 5-10 days earlier
- 5-10 days later
- 10-15 days later
- 15-20 days later
- 20+ days later

USGS[116]

Date of onset of spring runoff pulse. Reddish-brown circles indicate significant trends toward onsets more than 20 days earlier. Lighter circles indicate less advance of the onset. Blue circles indicate later onset. The changes depend on a number of factors in addition to temperature, including altitude and timing of snowfall.

Although there has been an overall increase in precipitation and no clear trend in drought for the nation as a whole, increasing temperatures have made droughts more severe and widespread than they would have otherwise been. Without the observed increase in precipitation, higher temperatures would have led to an increase in the area of the contiguous United States in severe to extreme drought, with some estimates of a 30 percent increase.[112] In the future, droughts are likely to become more frequent and severe in some regions.[68] The Southwest, in particular, is expected to experience increasing drought as changes in atmospheric circulation patterns cause the dry zone just outside the tropics to expand farther northward into the United States.[97]

Rising temperatures have also led to earlier melting of the snowpack in the western United States.[40] Because snowpack runoff is critical to the water resources in the western United States, changes in the timing and amount of runoff can exacerbate problems with already limited water supplies in the region.

Heat waves

A heat wave is a period of several days to weeks of abnormally hot weather, often with high humidity. During the 1930s, there was a high frequency of heat waves due to high daytime temperatures resulting in large part from an extended multi-year period of intense drought. By contrast, in the past

3 to 4 decades, there has been an increasing trend in high-humidity heat waves, which are characterized by the persistence of extremely high nighttime temperatures.[112]

As average temperatures continue to rise throughout this century, the frequency of cold extremes will decrease and the frequency and intensity of high temperature extremes will increase.[115] The number of days with high temperatures above

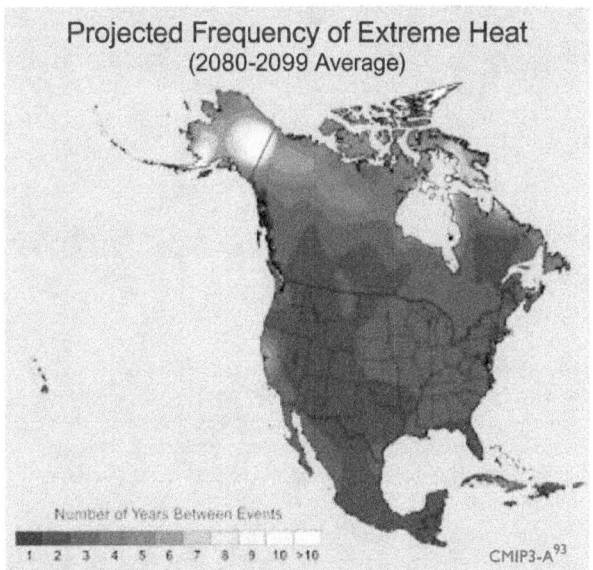

Projected Frequency of Extreme Heat
(2080-2099 Average)

Number of Years Between Events

1 2 3 4 5 6 7 8 9 10 >10

CMIP3-A[93]

Simulations for 2080-2099 indicate how currently rare extremes (a 1-in-20-year event) are projected to become more commonplace. A day so hot that it is currently experienced once every 20 years would occur every other year or more frequently by the end of the century under the higher emissions scenario.[91]

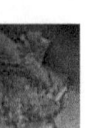

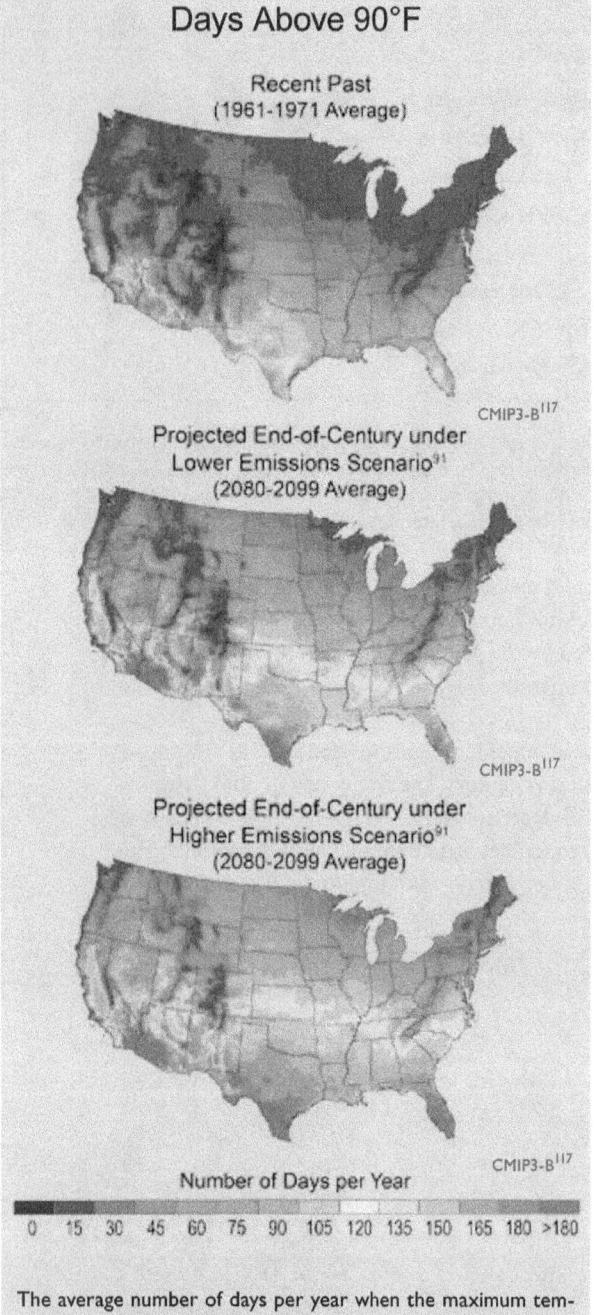

Days Above 90°F

Recent Past
(1961-1971 Average)

CMIP3-B[117]

Projected End-of-Century under Lower Emissions Scenario[91]
(2080-2099 Average)

CMIP3-B[117]

Projected End-of-Century under Higher Emissions Scenario[91]
(2080-2099 Average)

Number of Days per Year   CMIP3-B[117]

0 15 30 45 60 75 90 105 120 135 150 165 180 >180

The average number of days per year when the maximum temperature exceeded 90°F from 1961-1979 (top) and the projected number of days per year above 90°F by the 2080s and 2090s for lower emissions (middle) and higher emissions (bottom).[91] Much of the southern United States is projected to have more than twice as many days per year above 90°F by the end of this century.

90°F is projected to increase throughout the country as illustrated in the maps on the left. Parts of the South that currently have about 60 days per year with temperatures over 90°F are projected to experience 150 or more days a year above 90°F by the end of this century, under a higher emissions scenario.[91] There is higher confidence in the regional patterns than in results for any specific location (see *An Agenda for Climate Impacts Science* section).

With rising high temperatures, extreme heat waves that are currently considered rare will occur more frequently in the future. Recent studies using an ensemble of models show that events that now occur once every 20 years are projected to occur about every other year in much of the country by the end of this century. In addition to occurring more frequently, at the end of this century these very hot days are projected to be about 10°F hotter than they are today.[68]

The destructive energy of Atlantic hurricanes has increased in recent decades. The intensity of these storms is likely to increase in this century.

Of all the world's tropical storm and hurricane basins, the North Atlantic has been the most thoroughly monitored and studied. The advent of routine aircraft monitoring in the 1940s and the use of satellite observations since the 1960s have greatly aided monitoring of tropical storms and hurricanes. In addition, observations of tropical storm and hurricane strength made from island and mainland weather stations and from ships at sea began in the 1800s and continue today. Because of new and evolving observing techniques and technologies, scientists pay careful attention to ensuring consistency in tropical storm and hurricane records from the earliest manual observations to today's automated measurements. This is accomplished through collection, analysis, and cross-referencing of data from numerous sources and, where necessary, the application of adjustment techniques to account for differences in observing and reporting methodologies through time. Nevertheless, data uncertainty is larger in the early part of the record. Confidence in the tropical storm and hurricane record increases after 1900 and is greatest during the satellite era, from 1965 to the present.[112]

The total number of hurricanes and strongest hurricanes (Category 4 and 5) observed from 1881 through 2008 shows multi-decade periods of above average activity in the 1800s, the mid-1900s, and since 1995. The power and frequency of Atlantic hurricanes have increased substantially in recent decades.[112] There has been little change in the total number of landfalling hurricanes, in part because a variety of factors affect whether a hurricane will make landfall. These include large-scale steering winds, atmospheric stability, wind shear, and ocean heat content. This highlights the importance of understanding the broader changes occurring throughout the Atlantic Basin beyond the storms making landfall along the U.S. coast.[112]

Tropical storms and hurricanes develop and gain strength over warm ocean waters. As oceans warm, they provide a source of energy for hurricane growth. During the past 30 years, annual sea surface temperatures in the main Atlantic hurricane development region increased nearly 2°F. This warming coincided with an increase in the destructive energy (as defined by the Power Dissipation Index, a combination of intensity, duration, and frequency) of Atlantic tropical storms and hurricanes. The strongest hurricanes (Category 4 and 5) have, in particular, increased in intensity.[112] The graph below shows the strong correlation between hurricane power and sea surface temperature in the Atlantic and the overall increase in both during the past 30 years. Climate models project that hurricane intensity will continue to increase, though at a lesser rate than that observed in recent decades.[100]

New evidence has emerged recently for other temperature related linkages that can help

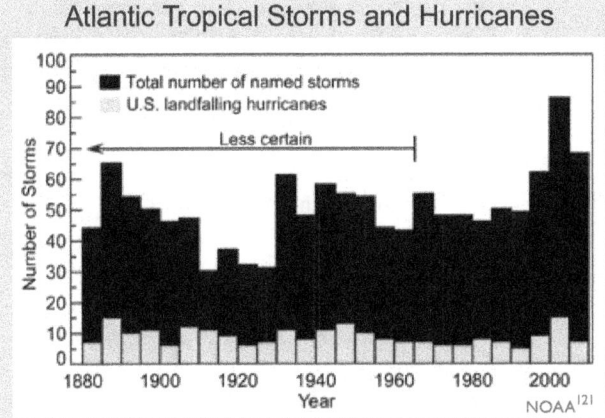

Top: Total numbers of North Atlantic named storms (tropical storms and hurricanes) (black) and total U.S. landfalling hurricanes (yellow) in 5-year periods based on annual data from 1881 to 2008. The bar for the last 5-year period is based on the assumption that the level of activity from 2006 to 2008 persists through 2010. In the era before satellites, indicated by the arrow above, the total number of named storms is less certain and has been adjusted upward to account for missing storms. Adjustments are based on relationships established during the satellite era between the number of observed storms and the number that would have been missed if satellite data had not been available.
Bottom: Total number of strongest (Category 4 and 5) North Atlantic basin hurricanes (purple) and strongest U.S. landfalling hurricanes (orange) in 5-year periods based on annual data from 1946 to 2008. The bar for the last 5-year period is based on the assumption that the level of activity from 2006 to 2008 persists through 2010. From 1946 to the mid-1960s, as indicated by the arrow above, hurricane intensity was measured primarily by aircraft reconnaissance. Data prior to aircraft reconnaissance are not shown due to the greater uncertainty in estimates of a hurricane's maximum intensity. Satellites have increased the reliability of hurricane intensity estimates since the mid-1960s.

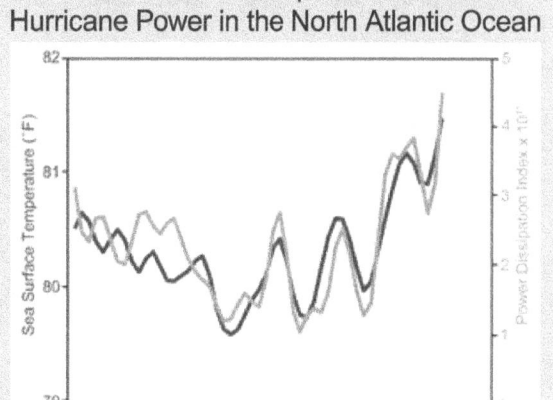

Observed sea surface temperature (blue) and the Power Dissipation Index (green), which combines frequency, intensity and duration for North Atlantic hurricanes.[120] Hurricane rainfall and wind speeds are likely to increase in response to human-caused warming. Analyses of model simulations suggest that for each 1.8°F increase in tropical sea surface temperatures, rainfall rates will increase by 6 to 18 percent.[68]

explain the increase in Atlantic hurricane activity. This includes the contrast in sea surface temperature between the main hurricane development region and the broader tropical ocean.[99,118,119] Other causes beyond the rise in ocean temperature, such as atmospheric stability and circulation, can also influence hurricane power. For these and other reasons, a confident assessment requires further study.[68]

Evidence of increasing hurricane strength in the Atlantic and other oceans with linkages to rising sea surface temperatures is also supported by satellite records dating back to 1981. An increase in the maximum wind speeds of the strongest hurricanes has been documented and linked to increasing sea surface temperatures.[122]

Projections are that sea surface temperatures in the main Atlantic hurricane development region will increase at even faster rates during the second half of this century under higher emissions scenarios. This highlights the need to better understand the relationship between increasing temperatures and hurricane intensity. As ocean temperatures continue to increase in the future, it is likely that hurricane rainfall and wind speeds will increase in response to human-caused warming.[68] Analyses

of model simulations suggest that for each 1.8°F increase in tropical sea surface temperatures, core rainfall rates will increase by 6 to 18 percent and the surface wind speeds of the strongest hurricanes will increase by about 1 to 8 percent.[114] Even without further coastal development, storm surge levels and hurricane damages are likely to increase because of increasing hurricane intensity coupled with sea-level rise, the latter being a virtually certain outcome of the warming global climate.[68]

In the eastern Pacific, the strongest hurricanes have become stronger since the 1980s, even while the total number of storms has decreased.

Although on average more hurricanes form in the eastern Pacific than the Atlantic each year, cool ocean waters along the U.S. West Coast and atmospheric steering patterns help protect the contiguous U.S. from landfalls. Threats to the Hawaiian Islands are greater, but landfalling storms are rare in comparison to those of the U.S. East and Gulf Coasts. Nevertheless, changes in hurricane intensity and frequency could influence the impact of landfalling Pacific hurricanes in the future.

The total number of tropical storms and hurricanes in the eastern Pacific on seasonal to multi-decade time periods is generally opposite to that observed in the Atlantic. For example, during El Niño events it is common for hurricanes in the Atlantic to be suppressed while the eastern Pacific is more active. This reflects the large-scale atmospheric circulation patterns that extend across both the Atlantic and the Pacific oceans.[123,124]

Within the past three decades the total number of tropical storms and hurricanes and their destructive energy have decreased in the eastern Pacific.[68,124] However, satellite observations have shown that like the Atlantic, the strongest hurricanes (the top 5 percent), have gotten stronger since the early 1980s.[122,125] As ocean temperatures rise, the strongest hurricanes are likely to increase in both the eastern Pacific and the Atlantic.[68]

Observed and Projected Sea Surface Temperature Change
Atlantic Hurricane Formation Region

Observed (black) and projected temperatures (blue = lower scenario; red = higher scenario) in the Atlantic hurricane formation region. Increased intensity of hurricanes is linked to rising sea surface temperatures in the region of the ocean where hurricanes form. The shaded areas show the likely ranges while the lines show the central projections from a set of climate models.

Sea level has risen along most of the U.S. coast over the past 50 years, and will rise more in the future.

Recent global sea-level rise has been caused by the warming-induced expansion of the oceans, accelerated melting of most of the world's glaciers, and loss of ice on the Greenland and Antarctic ice sheets.[37] There is strong evidence that global sea level is currently rising at an increased rate.[37,126] A warming global climate will cause further sea-level rise over this century and beyond.[90,105]

During the past 50 years, sea level has risen up to 8 inches or more along some coastal areas of the United States, and has fallen in other locations. The amount of relative sea-level rise experienced along different parts of the U.S. coast depends on the changes in elevation of the land that occur as a result of subsidence (sinking) or uplift (rising), as well as increases in global sea level due to warming. In addition, atmospheric and oceanic circulation, which will be affected by climate change, will influence regional sea level. Regional differences

in sea-level rise are also expected to be related to where the meltwater originates.[104]

Human-induced sea-level rise is occurring globally. Large parts of the Atlantic Coast and Gulf of Mexico Coast have experienced significantly higher rates of relative sea-level rise than the global average during the last 50 years, with the local differences mainly due to land subsidence.[127] Portions of the Northwest and Alaska coast have, on the other hand, experienced slightly falling sea level as a result of long-term uplift as a consequence of glacier melting and other geological processes.

Regional variations in relative sea-level rise are expected in the future. For example, assuming historical geological forces continue, a 2-foot rise in global sea level (which is within the range of recent estimates) by the end of this century would result in a relative sea-level rise of 2.3 feet at New York City, 2.9 feet at Hampton Roads, Virginia, 3.5 feet at Galveston, Texas, and 1 foot at Neah Bay in Washington state.[128]

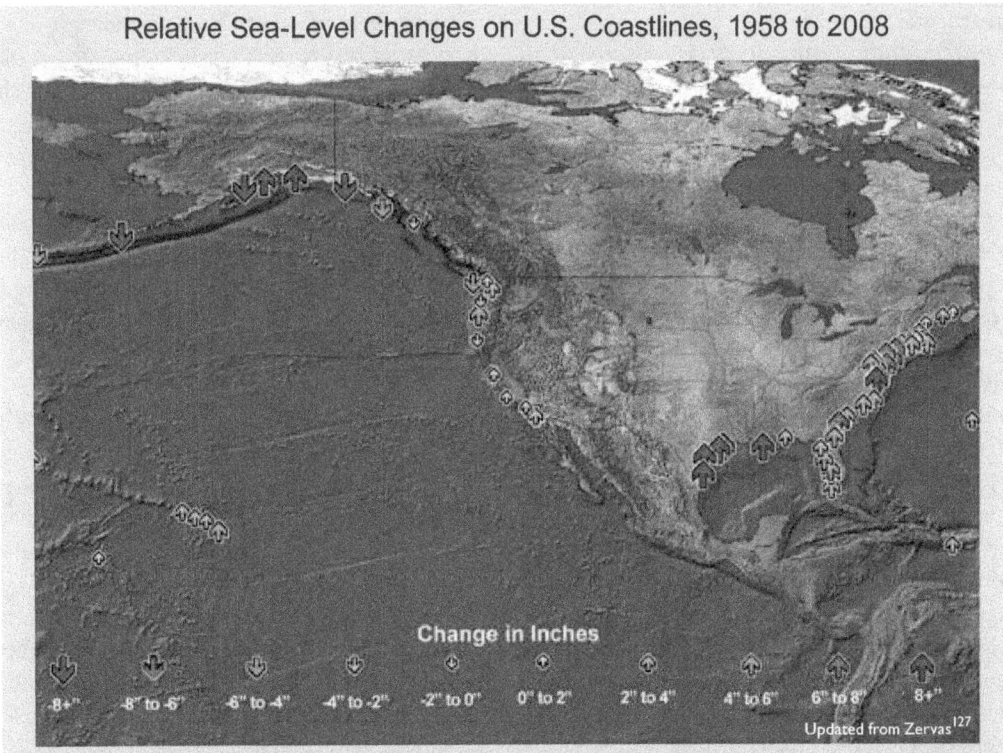

Relative Sea-Level Changes on U.S. Coastlines, 1958 to 2008

Change in Inches

-8+" -8" to -6" -6" to -4" -4" to -2" -2" to 0" 0" to 2" 2" to 4" 4" to 6" 6" to 8" 8+"

Updated from Zervas[127]

Observed changes in relative sea level from 1958 to 2008 for locations on the U.S. coast. Some areas along the Atlantic and Gulf coasts saw increases greater than 8 inches over the past 50 years.

Cold-season storm tracks are shifting northward and the strongest storms are likely to become stronger and more frequent.

Large-scale storm systems are the dominant weather phenomenon during the cold season in the United States. Although the analysis of these storms is complicated by a relatively short length of most observational records and by the highly variable nature of strong storms, some clear patterns have emerged.[112]

Storm tracks have shifted northward over the last 50 years as evidenced by a decrease in the frequency of storms in mid-latitude areas of the Northern Hemisphere, while high-latitude activity has increased. There is also evidence of an increase in the intensity of storms in both the mid- and high-latitude areas of the Northern Hemisphere, with greater confidence in the increases occurring in high latitudes.[112] The northward shift is projected to continue, and strong cold season storms are likely to become stronger and more frequent, with greater wind speeds and more extreme wave heights.[68]

Snowstorms

The northward shift in storm tracks is reflected in regional changes in the frequency of snowstorms. The South and lower Midwest saw reduced snowstorm frequency during the last century. In contrast, the Northeast and upper Midwest saw increases in snowstorms, although considerable decade-to-decade variations were present in all regions, influenced, for example, by the frequency of El Niño events.[112]

There is also evidence of an increase in lake-effect snowfall along and near the southern and eastern shores of the Great Lakes since 1950.[97] Lake-effect snow is produced by the strong flow of cold air across large areas of relatively warmer ice-free water. As the climate has warmed, ice coverage on the Great Lakes has fallen. The maximum seasonal coverage of Great Lakes ice decreased at a rate of 8.4 percent per decade from 1973 through 2008, amounting to a roughly 30 percent decrease in ice coverage (see *Midwest* region). This has created conditions conducive to greater evaporation of

Areas in New York state east of Lake Ontario received over 10 feet of lake-effect snow during a 10-day period in early February 2007.

moisture and thus heavier snowstorms. Among recent extreme lake-effect snow events was a February 2007 10-day storm total of over 10 feet of snow in western New York state. Climate models suggest that lake-effect snowfalls are likely to increase over the next few decades.[130] In the longer term, lake-effect snows are likely to decrease as temperatures continue to rise, with the precipitation then falling as rain.[129]

Tornadoes and severe thunderstorms

Reports of severe weather including tornadoes and severe thunderstorms have increased during the past 50 years. However, the increase in the number of reports is widely believed to be due to improvements in monitoring technologies such as Doppler radars combined with changes in population and increasing public awareness. When adjusted to account for these factors, there is no clear trend in the frequency or strength of tornadoes since the 1950s for the United States as a whole.[112]

The distribution by intensity for the strongest 10 percent of hail and wind reports is little changed, providing no evidence of an observed increase in the severity of events.[112] Climate models project future increases in the frequency of environmental conditions favorable to severe thunderstorms.[131] But the inability to adequately model the small-scale conditions involved in thunderstorm development remains a limiting factor in projecting the future character of severe thunderstorms and other small-scale weather phenomena.[68]

Arctic sea ice is declining rapidly and this is very likely to continue.

Sea ice is a very important part of the climate system. In addition to direct impacts on coastal areas of Alaska, it more broadly affects surface reflectivity, ocean currents, cloudiness, humidity, and the exchange of heat and moisture at the ocean's surface. Open ocean water is darker in color than sea ice, which causes it to absorb more of the Sun's heat, which increases the warming of the water even more.[40,132]

The most complete record of sea ice is provided by satellite observations of sea ice extent since the 1970s. Prior to that, aircraft, ship, and coastal observations in the Arctic make it possible to extend the record of Northern Hemisphere sea ice extent back to at least 1900, although there is a lower level of confidence in the data prior to 1953.[40]

Arctic sea ice extent has fallen at a rate of 3 to 4 percent per decade over the last three decades. End-of-summer Arctic sea ice has fallen at an even faster rate of more than 11 percent per decade in that time. The observed decline in Arctic sea ice has been more rapid than projected by climate models.[133] Year-to-year changes in sea ice extent and record low amounts are influenced by natural variations in atmospheric pressure and wind patterns.[134] However, clear linkages between rising greenhouse gas concentrations and declines in Arctic sea ice have been identified in the climate record as far back as the early 1990s.[61] The extreme loss in Arctic sea ice that occurred in 2007 would not have been possible without the long-term reductions that have coincided with a sustained increase in the atmospheric concentration of carbon dioxide and the rapid rise in global temperatures that have occurred since the mid-1970s.[135] Although the 2007 record low was not eclipsed in 2008, the 2008 sea ice extent is well below the long-term average, reflecting a continuation of the long-term decline in Arctic sea ice. In addition, the total volume of Arctic sea ice in 2008 was likely a record low because the ice was unusually thin.[136]

It is expected that declines in Arctic sea ice will continue in the coming decades with year-to-year fluctuations influenced by natural atmospheric variability. The overall rate of decline will be influenced mainly by the rate at which carbon dioxide and other greenhouse gas concentrations increase.[137]

Arctic Sea Ice
Annual Minimum
1979

NASA/GSFC[138]

2007

NASA/GSFC[138]

Arctic sea ice reaches its annual minimum in September. The satellite images above show September Arctic sea ice in 1979, the first year these data were available, and 2007.

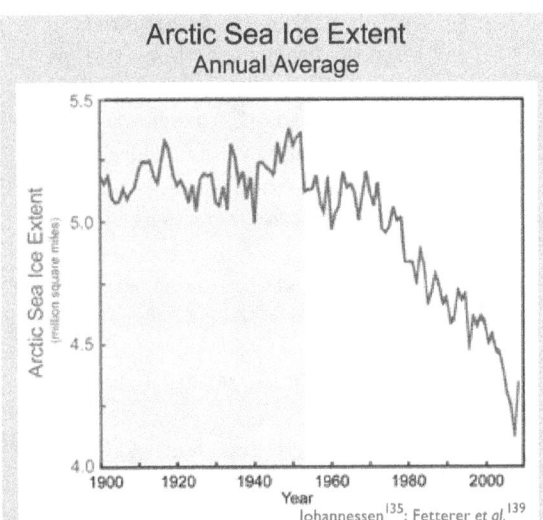

Arctic Sea Ice Extent
Annual Average

Johannessen[135]; Fetterer et al.[139]

Observations of annual average Arctic sea ice extent for the period 1900 to 2008. The gray shading indicates less confidence in the data before 1953.

U.S. Emission and Absorption of Heat-Trapping Gases

Since the industrial revolution, the United States has been the world's largest emitter of heat-trapping gases. With 4.5 percent of world's population, the United States is responsible for about 28 percent of the human-induced heat-trapping gases in the atmosphere today.[136] Although China has recently surpassed the United States in current total annual emissions, per capita emissions remain much higher in the United States. Carbon dioxide, the most important of the heat-trapping gases produced directly by human activities, is a cumulative problem because it has a long atmospheric lifetime. Roughly one-half of the carbon dioxide released from fossil fuel burning remains in the atmosphere after 100 years, and roughly one-fifth of it remains after 1,000 years.[90]

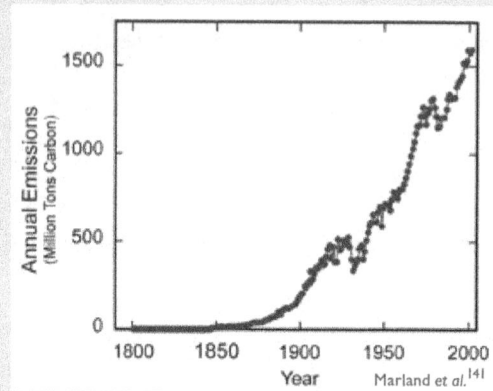

U.S. annual emissions of CO_2 from fossil-fuel use.[141]

U.S. carbon dioxide emissions grew dramatically over the past century. These emissions come almost entirely from burning fossil fuels. These sources of carbon dioxide are one side of the equation and on the other side are "sinks" that take up carbon dioxide. The growth of trees and other plants is an important natural carbon sink. In recent years, it is estimated that about 20 percent of U.S. carbon dioxide emissions have been offset by U.S. forest growth and other sinks (see figure below).[140] It is not known whether U.S. forests and other sinks will continue to take up roughly this amount of carbon dioxide in the future as climate change alters carbon release and uptake. For example, a warming-induced lengthening of the growing season would tend to increase carbon uptake. On the other hand, the increases in forest fires and in the decomposition rate of dead plant matter would decrease uptake, and might convert the carbon sink into a source.[140]

The amount of carbon released and taken up by natural sources varies considerably from year to year depending on climatic and other conditions. For example, fires release carbon dioxide, so years with many large fires result in more carbon release and less uptake as natural sinks (the vegetation) are lost. Similarly,

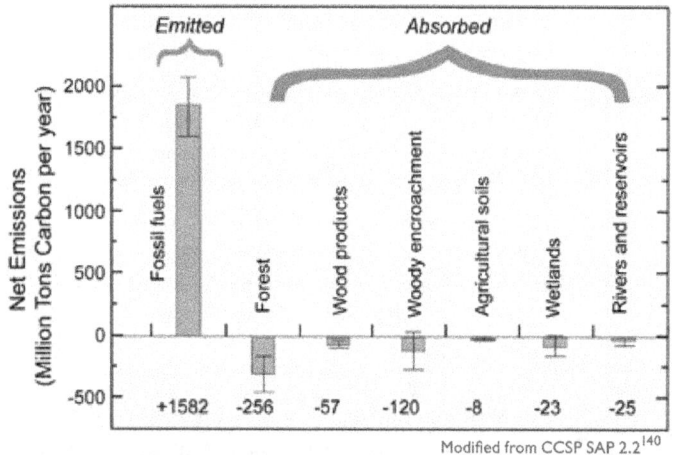

U.S. carbon dioxide emissions and uptake in millions of tons of carbon per year in 2003. The bar marked "Emitted" indicates the amount of carbon as carbon dioxide added to the atmosphere from U.S. emissions. The bars marked "Absorbed" indicate amounts of carbon as carbon dioxide removed from the atmosphere. The thin lines on each bar indicate estimates of uncertainty.

the trees destroyed by intense storms or droughts release carbon dioxide as they decompose, and the loss results in reduced strength of natural sinks until regrowth is well underway. For example, Hurricane Katrina killed or severely damaged over 320 million large trees. As these trees decompose over the next few years, they will release an amount of carbon dioxide equivalent to that taken up by all U.S. forests in a year.[112] The net change in carbon storage in the long run will depend on how much is taken up by the regrowth as well as how much was released by the original disturbance.

Water Resources

Key Sources

CCSP 3.3	CCSP 3.4	CCSP 4.3	CCSP 4.5	CCSP 4.6	CCSP 4.7
Extremes	Abrupt Climate Change	Impacts	Energy	Health	Transportation

CCSP 5.1	CCSP 5.2	IPCC	IPCC	IPCC
Data Uses & Limitations	Uncertainty	WG-1	WG-2	Water

Key Messages:

- Climate change has already altered, and will continue to alter, the water cycle, affecting where, when, and how much water is available for all uses.
- Floods and droughts are likely to become more common and more intense as regional and seasonal precipitation patterns change, and rainfall becomes more concentrated into heavy events (with longer, hotter dry periods in between).
- Precipitation and runoff are likely to increase in the Northeast and Midwest in winter and spring, and decrease in the West, especially the Southwest, in spring and summer.
- In areas where snowpack dominates, the timing of runoff will continue to shift to earlier in the spring and flows will be lower in late summer.
- Surface water quality and groundwater quantity will be affected by a changing climate.
- Climate change will place additional burdens on already stressed water systems.
- The past century is no longer a reasonable guide to the future for water management.

Changes in the water cycle, which are consistent with the warming observed over the past several decades, include:

- changes in precipitation patterns and intensity
- changes in the incidence of drought
- widespread melting of snow and ice
- increasing atmospheric water vapor
- increasing evaporation
- increasing water temperatures
- reductions in lake and river ice
- changes in soil moisture and runoff

For the future, marked regional differences are projected, with increases in annual precipitation, runoff, and soil moisture in much of the Midwest and Northeast, and declines in much of the West, especially the Southwest.

Skagit River and surrounding mountains in the Northwest

The impacts of climate change include too little water in some places, too much water in other places, and degraded water quality. Some locations are expected to be subject to all of these conditions during different times of the year. Water cycle changes are expected to continue and to adversely affect energy production and use, human health, transportation, agriculture, and ecosystems (see table on page 50).[142]

Climate change has already altered, and will continue to alter, the water cycle, affecting where, when, and how much water is available for all uses.

Substantial changes to the water cycle are expected as the planet warms because the movement of water in the atmosphere and oceans is one of the primary mechanisms for the redistribution of heat around the world. Evidence is mounting that human-induced climate change is already altering many of the existing patterns of precipitation in the United States, including when, where, how much, and what kind of precipitation falls.[68,142] A warmer climate increases evaporation of water from land and sea, and allows more moisture to be held in the atmosphere. For every 1°F rise in temperature, the water holding capacity of the atmosphere increases by about 4 percent.[49]

Projected Changes in the Water Cycle

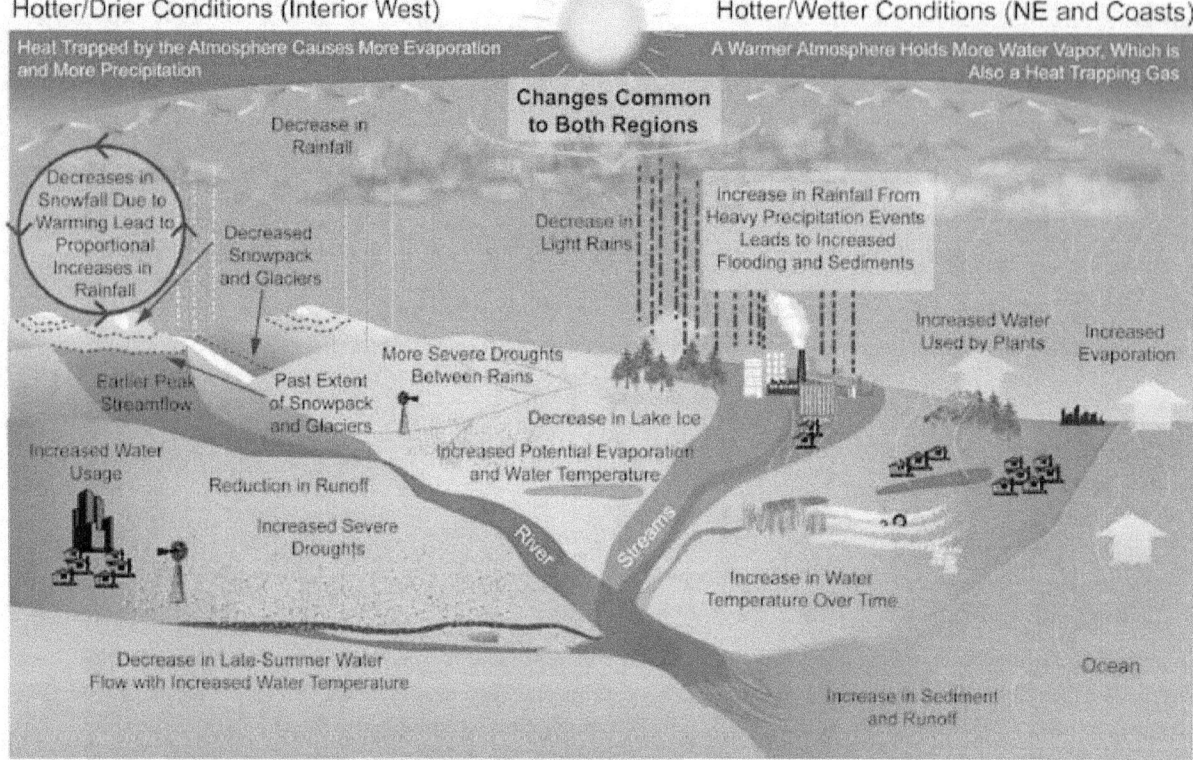

The water cycle exhibits many changes as the Earth warms. Wet and dry areas respond differently.

NOAA/NCDC

In addition, changes in atmospheric circulation will tend to move storm tracks northward with the result that dry areas will become drier and wet areas wetter. Hence, the arid Southwest is projected to experience longer and more severe droughts from the combination of increased evaporation and reductions in precipitation.[108]

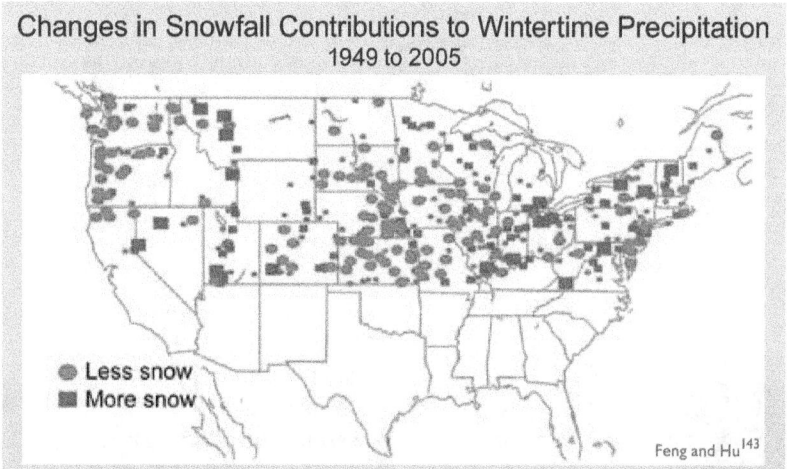

Changes in Snowfall Contributions to Wintertime Precipitation
1949 to 2005

● Less snow
■ More snow

Feng and Hu[143]

Trends in winter snow-to-total precipitation ratio from 1949 to 2005. Red circles indicate less snow, while blue squares indicate more snow. Large circles and squares indicate the most significant trends.[143] Areas south of 37°N latitude were excluded from the analysis because most of that area receives little snowfall. White areas above that line have inadequate data for this analysis.

The additional atmospheric moisture contributes to more overall precipitation in some areas, especially in much of the Northeast, Midwest, and Alaska. Over the past 50 years, precipitation and streamflow have increased in much of the Northeast and Midwest, with a reduction in drought duration and severity. Much of the Southeast and West has had reductions in precipitation and increases in drought severity and duration, especially in the Southwest.

In most areas of the country, the fraction of precipitation falling as rain versus snow has increased during the last 50 years. Despite this general shift from snow to rain, snowfalls

Observed Water-Related Changes During the Last Century[142]		
Observed Change	**Direction of Change**	**Region Affected**
One to four week earlier peak streamflow due to earlier warming-driven snowmelt	Earlier	West and Northeast
Proportion of precipitation falling as snow	Decreasing	West and Northeast
Duration and extent of snow cover	Decreasing	Most of the United States
Mountain snow water equivalent	Decreasing	West
Annual precipitation	Increasing	Most of the United States
Annual precipitation	Decreasing	Southwest
Frequency of heavy precipitation events	Increasing	Most of the United States
Runoff and streamflow	Decreasing	Colorado and Columbia River Basins
Streamflow	Increasing	Most of East
Amount of ice in mountain glaciers	Decreasing	U.S. western mountains, Alaska
Water temperature of lakes and streams	Increasing	Most of the United States
Ice cover on lakes and rivers	Decreasing	Great Lakes and Northeast
Periods of drought	Increasing	Parts of West and East
Salinization of surface waters	Increasing	Florida, Louisiana
Widespread thawing of permafrost	Increasing	Alaska

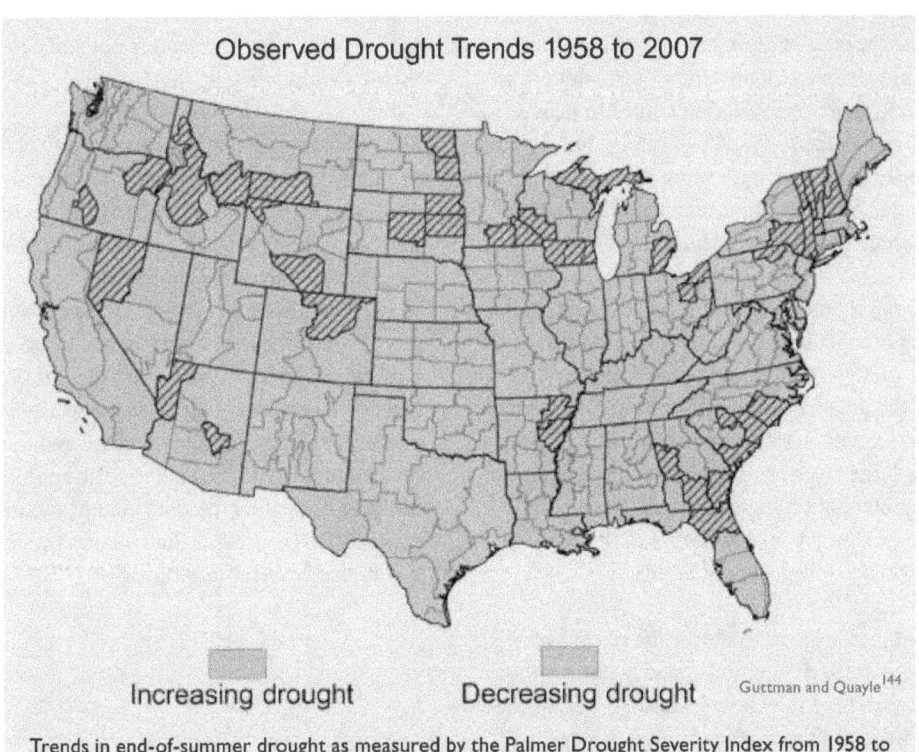

Observed Drought Trends 1958 to 2007

Increasing drought   Decreasing drought   Guttman and Quayle[144]

Trends in end-of-summer drought as measured by the Palmer Drought Severity Index from 1958 to 2007 in each of 344 U.S. climate divisions.[144] Hatching indicates significant trends.

along the downwind coasts of the Great Lakes have increased. Factors contributing to this increase include reduced ice cover due to warming, which lengthens the period of open water. In addition, cold air moving over relatively warm, open lake water induces strong evaporation, often causing heavy lake-effect snow. Heavy snowfall and snowstorm frequency have increased in many northern parts of the United States. In the South however, where temperatures are already marginal for heavy snowfall, climate warming has led to a reduction in heavy snowfall and snowstorm frequency. These trends suggest a northward shift in snowstorm occurrence.[68]

Floods and droughts are likely to become more common and more intense as regional and seasonal precipitation patterns change, and rainfall becomes more concentrated into heavy events (with longer, hotter dry periods in between).

While it sounds counterintuitive, a warmer world produces both wetter and drier conditions. Even though total global precipitation increases, the regional and seasonal distribution of precipitation changes, and more precipitation comes in heavier rains (which can cause flooding) rather than light events. In the past century, averaged over the United States, total precipitation has increased by about 7 percent, while the heaviest 1 percent of rain events increased by nearly 20 percent.[68] This has been especially noteworthy in the Northeast, where the annual number of days with very heavy precipitation has increased most in the past 50 years, as shown in the adjacent figure. Flooding often occurs when heavy precipitation persists for weeks to months in large river basins. Such extended periods of heavy precipitation have also been increasing over the past century, most notably in the past two to three decades in the United States.[112]

Observations also show that over the past several decades, extended dry periods have become more frequent in parts of the United States, especially the Southwest and the eastern United States.[146,147] Longer periods between rainfalls, combined with

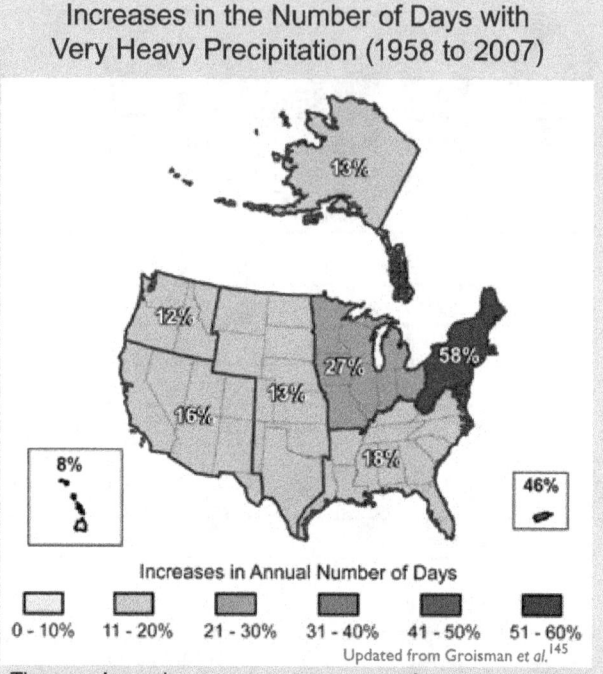

Increases in the Number of Days with Very Heavy Precipitation (1958 to 2007)

Increases in Annual Number of Days

| 0 - 10% | 11 - 20% | 21 - 30% | 31 - 40% | 41 - 50% | 51 - 60% |

Updated from Groisman et al.[145]

The map shows the percentage increases in the average number of days with very heavy precipitation (defined as the heaviest 1 percent of all events) from 1958 to 2007 for each region. There are clear trends toward more days with very heavy precipitation for the nation as a whole, and particularly in the Northeast and Midwest.

higher air temperatures, dry out soils and vegetation, causing drought.

For the future, precipitation intensity is projected to increase everywhere, with the largest increases occurring in areas in which average precipitation increases the most. For example, the Midwest and Northeast, where total precipitation is expected to increase the most, would also experience the largest increases in heavy precipitation events. The number of dry days between precipitation events is also projected to increase, especially in the more arid areas. Mid-continental areas and the Southwest are particularly threatened by future drought. The magnitude of the projected changes in extremes is expected to be greater than changes in averages, and hence detectable sooner.[49,68,90,142,148]

Precipitation and runoff are likely to increase in the Northeast and Midwest in winter and spring, and decrease in the West, especially the Southwest, in spring and summer.

Runoff, which accumulates as streamflow, is the amount of precipitation that is not evaporated, stored as snowpack or soil moisture, or filtered down to groundwater. The proportion of precipitation that runs off is determined by a variety of factors including temperature, wind speed, humidity, solar intensity at the ground, vegetation, and soil moisture. While runoff generally tracks precipitation, increases and decreases in precipitation do not necessarily lead to equal increases and decreases in runoff. For example, droughts cause soil moisture reductions that can reduce expected runoff until soil moisture is replenished. Conversely, water-saturated soils can generate floods with only moderate additional precipitation. During the last century, consistent increases in precipitation have been found in the Midwest and Northeast along with increased runoff.[149,150] Climate models consistently project that the East will experience increased runoff, while there will be substantial declines in the interior West, especially the Southwest. Projections for runoff in California and other parts of the West also show reductions, although less than in the interior West. In short, wet areas are projected to get wetter and dry areas drier. Climate models also consistently project heat-related summer soil moisture reductions in the middle of the continent.[115,142,146,149]

In areas where snowpack dominates, the timing of runoff will continue to shift to earlier in the spring and flows will be lower in late summer.

Large portions of the West and some areas in the Northeast rely on snowpack as a natural reservoir to hold winter precipitation until it later runs off as streamflow in spring, summer, and fall. Over the last 50

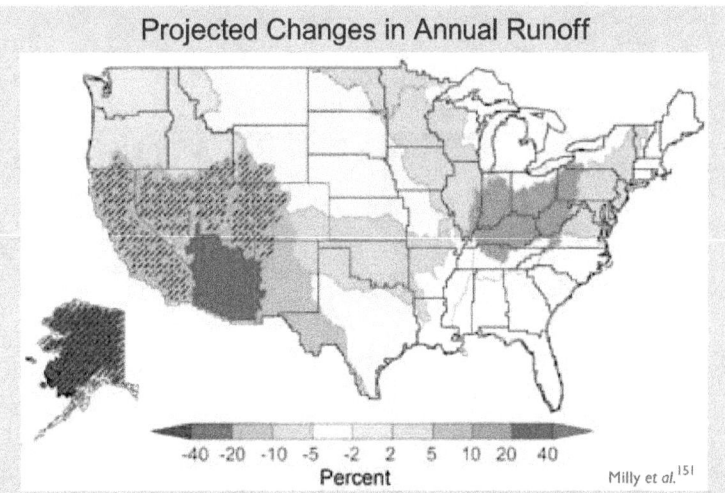

Projected Changes in Annual Runoff

Percent

Milly *et al.*[151]

Projected changes in median runoff for 2041-2060, relative to a 1901-1970 baseline, are mapped by water-resource region. Colors indicate percentage changes in runoff. Hatched areas indicate greater confidence due to strong agreement among model projections. White areas indicate divergence among model projections. Results are based on emissions in between the lower and higher emissions scenarios.[91]

years, there have been widespread temperature-related reductions in snowpack in the West, with the largest reductions occurring in lower elevation mountains in the Northwest and California where snowfall occurs at temperatures close to the freezing point.[142,153] The Northeast has also experienced snowpack reductions during a similar period. Observations indicate a transition to more rain and less snow in both the West and Northeast in the last 50 years.[143,154-156] Runoff in snowmelt-dominated areas is occurring up to 20 days earlier in the West, and up to 14 days earlier in the Northeast.[157,158] Future projections for most snowmelt-dominated basins in the West consistently indicate earlier spring

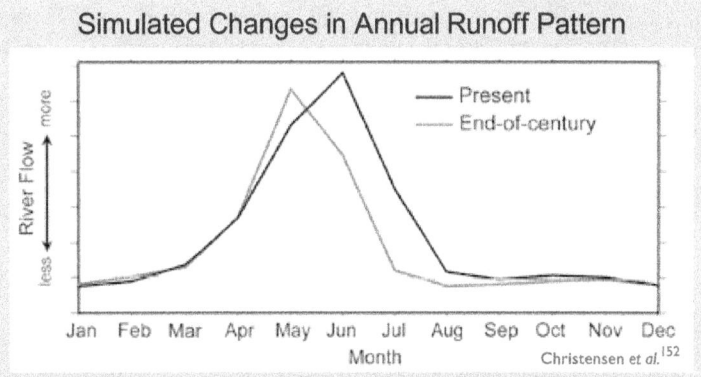

Simulated Changes in Annual Runoff Pattern

— Present
— End-of-century

River Flow
more
less

Jan Feb Mar Apr May Jun Jul Aug Sep Oct Nov Dec
Month

Christensen *et al.*[152]

General schematic of changes in the annual pattern of runoff for snowmelt-dominated streams. Compared to the historical pattern, runoff peak is projected to shift to earlier in the spring and late summer flows are expected to be lower. The above example is for the Green River, which is part of the Colorado River watershed.

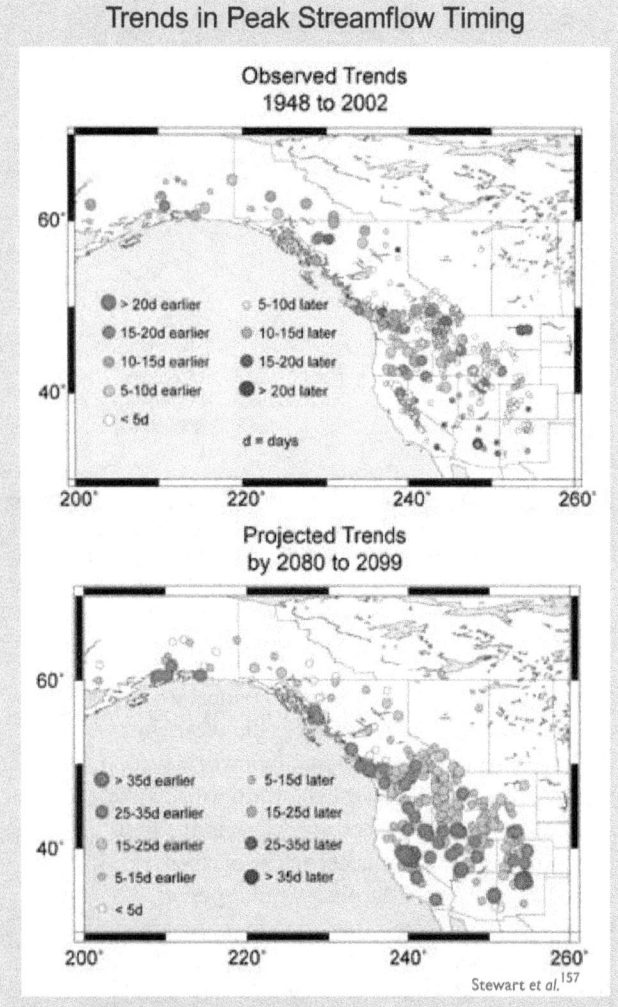

Trends in Peak Streamflow Timing

Observed Trends
1948 to 2002

Projected Trends
by 2080 to 2099

Stewart et al.[157]

Top map shows changes in runoff timing in snowmelt-driven streams from 1948 to 2002 with red circles indicating earlier runoff, and blue circles indicating later runoff. Bottom map shows projected changes in snowmelt-driven streams by 2080-2099, compared to 1951-1980, under a higher emissions scenario.[91]

runoff, in some cases up to 60 days earlier.[157,159] For the Northeast, projections indicate spring runoff will advance by up to 14 days.[150] Earlier runoff produces lower late-summer streamflows, which stress human and environmental systems through less water availability and higher water temperatures.[145] Scientific analyses to determine the causes of recent changes in snowpack, runoff timing, and increased winter temperatures have attributed these changes to human-caused climate change.[34,160,161]

Surface water quality and groundwater quantity will be affected by a changing climate.

Changes in water quality

Increased air temperatures lead to higher water temperatures, which have already been detected in many streams, especially during low-flow periods. In lakes and reservoirs, higher water temperatures lead to longer periods of summer stratification (when surface and bottom waters do not mix). Dissolved oxygen is reduced in lakes, reservoirs, and rivers at higher temperatures. Oxygen is an essential resource for many living things, and its availability is reduced at higher temperatures both because the amount that can be dissolved in water is lower and because respiration rates of living things are higher. Low oxygen stresses aquatic animals such as coldwater fish and the insects and crustaceans on which they feed.[142] Lower oxygen levels also decrease the self-purification capabilities of rivers.

The negative effects of water pollution, including sediments, nitrogen from agriculture, disease pathogens, pesticides, herbicides, salt, and thermal pollution, will be amplified by observed and projected increases in precipitation intensity and longer periods when streamflows are low.[146] The U.S. Environmental Protection Agency expects the number of waterways considered "impaired" by water pollution to increase.[162] Heavy downpours lead to increased sediment in runoff and outbreaks of waterborne diseases.[163,164] Increases in pollution carried to lakes, estuaries, and the coastal ocean, especially when coupled with increased temperature, can result in blooms of harmful algae and bacteria. However, pollution has the potential of being diluted in regions that experience increased streamflow.

Water-quality changes during the last century were probably due to causes other than climate change, primarily changes in pollutants.[149]

Changes in groundwater

Many parts of the United States are heavily dependent on groundwater for drinking, residential, and agricultural water supplies.[164] How climate change will affect groundwater is not well known,

Heavy rain can cause sediments to become suspended in water, reducing its quality, as seen in the brown swath above in New York City's Ashokan reservoir following Hurricane Floyd in September 1999.

but increased water demands by society in regions that already rely on groundwater will clearly stress this resource, which is often drawn down faster than it can be recharged.[164] In many locations, groundwater is closely connected to surface water and thus trends in surface water supplies over time affect groundwater. Changes in the water cycle that reduce precipitation or increase evaporation and runoff would reduce the amount of water available for recharge. Changes in vegetation and soils that occur as temperature changes or due to fire or pest outbreaks are also likely to affect recharge by altering evaporation and infiltration rates. More frequent and larger floods are likely to increase groundwater recharge in semi-arid and arid areas,

where most recharge occurs through dry stream-beds after heavy rainfalls and floods.[142]

Sea-level rise is expected to increase saltwater intrusion into coastal freshwater aquifers, making some unusable without desalination.[146] Increased evaporation or reduced recharge into coastal aquifers exacerbates saltwater intrusion. Shallow groundwater aquifers that exchange water with streams are likely to be the most sensitive part of the groundwater system to climate change. Small reductions in groundwater levels can lead to large reductions in streamflow and increases in groundwater levels can increase streamflow.[165] Further, the interface between streams and groundwater is an important site for pollution removal by microorganisms. Their activity will change in response to increased temperature and increased or decreased streamflow as climate changes, and this will affect water quality. Like water quality, research on the impacts of climate change on groundwater has been minimal.[149]

Climate change will place additional burdens on already stressed water systems.

In many places, the nation's water systems are already taxed due to aging infrastructure, population increases, and competition among water needs for farming, municipalities, hydropower, recreation, and ecosystems.[167-169] Climate change will add another factor to existing water management challenges, thus increasing vulnerability.[170] The U.S. Bureau of Reclamation has identified many areas in the West that are already at risk for serious conflict over water, even in the absence of climate change[171] (see figure next page).

Adapting to gradual changes, such as changes in average amounts of precipitation, is less difficult than adapting to changes in extremes. Where extreme events, such as droughts or floods, become more intense or more frequent with climate change, the economic and social costs of these events will increase.[172] Water systems have life spans of many years and are designed with spare

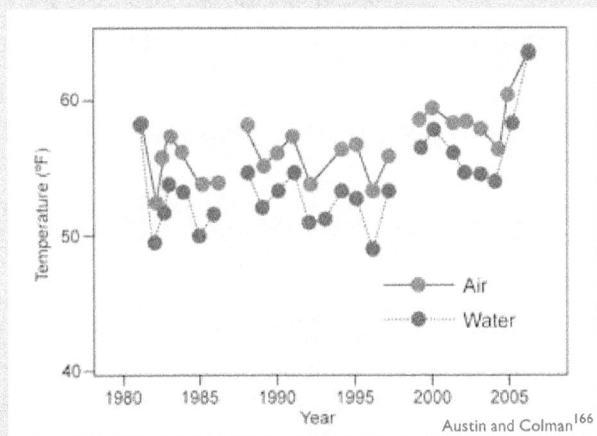

Lake Superior Summer Air and Water Temperatures
1979 to 2006

Austin and Colman[166]

The recent large jump in summer water temperature is related to the recent large reduction in ice cover (see *Midwest* region).

capacity. These systems are thus able to cope with small changes in average conditions.[172] Water resource planning today considers a broad range of stresses and hence adaptation to climate change will be one factor among many in deciding what actions will be taken to minimize vulnerability.[172-174]

Rapid regional population growth

The U.S. population is estimated to have grown to more than 300 million people, nearly a 7 percent increase since the 2000 Census. Current Census Bureau projections are for this growth rate to continue, with the national population projected to reach 350 million by 2025 and 420 million by 2050. The highest rates of population growth to 2025 are projected to occur in areas such as the Southwest that are at risk for reductions in water supplies due to climate change.[167]

Aging water infrastructure

The nation's drinking water and wastewater infrastructure is aging. In older cities, some buried water mains are over 100 years old and breaks of these lines are a significant problem. Sewer overflows resulting in the discharge of untreated wastewater also occur frequently. Heavier downpours will exacerbate existing problems in many cities, especially where stormwater catchments and sewers are combined.

Damage to the city water system in Asheville, North Carolina, due to heavy rain in 2004.

Drinking water and sewer infrastructure is very expensive to install and maintain. Climate change will present a new set of challenges for designing upgrades to the nation's water delivery and sewage removal infrastructure.[168]

Potential Water Supply Conflicts by 2025

Indian Lands and Native Entities

Water Supply Issue Areas

Unmet Rural Water Needs

Conflict Potential-- Moderate

Conflict Potential-- Substantial

Conflict Potential-- Highly Likely

USBR[171]

The map shows regions in the West where water supply conflicts are likely to occur by 2025 based on a combination of factors including population trends and potential endangered species' needs for water. The red zones are where the conflicts are most likely to occur. This analysis does not factor in the effects of climate change, which is expected to exacerbate many of these already-identified issues.[171]

Existing water disputes across the country

Many locations in the United States are already undergoing water stress. The Great Lakes states are establishing an interstate compact to protect against reductions in lake levels and potential water exports. Georgia, Alabama, and Florida are in a dispute over water for drinking, recreation, farming, environmental purposes, and hydropower in the Apalachicola–Chattahoochee–Flint River system.[175,176]

The State Water Project in California is facing a variety of problems in the Sacramento Delta, including endangered species, saltwater intrusion, and potential loss of islands due to flood- or earthquake-caused levee failures.[177-182] A dispute over endangered fish in the Rio Grande has been ongoing for many years.[183] The Klamath River in Oregon and California has been the location of a multi-year disagreement over native fish, hydropower, and farming.[184,185] The Colorado River has been the site of numerous interstate quarrels over the last century.[186,187] Large, unquantified Native

American water rights challenge existing uses in the West (see *Southwest* region).[188] By changing the existing patterns of precipitation and runoff, climate change will add another stress to existing problems.

Changing water demands

Water demands are expected to change with increased temperatures. Evaporation is projected to increase over most of the United States as temperatures rise. Higher temperatures and longer dry periods are expected to lead to increased water demand for irrigation. This may be partially offset by more efficient use of water by plants due to rising atmospheric carbon dioxide. Higher temperatures are projected to increase cooling water withdrawals by electrical generating stations. In addition, greater cooling requirements in summer will increase electricity use, which in turn will require more cooling water for power plants. Industrial and municipal demands are expected to increase slightly.[146]

The past century is no longer a reasonable guide to the future for water management.

Water planning and management have been based on historical fluctuations in records of stream flows, lake levels, precipitation, temperature, and water demands. All aspects of water management including reservoir sizing, reservoir flood operations, maximum urban stormwater runoff amounts, and projected water demands have been based on these records. Water managers have proven adept at balancing supplies and demand through the significant climate variability of the past century.[142] Because climate change will significantly modify many aspects of the water cycle, the assumption of an unchanging climate is no longer appropriate for many aspects of water planning. Past assumptions derived from the historical record about supply and demand will need to be revisited for existing and proposed water projects.[142,151,174]

Drought studies that consider the past 1,200 years indicate that in the West, the last

century was significantly wetter than most other centuries. Multi-decade "megadroughts" in the years 900 to 1300 were substantially worse than the worst droughts of the last century, including the Dust Bowl era. The causes of these events are only partially known; if they were to reoccur, they would clearly stress water management, even in the absence of climate change (see figure below).[97,149,189]

The intersection of substantial changes in the water cycle with multiple stresses such as population growth and competition for water supplies means that water planning will be doubly challenging. The ability to modify operational rules and water allocations is likely to be critical for the protection of infrastructure, for public safety, to ensure reliability of water delivery, and to protect the environment. There are, however, many institutional and legal barriers to such changes in both the short and long term.[190] Four examples:

- The allocation of the water in many interstate rivers is governed by compacts, international treaties, federal laws, court decrees, and other agreements that are difficult to modify.

- Reservoir operations are governed by "rule curves" that require a certain amount of space to be saved in a reservoir at certain times of

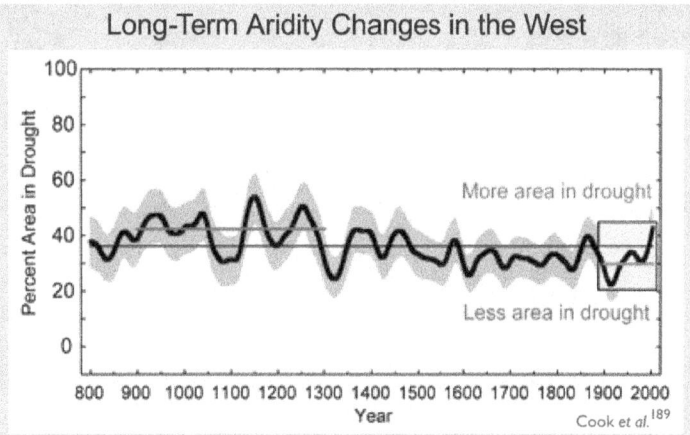

Long-Term Aridity Changes in the West

Cook et al.[189]

The black line shows the percentage of the area affected by drought (Palmer Drought Severity Index less than −1) in the West over the past 1,200 years. The red line indicates the average drought area in the years 900 to 1300. The blue horizontal line in the yellow box indicates the average during the period from 1900 to 2000, illustrating that the most recent period, during which population and water infrastructure grew rapidly in the West, was wetter than the long-term average (thin horizontal black line).[189] Droughts shown in the period 1100-1300 significantly exceed those that have occurred over the past 100 years.

year to capture a potential flood. Developed by the U.S. Army Corps of Engineers based on historical flood data, many of these rule curves have never been modified, and modifications might require Environmental Impact Statements.[151]

Highlights of Water-Related Impacts by Sector	
Sector	**Examples of Impacts**
Human Health	Heavy downpours increase incidence of waterborne disease and floods, resulting in potential hazards to human life and health.[163]
Energy Supply and Use	Hydropower production is reduced due to low flows in some regions. Power generation is reduced in fossil fuel and nuclear plants due to increased water temperatures and reduced cooling water availability.[191]
Transportation	Floods and droughts disrupt transportation. Heavy downpours affect harbor infrastructure and inland waterways. Declining Great Lakes levels reduce freight capacity.[192]
Agriculture and Forests	Intense precipitation can delay spring planting and damage crops. Earlier spring snowmelt leads to increased number of forest fires.[193]
Ecosystems	Coldwater fish threatened by rising water temperatures. Some warmwater fish will expand ranges.[70]

- In most parts of the West, water is allocated based on a "first in time means first in right" system, and because agriculture was developed before cities were established, large volumes of water typically are allocated to agriculture. Transferring agricultural rights to municipalities, even for short periods during drought, can involve substantial expense and time and can be socially divisive.

- Conserving water does not necessarily lead to a right to that saved water, thus creating a disincentive for conservation.

Total U.S. water diversions peaked in the 1980s, which implies that expanding supplies in many areas to meet new needs are unlikely to be a viable option, especially in arid areas likely to experience less precipitation. However, over the last 30 years, per capita water use has decreased significantly (due, for example, to more efficient technologies such as drip irrigation) and it is anticipated that per capita use will continue to decrease, thus easing stress.[149]

Adaptation: New York City Begins Planning for Climate Change

The New York City Department of Environmental Protection (DEP), the agency in charge of providing the city's drinking water and wastewater treatment, is beginning to alter its planning to take into account the effects of climate change – sea-level rise, higher temperatures, increases in extreme events, and changing precipitation patterns – on the city's water systems. In partnership with Columbia University, DEP is evaluating climate change projections, impacts, indicators, and adaptation and mitigation strategies.

City planners have begun to address these issues by defining risks using probabilistic climate scenarios and considering potential adaptations that relate to operations/management, infrastructure, and policy. For example, DEP is examining the feasibility of relocating critical control systems to higher floors in low-lying buildings or to higher ground, building flood walls, and modifying design criteria to reflect changing hydrologic processes.

Important near-term goals of the overall effort include updating the existing 100-year flood elevations using climate model projections and identifying additional monitoring stations needed to track changes. DEP will also establish a system for reporting the impacts of extreme weather events on the City's watershed and infrastructure. In the immediate future, DEP will evaluate flood protection measures for three existing water pollution control plants that are scheduled for renovation.[194]

Spotlight on the Colorado River

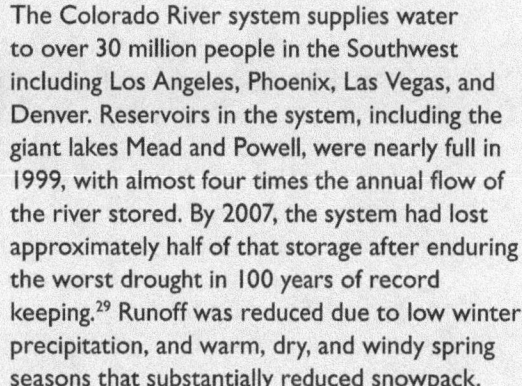

The Colorado River system supplies water to over 30 million people in the Southwest including Los Angeles, Phoenix, Las Vegas, and Denver. Reservoirs in the system, including the giant lakes Mead and Powell, were nearly full in 1999, with almost four times the annual flow of the river stored. By 2007, the system had lost approximately half of that storage after enduring the worst drought in 100 years of record keeping.[29] Runoff was reduced due to low winter precipitation, and warm, dry, and windy spring seasons that substantially reduced snowpack.

Numerous studies over the last 30 years have indicated that the river is likely to experience reductions in runoff due to climate change. In addition, diversions from the river to meet the needs of cities and agriculture are approaching its average flow. Under current conditions, even without climate change, large year-to-year fluctuations in reservoir storage are possible.[152] If reductions in flow projected to accompany global climate change occur, water managers will be challenged to satisfy all existing demands, let alone the increasing demands of a rapidly growing population.[167,195]

Efforts are underway to address these challenges. In 2005, the Department of Interior's Bureau of Reclamation began a process to formalize operating rules for lakes Mead and Powell during times of low flows and to apportion limited water among the states.[196]

June 29, 2002

December 23, 2003

Matching photographs taken 18 months apart during the most serious period of recent drought show a significant decrease in Lake Powell.

Change in Water Volume of Lakes Mead and Powell

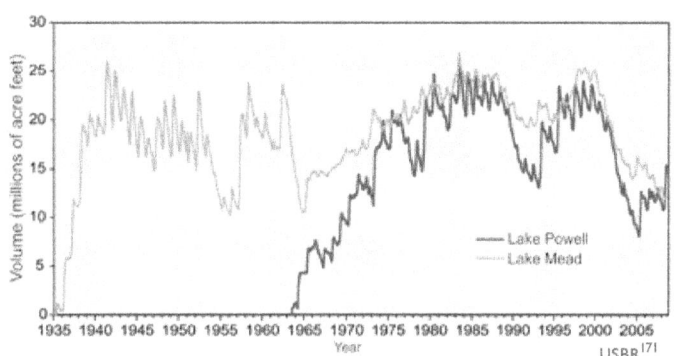

USBR[171]

The filling of Lake Mead (green) was initiated in 1935, and that of Lake Powell (blue) in 1963. In 1999, the lakes were nearly full, but by 2007, the lakes had lost nearly half of their storage water after the worst drought in 100 years.

Water and Energy Connections

Water and energy are tightly interconnected; water systems use large amounts of energy, and energy systems use large amounts of water. Both are expected to be under increasing pressure in the future and both will be affected by a changing climate. In the energy sector, water is used directly for hydropower, and cooling water is critical for nearly all other forms of electrical power generation. Withdrawals of freshwater used to cool power plants that use heat to generate electricity are very large, nearly equaling the water withdrawn for irrigation. Water consumption by power plants is about 20 percent of all non-agricultural uses, or half that of all domestic use.[197]

In the water sector, two very unusual attributes of water, significant weight due to its relatively high density, and high heat capacity, make water use energy intensive. Large amounts of energy are needed for pumping, heating, and treating drinking water and wastewater. Water supply and treatment consumes roughly 4 percent of the nation's power supply, and electricity accounts for about 75 percent of the cost of municipal water processing and transport. In California, 30 percent of all non-power plant natural gas is used for water-related activities.[198,199] The energy required to provide water depends on its source (groundwater, surface water, desalinated water, treated wastewater, or recycled water), the distance the water is conveyed, the amount of water moved, and the local topography. Surface water often requires more treatment than groundwater. Desalination requires large amounts of energy to produce freshwater. Treated wastewater and recycled water (used primarily for agriculture and industry) require energy for treatment, but little energy for supply and conveyance. Conserving water has the dual benefit of conserving energy and potentially reducing greenhouse gas emissions if fossil fuels are the predominant source of that energy.

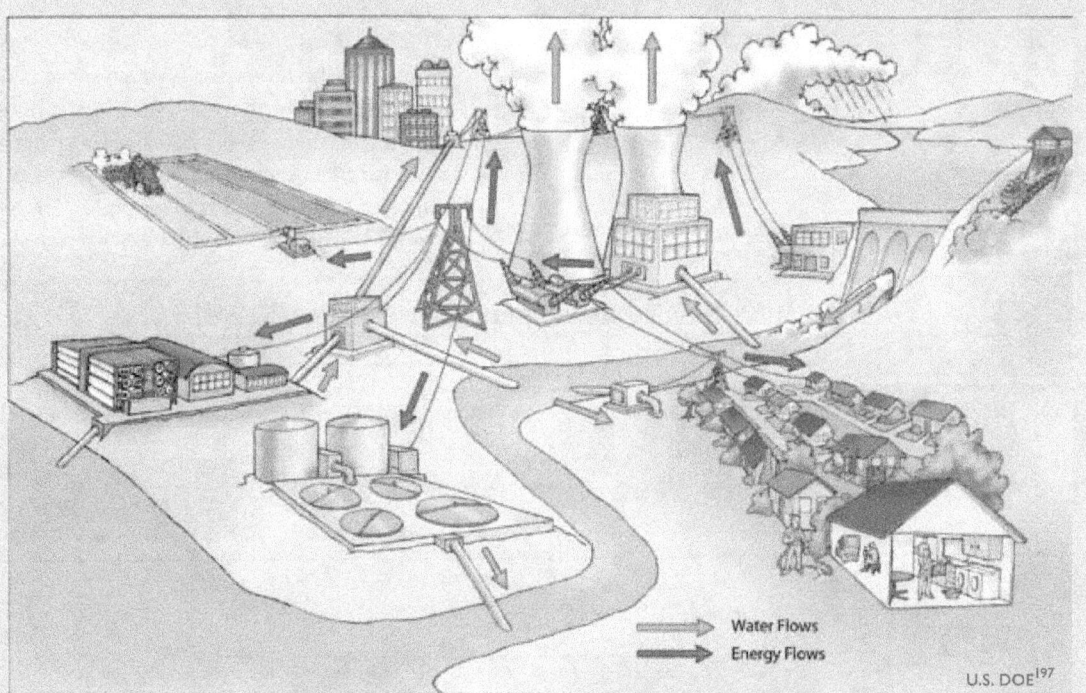

Water Flows
Energy Flows

U.S. DOE[197]

Water and energy are intimately connected. Water is used by the power generation sector for cooling, and energy is used by the water sector for pumping, drinking water treatment, and wastewater treatment. Without energy, there would be limited water distribution, and without water, there would be limited energy production.

Energy Supply and Use

Key Messages:

- Warming will be accompanied by decreases in demand for heating energy and increases in demand for cooling energy. The latter will result in significant increases in electricity use and higher peak demand in most regions.
- Energy production is likely to be constrained by rising temperatures and limited water supplies in many regions.
- Energy production and delivery systems are exposed to sea-level rise and extreme weather events in vulnerable regions.
- Climate change is likely to affect some renewable energy sources across the nation, such as hydropower production in regions subject to changing patterns of precipitation or snowmelt.

Energy is at the heart of the global warming challenge.[3] It is humanity's production and use of energy that is the primary cause of global warming, and in turn, climate change will eventually affect our production and use of energy. The vast majority of U.S. greenhouse gas emissions, about 87 percent, come from energy production and use.[200]

At the same time, other U.S. trends are increasing energy use: population shifts to the South, especially the Southwest, where air conditioning use is high, an increase in the square footage built per person, increased electrification of the residential and commercial sectors, and increased market penetration of air conditioning.[201]

Many of the effects of climate change on energy production and use in the United States are not well studied. Some of the effects of climate change, however, have clear implications for

energy production and use. For instance, rising temperatures are expected to increase energy requirements for cooling and reduce energy requirements for heating.[164,201] Changes in precipitation have the potential to affect prospects for hydropower, positively or negatively.[201] Increases in hurricane intensity are likely to cause further disruptions to oil and gas operations in the Gulf, like those experienced in 2005 with Hurricane Katrina and in 2008 with Hurricane Ike.[201] Concerns about climate

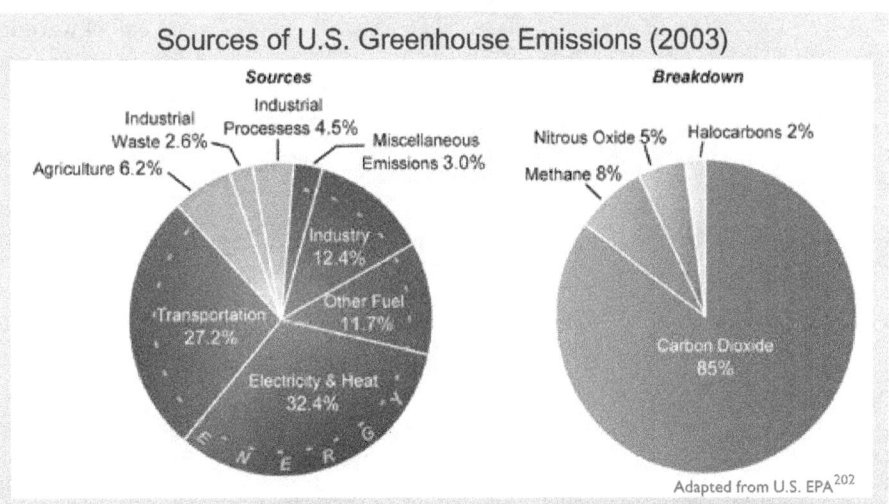

Sources of U.S. Greenhouse Emissions (2003)

Sources
- Industrial Waste 2.6%
- Agriculture 6.2%
- Industrial Processess 4.5%
- Miscellaneous Emissions 3.0%
- Industry 12.4%
- Other Fuel 11.7%
- Transportation 27.2%
- Electricity & Heat 32.4%

Breakdown
- Nitrous Oxide 5%
- Halocarbons 2%
- Methane 8%
- Carbon Dioxide 85%

Adapted from U.S. EPA[202]

About 87 percent of U.S. greenhouse gas emissions come from energy production and use, as shown in the left pie chart. The right pie chart breaks down these emissions by greenhouse gas.

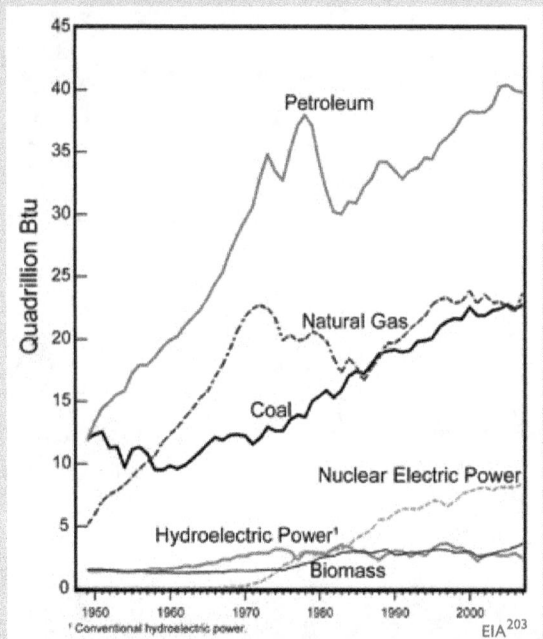

Primary Energy Consumption by Major Source (1949 to 2007)

U.S. energy supply is dominated by fossil fuels. Petroleum, the top source of energy shown above, is primarily used for transportation (70 percent of oil use). Natural gas is used in roughly equal parts to generate electricity, power industrial processes, and heat water and buildings. Coal is primarily used to generate electricity (91 percent of coal use). Nuclear power is used entirely for electricity generation.

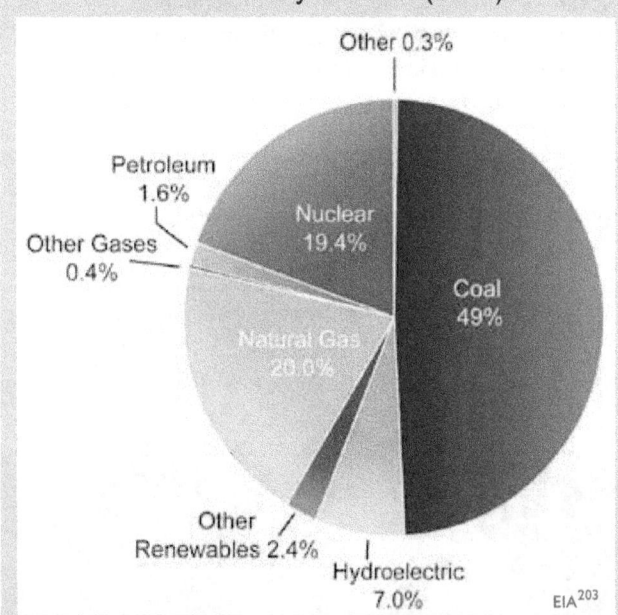

U.S. Electricity Sources (2007)

Coal, natural gas, and nuclear power plants together account for about 90 percent of current U.S. electricity production.

change impacts will almost certainly alter perceptions and valuations of energy technology alternatives. These effects are very likely to be relevant for energy policies, decisions, and institutions in the United States, affecting courses of action and appropriate strategies for risk management.[201]

The overall scale of the national energy economy is very large, and the energy industry has both the financial and the managerial resources to be adaptive. Impacts due to climate change are likely to be most apparent at sub-national scales, such as regional effects of extreme weather events and reduced water availability, and effects of increased cooling demands on especially vulnerable places and populations.[204]

Warming will be accompanied by decreases in demand for heating energy and increases in demand for cooling energy. The latter will result in significant increases in electricity use and higher peak demand in most regions.

Research on the effects of climate change on energy production and use has largely been limited to impacts on energy use in buildings. These studies have considered effects of global warming on energy requirements for heating and cooling in buildings in the United States.[205] They find that the demand for cooling energy increases from 5 to 20 percent per 1.8°F of warming, and the demand for heating energy drops by 3 to 15 percent per 1.8°F of warming.[205] These ranges reflect different assumptions about factors such as the rate of market penetration of improved building equipment technologies.[205]

Studies project that temperature increases due to global warming are very likely to increase peak demand for electricity in most regions of the country.[205] An increase in peak demand can lead to a disproportionate increase in energy infrastructure investment.[205]

Since nearly all of the cooling of buildings is provided by electricity use, whereas the vast majority of the heating of buildings is provided by natural gas and fuel oil,[201,206] the projected changes imply increased demands for electricity. This is especially the case where climate change would result in significant increases in the heat index in summer, and where relatively little space cooling has been needed in the past, but demands are likely to increase in the future.[205] The increase in electricity demand is likely to be accelerated by population movements to the South and South-west, which are regions of especially high per capita electricity use, due to demands for cooling in commercial buildings and households.[205] Because nearly half of the nation's electricity is currently generated from coal, these factors have the potential to increase total national carbon dioxide emissions in the absence of improved energy efficiency, development of non-carbon energy sources, and/or carbon capture and storage.[205]

Other effects of climate change on energy consumption are less clear, because little research has been done.[205] For instance, in addition to cooling, air conditioners also remove moisture from the air; thus the increase in humidity projected to accompany global warming is likely to increase electricity consumption by air conditioners even further.[205] As other examples, warming would increase the use of air conditioners in highway vehicles, and water scarcity in some regions has the potential to increase energy demands for water pumping. It is important to improve the information available about these other kinds of effects.

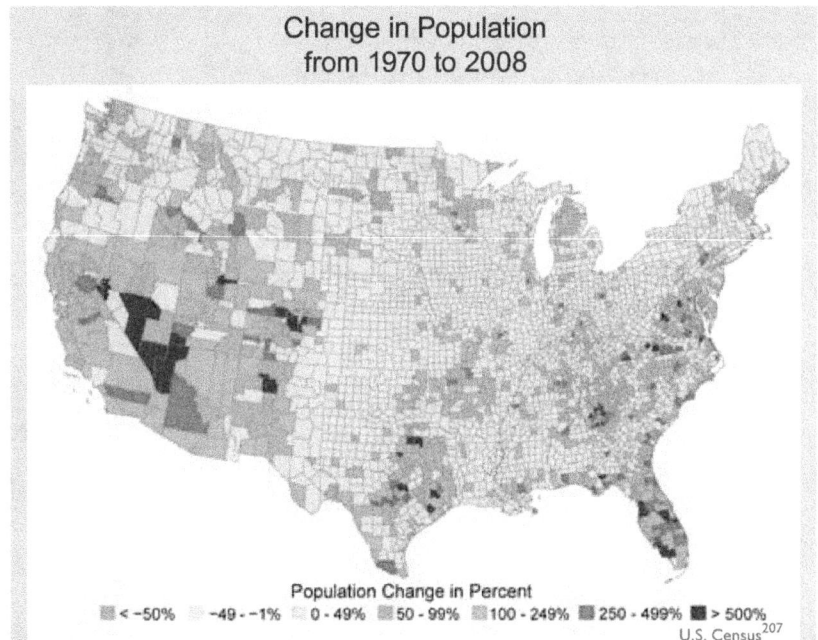

Change in Population from 1970 to 2008

Population Change in Percent

■ < -50%   -49 - -1%   0 - 49%   ■ 50 - 99%   ■ 100 - 249%   ■ 250 - 499%   ■ > 500%

U.S. Census[207]

The map above, showing percentage changes in county population between 1970 and 2008, graphically illustrates the large increases in places that require air conditioning. Areas with very large increases are shown in orange, red, and maroon. Some places had enormous growth, in the hundreds of thousands of people. For example, counties in the vicinity of South Florida, Atlanta, Los Angeles, Phoenix, Las Vegas, Denver, Dallas, and Houston all had very large increases.

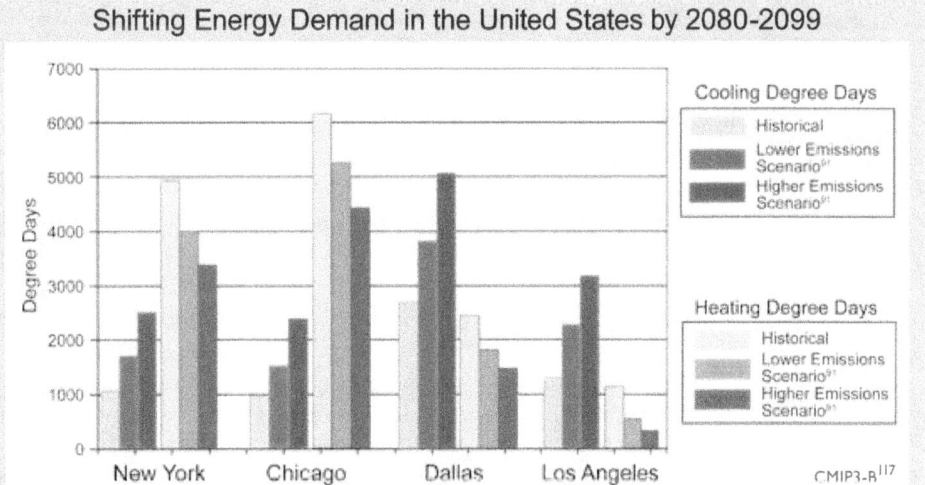

Shifting Energy Demand in the United States by 2080-2099

Cooling Degree Days
- Historical
- Lower Emissions Scenario[91]
- Higher Emissions Scenario[91]

Heating Degree Days
- Historical
- Lower Emissions Scenario[91]
- Higher Emissions Scenario[91]

CMIP3-B[117]

"Degree days" are a way of measuring the energy needed for heating and cooling by adding up how many degrees hotter or colder each day's average temperature is from 65°F over the course of a year. Colder locations have high numbers of heating degree days and low numbers of cooling degree days, while hotter locations have high numbers of cooling degree days and low numbers of heating degree days. Nationally, the demand for energy will increase in summer and decrease in winter. Cooling uses electricity while heating uses a combination of energy sources, so the overall effect nationally and in most regions will be an increased need for electricity. The projections shown in the chart are for late this century. The higher emissions scenario[91] used here is referred to as "even higher" on page 23.

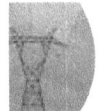

Energy production is likely to be constrained by rising temperatures and limited water supplies in many regions.

In some regions, reductions in water supply due to decreases in precipitation and/or water from melting snowpack are likely to be significant, increasing the competition for water among various sectors including energy production (see *Water Resources* sector).[191,208]

The production of energy from fossil fuels (coal, oil, and natural gas) is inextricably linked to the availability of adequate and sustainable supplies of water.[191,208] While providing the United States with the majority of its annual energy needs, fossil fuels also place a high demand on the nation's water resources in terms of both quantity and quality impacts.[191,208] Generation of electricity in thermal power plants (coal, nuclear, gas, or oil) is water intensive. Power plants rank only slightly behind irrigation in terms of freshwater withdrawals in the United States.[191]

There is a high likelihood that water shortages will limit power plant electricity production in many regions. Future water constraints on electricity production in thermal power plants are projected for Arizona, Utah, Texas, Louisiana, Georgia, Alabama, Florida, California, Oregon, and Washington state by 2025.[191] Additional parts of the United States could face similar constraints as a result of drought, growing populations, and increasing demand for water for various uses, at least seasonally.[209] Situations where the development of new power plants is being slowed down or halted due to inadequate cooling water are becoming more frequent throughout the nation.[191]

The issue of competition among various water uses is dealt with in more detail in the *Water Resources* sector. In connection with these issues and other regional water scarcity impacts, energy is likely to be needed to move and manage water. This is one of many examples of interactions among the impacts of climate change on various sectors that, in this case, affects energy requirements.

Nuclear, coal, and natural gas power plants require large amounts of water for cooling.[191]

In addition to the problem of water availability, there are issues related to an increase in water temperature. Use of warmer water reduces the efficiency of thermal power plant cooling technologies. And, warmer water discharged from power plants can alter species composition in aquatic ecosystems.[210] Large coal and nuclear plants have been limited in their operations by reduced river levels caused by higher temperatures and thermal limits on water discharge.[191]

The efficiency of thermal power plants, fossil or nuclear, is sensitive to ambient air and water temperatures; higher temperatures reduce power outputs by affecting the efficiency of cooling.[191] Although this effect is not large in percentage terms, even a relatively small change could have significant implications for total national electric power supply.[191] For example, an average reduction of 1 percent in electricity generated by thermal power plants nationwide would mean a loss of 25 billion kilowatt-hours per year,[211] about the amount of electricity consumed by 2 million Americans, a loss that would need to be supplied in some other way or offset through measures that improve energy efficiency.

Energy production and delivery systems are exposed to sea-level rise and extreme weather events in vulnerable regions.

Sea-level rise

A significant fraction of America's energy infrastructure is located near the coasts, from power plants, to oil refineries, to facilities that receive oil and gas deliveries.[191] Rising sea levels are likely to lead to direct losses, such as equipment damage from flooding or erosion, and indirect effects, such as the costs of raising vulnerable assets to higher levels or building new facilities farther inland, increasing transportation costs.[191] The U.S. East Coast and Gulf Coast have been identified as particularly vulnerable to sea-level rise because the land is relatively flat and also sinking in many places.[191]

Extreme events

Observed and projected increases in a variety of extreme events will have significant impacts on the energy sector. As witnessed in 2005, hurricanes can have a debilitating impact on energy infrastructure. Direct losses to the energy industry in 2005 are estimated at $15 billion,[191] with millions more in restoration and recovery costs. As one example, the Yscloskey Gas Processing Plant (located on the Louisiana coast) was forced to close for six months following Hurricane Katrina, resulting in lost revenues to the plant's owners and employees, and higher prices to consumers, as gas had to be procured from other sources.[191]

The impacts of an increase in severe weather are not limited to hurricane-prone areas. For example, rail transportation lines, which carry approximately two-thirds of the coal to the nation's power plants,[212] often follow riverbeds, especially in the Appalachian region.[191] More intense rainstorms, which have been observed and projected,[68,112] can lead to rivers flooding, which can "wash out" or degrade nearby railbeds and roadbeds.[191] This is also a problem in the Midwest, which experienced major flooding of the Mississippi River in 1993 and 2008.[213]

Development of new energy facilities could be restricted by siting concerns related to sea-level rise, exposure to extreme events, and increased capital costs resulting from a need to provide greater protection from extreme events.[191]

The electricity grid is also vulnerable to climate change effects, from temperature changes to severe weather events.[191] The most familiar example is

Regional Spotlight: Gulf Coast Oil and Gas

The Gulf Coast is home to the U.S. oil and gas industries, representing nearly 30 percent of the nation's crude oil production and approximately 20 percent of its natural gas production. One-third of the national refining and processing capacity lies on coastal plains adjacent to the Gulf of Mexico. Several thousand offshore drilling platforms, dozens of refineries, and thousands of miles of pipelines are vulnerable to damage and disruption due to sea-level rise and the high winds and storm surge associated with hurricanes and other tropical storms. For example, hurricanes Katrina and Rita halted all oil and gas production from the Gulf, disrupted nearly 20 percent of the nation's refinery capacity, and closed many oil and gas pipelines.[214] Relative sea-level rise in parts of the Gulf Coast region (Louisiana and East Texas) is projected to be as high as 2 to 4 feet by 2050 to 2100, due to the combination of global sea-level rise caused by warming oceans and melting ice and local land sinking.[215] Combined with onshore and offshore storm activity, this would represent an increased threat to this regional energy infrastructure. Some adaptations to these risks are beginning to emerge (see Adaptation box, page 58).

Offshore oil production is particularly susceptible to extreme weather events. Hurricane Ivan in 2004 destroyed seven platforms in the Gulf of Mexico, significantly damaged 24 platforms, and damaged 102 pipelines. Hurricanes Katrina and Rita in 2005 destroyed more than 100 platforms and damaged 558 pipelines. For example, Chevron's $250 million "Typhoon" platform was damaged beyond repair. Plans are being made to sink its remains to the seafloor.

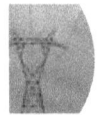

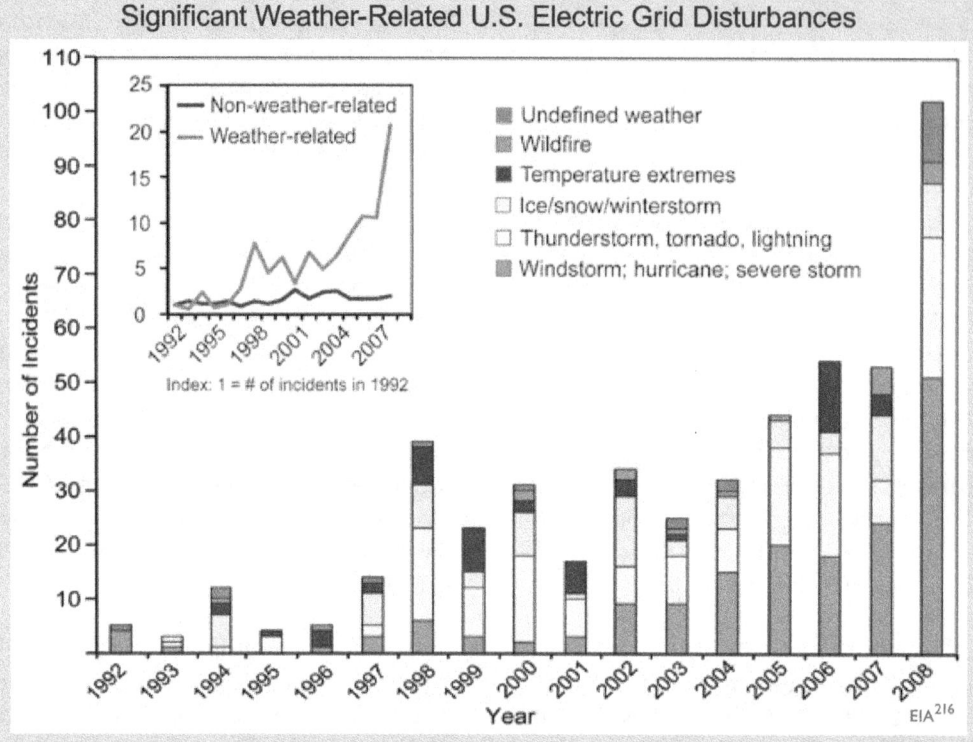

Significant Weather-Related U.S. Electric Grid Disturbances

Legend:
- Non-weather-related
- Weather-related

Index: 1 = # of incidents in 1992

- Undefined weather
- Wildfire
- Temperature extremes
- Ice/snow/winterstorm
- Thunderstorm, tornado, lightning
- Windstorm; hurricane; severe storm

EIA[216]

The number of incidents caused by extreme weather has increased tenfold since 1992. The portion of all events that are caused by weather-related phenomena has more than tripled from about 20 percent in the early 1990s to about 65 percent in recent years. The weather-related events are more severe, with an average of about 180,000 customers affected per event compared to about 100,000 for non-weather-related events (and 50,000 excluding the massive blackout of August 2003).[201] The data shown include disturbances that occurred on the nation's large-scale "bulk" electric transmission systems. Most outages occur in local distribution networks and are not included in the graph. Although the figure does not demonstrate a cause-effect relationship between climate change and grid disruption, it does suggest that weather and climate extremes often have important effects on grid disruptions. We do know that more frequent weather and climate extremes are likely in the future,[68] which poses unknown new risks for the electric grid.

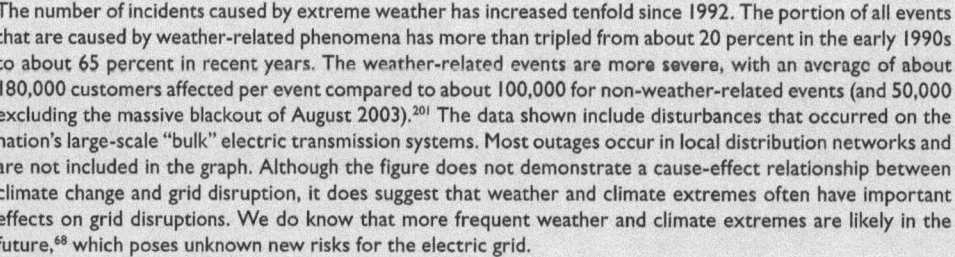

Adaptation: Addressing Oil Infrastructure Vulnerabilities in the Gulf Coast

Port Fourchon, Louisiana, supports 75 percent of deepwater oil and gas production in the Gulf of Mexico, and its role in supporting oil production in the region is increasing. The Louisiana Offshore Oil Port, located about 20 miles offshore, links daily imports of 1 million barrels of oil and production of 300,000 barrels in the Gulf of Mexico to 50 percent of national refining capacity. One road, Louisiana Highway 1, connects Port Fourchon with the nation. It transports machinery, supplies, and workers and is the evacuation route for onshore and offshore workers. Responding to threats of storm surge and flooding, related in part to concerns about climate change, Louisiana is currently upgrading Highway 1, including elevating it above the 500-year flood level and building a higher bridge over Bayou LaFourche and the Boudreaux Canal.[217]

Regional Spotlight: Florida's Energy Infrastructure

Florida's energy infrastructure is particularly vulnerable to sea-level rise and storm impacts. Most of the petroleum products consumed in Florida are delivered by barge to three ports, two on the east coast and one on the west coast. The interdependencies of natural gas distribution, transportation fuel distribution and delivery, and electrical generation and distribution were found to be major issues in Florida's recovery from recent major hurricanes.[191]

effects of severe weather events on power lines, such as from ice storms, thunderstorms, and hurricanes. In the summer heat wave of 2006, for example, electric power transformers failed in several areas (including St. Louis, Missouri, and Queens, New York) due to high temperatures, causing interruptions of electric power supply. It is not yet possible to project effects of climate change on the grid, because so many of the effects would be more localized than current climate change models can depict; but, weather-related grid disturbances are recognized as a challenge for strategic planning and risk management.

Climate change is likely to affect some renewable energy sources across the nation, such as hydropower production in regions subject to changing patterns of precipitation or snowmelt.

Renewable sources currently account for about 9 percent of electricity production in the United States.[203] Hydroelectric power is by far the largest renewable contributor to electricity generation,[191] accounting for about 7 percent of total U.S. electricity.[218] Like many things discussed in this report, renewable energy resources have strong interrelationships with climate change; using renewable energy can reduce the magnitude of climate change, while climate change can affect the prospects for using some renewable energy sources.

Hydropower is a major source of electricity in some regions of the United States, notably in the Northwest.[191] It is likely to be significantly affected by climate change in regions subject to reduced precipitation and/or water from melting snowpack. Significant changes are already being detected in the timing and amount of streamflows in many western rivers,[164] consistent with the predicted effects of global warming. More precipitation coming as rain rather than snow, reduced snowpack, earlier peak runoff, and related effects are beginning to affect hydropower availability.[164]

Hydroelectric generation is very sensitive to changes in precipitation and river discharge. For example, every 1 percent decrease in precipitation results in a 2 to 3 percent drop in streamflow;[219] every 1 percent decrease in streamflow in the Colorado River Basin results in a 3 percent drop in power generation.[191] Such magnifying sensitivities occur because water flows through multiple power plants in a river basin.[191]

Climate impacts on hydropower occur when either the total amount or the timing of runoff is altered, such as when natural water storage in snowpack and glaciers is reduced under hotter conditions. Glaciers, snowpack, and their associated runoff are already declining in the West, and larger declines are projected.[164]

Hydropower operations are also affected by changes to air temperatures, humidity, or wind patterns due to climate change.[191] These variables cause changes in water quantity and quality, including water temperature. Warmer air and water generally increase the evaporation of water from the surface

of reservoirs, reducing the amount of water available for power production and other uses. Huge reservoirs with large surface areas, located in arid, sunny parts of the country, such as Lake Mead (located on Arizona-Nevada border on the Colorado River), are particularly susceptible to increased evaporation due to warming, meaning less water will be available for all uses, including hydropower.[191] And, where hydropower dams flow into waterways that

Hydroelectric dam in the Northwest

support trout, salmon or other coldwater fisheries, warming of reservoir releases might have detrimental consequences that require changes in operations that reduce power production.[191] Such impacts will increasingly translate into competition for water resources.

Climate change is also likely to affect other renewable energy sources. For example, changing cloud cover affects solar energy resources, changes in winds affect wind power, and temperature and water availability affect biomass production (particularly related to water requirements for biofuels).[191] The limited research to date on these important issues does not support firm conclusions about where such impacts would occur and how significant they would be.[205] This is an area that calls for much more study (see *An Agenda for Climate Impacts Science* section, Recommendation 2).

Regional Spotlight: Energy Impacts of Alaska's Rapid Warming

Significant impacts of warming on the energy sector can already be observed in Alaska, where temperatures have risen about twice as much as the rest of the nation. In Alaska, frozen ground and ice roads are an important means of winter travel, and warming has resulted in a much shorter cold season. Impacts on the oil and natural gas industries on Alaska's North Slope have been one of the results. For example, the season during which oil and gas exploration and extraction equipment can be operated on the tundra has been shortened due to warming. In addition, the thawing of permafrost, on which buildings, pipelines, airfields, and coastal installations supporting oil and gas development are located, adversely affects these structures and increases the cost of maintaining them.[191]

Different energy impacts are expected in the marine environment as sea ice continues to retreat and thin. These trends are expected to improve shipping accessibility, including oil and gas transport by sea, around the margins of the Arctic Basin, at least in the summer. The improved accessibility, however, will not be uniform throughout the different regions. Offshore oil exploration and extraction might benefit from less extensive and thinner sea ice, although equipment will have to be designed to withstand increased wave forces and ice movement.[191,220]

Transportation

Key Messages:

- Sea-level rise and storm surge will increase the risk of major coastal impacts, including both temporary and permanent flooding of airports, roads, rail lines, and tunnels.
- Flooding from increasingly intense downpours will increase the risk of disruptions and delays in air, rail, and road transportation, and damage from mudslides in some areas.
- The increase in extreme heat will limit some transportation operations and cause pavement and track damage. Decreased extreme cold will provide some benefits such as reduced snow and ice removal costs.
- Increased intensity of strong hurricanes would lead to more evacuations, infrastructure damage and failure, and transportation interruptions.
- Arctic warming will continue to reduce sea ice, lengthening the ocean transport season, but also resulting in greater coastal erosion due to waves. Permafrost thaw in Alaska will damage infrastructure. The ice road season will become shorter.

Key Sources

CCSP 1.2 Past Climate | CCSP 3.3 Extremes | CCSP 4.3 Impacts | CCSP 4.7 Transportation

IPCC WG-1 | IPCC WG-2 | NRC Transportation Impacts | ACIA Arctic Impacts | NAST U.S. Impacts

The U.S. transport sector is a significant source of greenhouse gases, accounting for 27 percent of U.S. emissions.[221] While it is widely recognized that emissions from transportation have a major impact on climate, climate change will also have a major impact on transportation.

Climate change impacts pose significant challenges to our nation's multi-modal transportation system and cause disruptions in other sectors across the economy. For example, major flooding in the Midwest in 1993 and 2008 restricted regional travel of all types, and disrupted freight and rail shipments across the country, such as those bringing coal to power plants and chlorine to water treatment systems. The U.S. transportation network is vital to the nation's economy, safety, and quality of life.

Extreme events present major challenges for transportation, and such events are becoming more frequent and intense. Historical weather patterns are no longer a reliable predictor of the future.[222] Transportation planners have not typically accounted for climate change in their long-term planning and project development. The longevity of transportation infrastructure, the long-term nature of climate change, and the potential impacts identified by recent studies warrant serious attention to climate change in planning new or rehabilitated transportation systems.[223]

Buildings and debris float up against a railroad bridge on the Cedar River during record flooding in June 2008, in Cedar Rapids, Iowa.

The strategic examination of national, regional, state, and local networks is an important step toward understanding the risks posed by climate change. A range of adaptation responses can be employed to reduce risks through redesign or relocation of infrastructure, increased redundancy of critical services, and operational improvements. Adapting to climate change is an evolutionary process. Through adoption of longer planning horizons, risk management, and adaptive responses, vulnerable transportation infrastructure can be made more resilient.[215]

Sea-level rise and storm surge will increase the risk of major coastal impacts, including both temporary and permanent flooding of airports, roads, rail lines, and tunnels.

Sea-level rise

Transportation infrastructure in U.S. coastal areas is increasingly vulnerable to sea-level rise. Given the high population density near the coasts, the potential exposure of transportation infrastructure to flooding is immense. Population swells in these areas during the summer months because beaches are very important tourist destinations.[222]

In the Gulf Coast area alone, an estimated 2,400 miles of major roadway and 246 miles of freight rail lines are at risk of permanent flooding within 50 to 100 years as global warming and land subsidence (sinking) combine to produce an anticipated relative sea-level rise in the range of 4 feet.[217] Since the Gulf Coast region's transportation network is interdependent and relies on minor roads and other low-lying infrastructure, the risks of service disruptions due to sea-level rise are likely to be even greater.[217]

Coastal areas are also major centers of economic activity. Six of the nation's top 10 freight gateways (measured by the value of shipments) will be threatened by sea-level rise.[222] Seven of the 10 largest ports (by tons of traffic) are located on the Gulf Coast.[222] The region is also home to the U.S. oil and gas industry, with its offshore drilling platforms, refineries, and pipelines. Roughly two-thirds of all U.S. oil imports are transported through this region[224] (see *Energy* sector). Sea-level rise would potentially affect commercial transportation activity valued in the hundreds of billions of dollars annually through inundation of area roads, railroads, airports, seaports, and pipelines.[217]

Storm surge

More intense storms, especially when coupled with sea-level rise, will result in far-reaching and damaging storm surges. An estimated 60,000 miles of coastal highway are already exposed to periodic flooding from coastal storms and high waves.[222] Some of these highways currently serve as evacuation routes during hurricanes and other coastal storms, and these routes could become seriously compromised in the future.

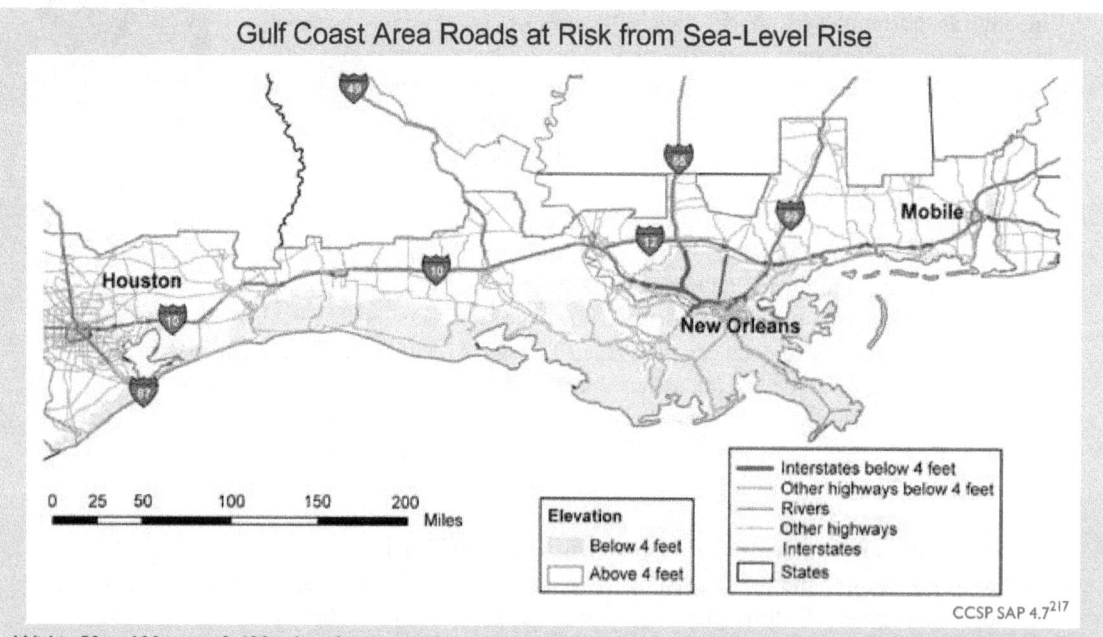

Gulf Coast Area Roads at Risk from Sea-Level Rise

Elevation	—— Interstates below 4 feet
	---- Other highways below 4 feet
▨ Below 4 feet	—— Rivers
☐ Above 4 feet	---- Other highways
	—— Interstates
	☐ States

CCSP SAP 4.7[217]

Within 50 to 100 years, 2,400 miles of major roadway are projected to be inundated by sea-level rise in the Gulf Coast region. The map shows roadways at risk in the event of a sea-level rise of about 4 feet, within the range of projections for this region in this century under medium- and high-emissions scenarios.[91] In total, 24 percent of interstate highway miles and 28 percent of secondary road miles in the Gulf Coast region are at elevations below 4 feet.[217]

Regional Spotlight: Gulf Coast

Sea-level rise, combined with high rates of subsidence in some areas, will make much of the existing infrastructure more prone to frequent or permanent inundation; 27 percent of the major roads, 9 percent of the rail lines, and 72 percent of the ports in the area shown on the map on the previous page are built on land at or below 4 feet in elevation, a level within the range of projections for relative sea-level rise in this region in this century. Increased storm intensity may lead to increased service disruption and infrastructure damage. More than half of the area's major highways (64 percent of interstates, 57 percent of arterials), almost half of the rail miles, 29 airports, and virtually all of the ports, are below 23 feet in elevation and subject to flooding and damage due to hurricane storm surge. These factors merit consideration in today's transportation decisions and planning processes.[217]

Coastal areas are projected to experience continued development pressures as both retirement and tourist destinations. Many of the most populous counties of the Gulf Coast, which already experience the effects of tropical storms, are expected to grow rapidly in the coming decades.[222] This growth will generate demand for more transportation infrastructure and services, challenging transportation planners to meet the demand, address current and future flooding, and plan for future conditions.[223]

Land

More frequent inundation and interruptions in travel on coastal and low-lying roadways and rail lines due to storm surge are projected, potentially requiring changes to minimize disruptions. More frequent evacuations due to severe storm surges are also likely. Across the United States, many coastal cities have subways, tunnels, parking lots, and other transportation infrastructure below

ground. Underground tunnels and other low-lying infrastructure will experience more frequent and severe flooding. Higher sea levels and storm surges will also erode road base and undermine bridge supports. The loss of coastal wetlands and barrier islands will lead to further coastal erosion due to the loss of natural protection from wave action.

Water

Impacts on harbor infrastructure from wave damage and storm surges are projected to increase. Changes will be required in harbor and port facilities to accommodate higher tides and storm surges. There will be reduced clearance under some waterway bridges for boat traffic. Changes in the navigability of channels are expected; some will become more accessible (and extend farther inland) because of deeper waters, while others will be restricted because of changes in sedimentation rates and sandbar locations. In some areas, waterway systems will become part of open water as barrier islands disappear. Some channels are likely to have to be dredged more frequently as has been done across large open-water bodies in Texas.[222]

Regional Spotlight: New York Metropolitan Area

With the potential for significant sea-level rise estimated under continued high levels of emissions, the combined effects of sea-level rise and storm surge are projected to increase the frequency of flooding. What is currently called a 100-year storm is projected to occur as often as every 10 years by late this century. Portions of lower Manhattan and coastal areas of Brooklyn, Queens, Staten Island, and Nassau County, would experience a marked increase in flooding frequency. Much of the critical transportation infrastructure, including tunnels, subways, and airports, lies well within the range of projected storm surge and would be flooded during such events.[222,225,369]

Air

Airports in coastal cities are often located adjacent to rivers, estuaries, or open ocean. Airport runways in coastal areas face inundation unless effective protective measures are taken. There is the potential for closure or restrictions for several of the nation's busiest airports that lie in coastal zones, affecting service to the highest density populations in the United States.

Flooding from increasingly intense downpours will increase the risk of disruptions and delays in air, rail, and road transportation, and damage from mudslides in some areas.

Heavy downpours have already increased substantially in the United States; the heaviest 1 percent of precipitation events increased by 20 percent, while total precipitation increased by only 7 percent over the past century.[112] Such intense precipitation is likely to increase the frequency and severity of events such as the Great Flood of 1993, which caused catastrophic flooding along 500 miles of the Mississippi and Missouri river system, paralyzing surface transportation systems, including rail, truck, and marine traffic. Major east-west traffic was halted for roughly six weeks in an area stretching from St. Louis, Missouri, west to Kansas City, Missouri and north to Chicago, Illinois, affecting one-quarter of all U.S. freight, which either originated or terminated in the flood-affected region.[222]

The June 2008 Midwest flood was the second record-breaking flood in the past 15 years. Dozens of levees were breached or overtopped in Iowa, Illinois, and Missouri, flooding huge areas, including nine square miles in and around Cedar Rapids, Iowa. Numerous highway and rail bridges were impassable due to flooding of approaches and transport was shut down along many stretches of highway, rail lines, and normally navigable waterways.

Planners have generally relied on weather extremes of the past as a guide to the future, planning, for example, for a "100-year flood," which is now likely to come more frequently as a result of

climate change. Historical analysis of weather data has thus become less reliable as a forecasting tool. The accelerating changes in climate make it more difficult to predict the frequency and intensity of weather events that can affect transportation.[222]

Land

The increase in heavy precipitation will inevitably cause increases in weather-related accidents, delays, and traffic disruptions in a network already challenged by increasing congestion.[215] There will be increased flooding of evacuation routes, and construction activities will be disrupted. Changes in rain, snowfall, and seasonal flooding will impact safety and maintenance operations on the nation's roads and railways. For example, if more precipitation falls as rain rather than snow in winter and spring, there will be an increased risk of landslides, slope failures, and floods from the runoff, causing road closures as well as the need for road repair and reconstruction[222] (see *Water Resources* sector).

Increased flooding of roadways, rail lines, and underground tunnels is expected. Drainage systems will be overloaded more frequently and severely, causing backups and street flooding. Areas where flooding is already common will face more frequent and severe problems. For example, Louisiana Highway 1, a critical link in the transport of oil from the Gulf of Mexico, has recently experienced increased flooding, prompting authorities to elevate the road (see Adaptation Box page 58).[217] Increases in road washouts, damage to railbed support structures, and landslides and mudslides that damage roads and other infrastructure are expected. If soil moisture levels become too high, the structural integrity of roads, bridges, and tunnels, which in some cases are already under age-related stress and in need of repair, could be compromised. Standing water will have adverse impacts on road base. For example, damage due to long term submersion of roadways in Louisiana was estimated to be $50 million for just 200 miles of state-owned highway. The Louisiana Department of Transportation and Development noted that a total of 1,800 miles of roads were under water for long periods, requiring costly repairs.[217] Pipelines are likely to be damaged because intense precipitation can cause the ground to sink underneath the pipeline; in shallow river-

Completion of a road around the 42-square mile island of Kosrae in the U.S.-affiliated Federated States of Micronesia provides a good example of adaptation to climate change. A road around the island's perimeter existed, except for a 10-mile gap. Filling this gap would provide all-weather land access to a remote village and allow easier access to the island's interior.

In planning this new section of road, authorities decided to "climate-proof" it against projected increases in heavy downpours and sea-level rise. This led to the section of road being placed higher above sea level and with an improved drainage system to handle the projected heavier rainfall. While there were additional capital costs for incorporating this drainage system, the accumulated costs, including repairs and maintenance, would be lower after about 15 years, equating to a good rate of return on investment. Adding this improved drainage system to roads that are already built is more expensive than on new construction, but still has been found to be cost effective.[226]

beds, pipelines are more exposed to the elements and can be subject to scouring and shifting due to heavy precipitation.[217]

Water

Facilities on land at ports and harbors will be vulnerable to short term flooding from heavy downpours, interrupting shipping service. Changes in silt and debris buildup resulting from extreme precipitation events will affect channel depth, increasing dredging costs. The need to expand stormwater treatment facilities, which can be a significant expense for container and other terminals with large impermeable surfaces, will increase.

Air

Increased delays due to heavy downpours are likely to affect operations, causing increasing flight delays and cancellations.[222] Stormwater runoff that exceeds the capacity of collection and drainage systems will cause flooding, delays, and airport closings. Heavy downpours will affect the structural integrity of airport facilities, such as through flood damage to runways and other infrastructure. All of these impacts have implications for emergency evacuation planning, facility maintenance, and safety.[222]

The increase in extreme heat will limit some transportation operations and cause pavement and track damage. Decreased extreme cold will provide some benefits such as reduced snow and ice removal costs.

Land

Longer periods of extreme heat in summer can damage roads in several ways, including softening of asphalt that leads to rutting from heavy traffic.[164] Sustained air temperature over 90°F is a significant threshold for such problems (see maps page 34). Extreme heat can cause deformities in rail tracks, at minimum resulting in speed restrictions and, at worst, causing derailments. Air temperatures above 100°F can lead to equipment failure (see maps page 90). Extreme heat also causes thermal expansion of bridge joints, adversely affecting bridge operations and increasing maintenance costs. Vehicle overheating and tire deterioration are additional concerns.[222] Higher temperatures will also increase refrigeration needs for goods during transport, particularly in the South, raising transportation costs.[217]

Increases in very hot days and heat waves are expected to limit construction activities due to health and safety concerns for highway workers. Guid-

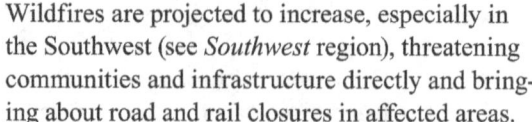

An example of intense precipitation affecting transportation infrastructure was the record-breaking 24-hour rainstorm in July 1996, which resulted in flash flooding in Chicago and its suburbs, with major impacts. Extensive travel delays occurred on metropolitan highways and railroads, and streets and bridges were damaged. Commuters were unable to reach Chicago for up to three days, and more than 300 freight trains were delayed or rerouted.[222]

The June 2008 Midwest floods caused I-80 in eastern Iowa to be closed for more than five days, disrupting major east-west shipping routes for trucks and the east-west rail lines through Iowa. These floods exemplify the kind of extreme precipitation events and their direct impacts on transportation that are likely to become more frequent in a warming world. These extremes create new and more difficult problems that must be addressed in the design, construction, rehabilitation, and operation of the nation's transportation infrastructure.

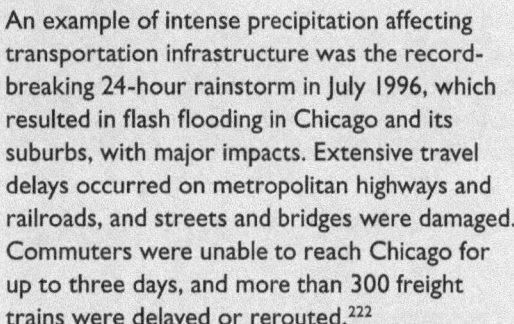

ance from the U.S. Occupational Safety and Health Administration states that concern for heat stress for moderate to heavy work begins at about 80°F as measured by an index that combines temperature, wind, humidity, and direct sunlight. For dry climates, such as Phoenix and Denver, National Weather Service heat indices above 90°F might allow work to proceed, while higher humidity areas such as New Orleans or Miami should consider 80 to 85°F as an initial level for work restrictions.[227] These trends and associated impacts will be exacerbated in many places by urban heat island effects (see *Human Health* and *Society* sectors).

Wildfires are projected to increase, especially in the Southwest (see *Southwest* region), threatening communities and infrastructure directly and bringing about road and rail closures in affected areas.

In many northern states, warmer winters will bring about reductions in snow and ice removal costs, lessen adverse environmental impacts from the use of salt and chemicals on roads and bridges, extend the construction season, and improve the mobility and safety of passenger and freight travel through reduced winter hazards. On the other hand, more freeze-thaw conditions are projected to occur in northern states, creating frost heaves and potholes on road and bridge surfaces and resulting in load restrictions on certain roads to minimize the damage. With the expected earlier onset of seasonal warming, the period of springtime load restrictions might be reduced in some areas, but it is likely to expand in others with shorter winters but longer thaw seasons. Longer construction seasons will be a benefit in colder locations.[222]

Water

Warming is projected to mean a longer shipping season but lower water levels for the Great Lakes and St. Lawrence Seaway. Higher temperatures, reduced lake ice, and increased evaporation are expected to combine to produce lower water levels as climate warming proceeds (see *Midwest* region). With lower lake levels, ships will be unable to carry as much cargo and hence shipping costs will increase. A recent study, for example, found that the projected reduction in Great Lakes water levels would result in an estimated 13 to 29 percent increase in shipping costs for Canadian commercial navigation by 2050, all else remaining equal.[222]

If low water levels become more common because of drier conditions due to climate change, this could create problems for river traffic, reminiscent of the stranding of more than 4,000 barges on the Mississippi River during the drought in 1988. Freight movements in the region could be seriously impaired, and extensive dredging could be required to keep shipping channels open. On the other hand, a longer shipping season afforded by a warmer climate could offset some of the resulting adverse economic effects.

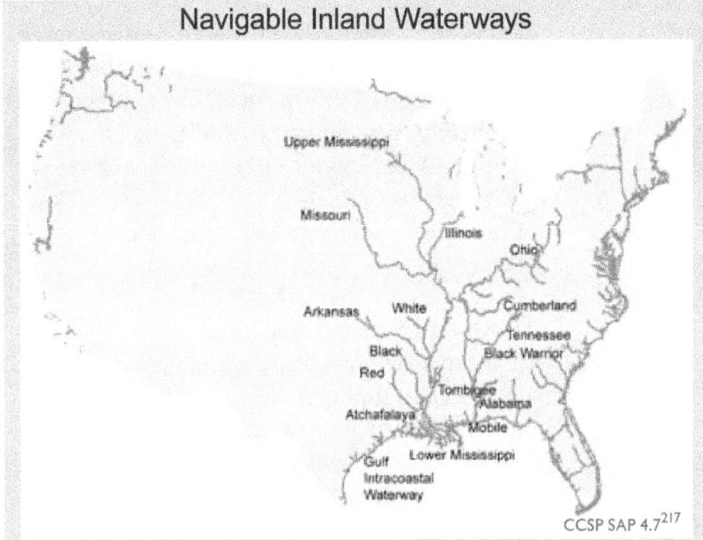

Navigable Inland Waterways

Upper Mississippi
Missouri
Illinois
Ohio
Arkansas   White   Cumberland
Black   Tennessee
Red   Black Warrior
Tombigbee
Alabama
Atchafalaya   Mobile
Gulf   Lower Mississippi
Intracoastal
Waterway

CCSP SAP 4.7[217]

Inland waterways are an important part of the transportation network in various parts of the United States. For example, these waterways provide 20 states with access to the Gulf of Mexico.[217] As conditions become drier, these main transportation pathways are likely to be adversely affected by the resulting lower water levels, creating problems for river traffic. Names of navigable rivers are shown above.

In cold areas, the projected decrease in very cold days will mean less ice accumulation on vessels, decks, riggings, and docks; less ice fog; and fewer ice jams in ports.[222]

Air

Rising temperatures will affect airport ground facilities, runways in particular, in much the same way they affect roads. Airports in some areas are likely to benefit from reduction in the cost of snow and ice removal and the impacts of salt and chemical use, though some locations have seen increases in snowfall. Airlines could benefit from reduced need to de-ice planes.

More heat extremes will create added operational difficulties, for example, causing greater energy consumption by planes on the ground. Extreme heat also affects aircraft lift; because hotter air is less dense, it reduces the lift produced by the wing and the thrust produced by the engine – problems exacerbated at high altitudes and high temperatures. As a result, planes need to take off faster, and if runways are not sufficiently long for aircraft to build up enough speed to generate lift, aircraft weight must be reduced. Thus, increases in extreme heat will result in payload restrictions, could cause flight cancellations and service disruptions

at affected airports, and could require some airports to lengthen runways. Recent hot summers have seen flights cancelled due to heat, especially in high altitude locations. Economic losses are expected at affected airports. A recent illustrative analysis projects a 17 percent reduction in freight carrying capacity for a single Boeing 747 at the Denver airport by 2030 and a 9 percent reduction at the Phoenix airport due to increased temperature and water vapor.[222]

Drought

Rising air temperatures increase evaporation, contributing to dry conditions, especially when accompanied by decreasing precipitation. Even where total annual precipitation does not decrease, precipitation is projected to become less frequent in many parts of the country.[68] Drought is expected to be an increasing problem in some regions; this, in turn, has impacts on transportation. For example, increased susceptibility to wildfires during droughts could threaten roads and other transportation infrastructure directly, or cause road closures due to fire threat or reduced visibility such as has occurred in Florida and California in recent years. There is also increased susceptibility to mudslides in areas deforested by wildfires. Airports could suffer from decreased visibility due to wildfires. River transport is seriously affected by drought, with reductions in the routes available, shipping season, and cargo carrying capacity.

Increased intensity of strong hurricanes would lead to more evacuations, infrastructure damage and failure, and transportation interruptions.

More intense hurricanes in some regions are a projected effect of climate change. Three aspects of tropical storms are relevant to transportation: precipitation, winds, and wind-induced storm surge. Stronger hurricanes have longer periods of intense precipitation, higher wind speeds (damage increases exponentially with wind speed[228]),

and higher storm surge and waves. Transportation planners, designers, and operators may need to adopt probabilistic approaches to developing transportation projects rather than relying on standards and the deterministic approaches of the past. The uncertainty associated with projecting impacts over a 50- to 100-year time period makes risk management a reasonable approach for realistically incorporating climate change into decision making and investment.[215]

Land

There will be a greater probability of infrastructure failures such as highway and rail bridge decks being displaced and railroad tracks being washed away. Storms leave debris on roads and rail lines, which can damage the infrastructure and interrupt travel and shipments of goods. In Louisiana, the Department of Transportation and

Development spent $74 million for debris removal alone in the wake of hurricanes Katrina and Rita. The Mississippi Department of Transportation expected to spend in excess of $1 billion to replace the Biloxi and Bay St. Louis bridges, repair other portions of roadway, and remove debris. As of June 2007, more than $672 million had been spent.

There will be more frequent and potentially more extensive emergency evacuations. Damage to signs, lighting fixtures, and supports will increase. The lifetime of highways that have been exposed to flooding is expected to decrease. Road and rail infrastructure for passenger and freight services are likely to face increased flooding by strong hurricanes. In the Gulf Coast, more than one-third of the rail miles are likely to flood when subjected to a storm surge of 18 feet.[217]

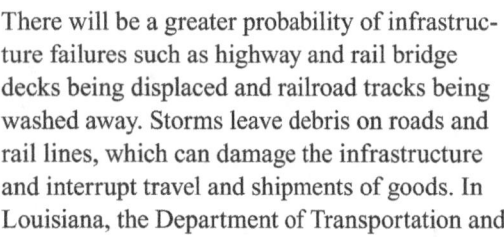

Spotlight on Hurricane Katrina

Hurricane Katrina was one of the most destructive and expensive natural disasters in U.S. history, claiming more than 1,800 lives and causing an estimated $134 billion in damage.[217,229] It also seriously disrupted transportation systems as key highway and railroad bridges were heavily damaged or destroyed, necessitating rerouting of traffic and placing increased strain on other routes, particularly other rail lines. Replacement of major infrastructure took from months to years. The CSX Gulf Coast line was re-opened after five months and $250 million in reconstruction costs, while the Biloxi-Ocean Springs Bridge took more than two years to reopen. Barge shipping was halted, as was grain export out of the Port of New Orleans, the nation's largest site of grain exports. The extensive oil and gas pipeline network was shut down by the loss of electrical power, producing shortages of natural gas and petroleum products. Total recovery costs for the roads, bridges, and utilities as well as debris removal have been estimated at $15 billion to $18 billion.[217]

Redundancies in the transportation system, as well as the storm timing and track, helped keep the storm from having major or long-lasting impacts on national-level freight flows. For example, truck traffic was diverted from the collapsed bridge that carries highway I-10 over Lake Pontchartrain to highway I-12, which parallels I-10 well north of the Gulf Coast. The primary north-south highways that connect the Gulf Coast with major inland transportation hubs were not damaged and were open for nearly full commercial freight movement within days. The railroads were able to route some traffic not bound directly for New Orleans through Memphis and other Midwest rail hubs. While a disaster of historic proportions, the effects of Hurricane Katrina could have been even worse if not for the redundancy and resilience of the transportation network in the area.

Hurricane Katrina damage to bridge

Water

All aspects of shipping are disrupted by major storms. For example, freight shipments need to be diverted from the storm region. Activities at offshore drilling sites and coastal pumping facilities are generally suspended and extensive damage to these facilities can occur, as was amply demonstrated during the 2005 hurricane season. Refineries and pipelines are also vulnerable to damage and disruption due to the high winds and storm surge associated with hurricanes and other tropical storms (see *Energy* sector). Barges that are unable to get to safe harbors can be destroyed or severely damaged. Waves and storm surge will damage harbor infrastructure such as cranes, docks, and other terminal facilities. There are implications for emergency evacuation planning, facility maintenance, and safety management.

Air

More frequent interruptions in air service and airport closures can be expected. Airport facilities including terminals, navigational equipment, perimeter fencing, and signs are likely to sustain increased wind damage. Airports are frequently located in low-lying areas and can be expected to flood with more intense storms. As a response to this vulnerability, some airports, such as LaGuardia in New York City, are already protected by levees. Eight airports in the Gulf Coast region of Louisiana and Texas are located in historical 100-year flood plains; the 100-year flood events will be more frequent in the future, creating the likelihood of serious costs and disruption.[217]

Arctic warming will continue to reduce sea ice, lengthening the ocean transport season, but also resulting in greater coastal erosion due to waves. Permafrost thaw in Alaska will damage infrastructure. The ice road season will become shorter.

Special issues in Alaska

Warming has been most rapid in high northern regions. As a result, Alaska is warming at twice the rate of the rest of the nation, bringing both major opportunities and major challenges. Alaska's transportation infrastructure differs sharply from that of the lower 48 states. Although Alaska is twice the size of Texas, its population and road mileage are more like Vermont's. Only 30 percent of Alaska's roads are paved. Air travel is much more common than in other states. Alaska has 84 commercial airports and more than 3,000 airstrips, many of which are the only means of transport for rural communities. Unlike other states, over much of Alaska, the land is generally more accessible in winter, when the ground is frozen and ice roads and bridges formed by frozen rivers are available.

Sea ice decline

The striking thinning and downward trend in the extent of Arctic sea ice is regarded as a considerable opportunity for shippers. Continued reduction in sea ice should result in opening of additional ice-free ports, improved access to ports and natural resources in remote areas, and longer shipping seasons, but it is likely to increase erosion rates on land as well, raising costs for maintaining ports and other transportation infrastructure.[132,220]

Later this century and beyond, shippers are looking forward to new Arctic shipping routes, including the fabled Northwest Passage, which could provide significant costs savings in shipping times and distances. However, the next few decades are likely to be very unpredictable for shipping through these new routes. The past three decades have seen very high year-to-year variability of sea ice extent in the Canadian Arctic, despite the overall decrease in September sea ice extent. The loss of sea ice from the shipping channels of the Canadian Archipelago might actually allow more frequent intrusions of icebergs, which would continue to impede shipping through the Northwest Passage.

Lack of sea ice, especially on the northern shores of Alaska, creates conditions whereby storms produce waves that cause serious coastal erosion.[137,219] Already a number of small towns, roads, and airports are threatened by retreating coastlines, necessitating the planned relocation of these communities (see *Alaska* region).[132,220]

Thawing ground

The challenges warming presents for transportation on land are considerable.[164] For highways, thawing of permafrost causes settling of the roadbed and

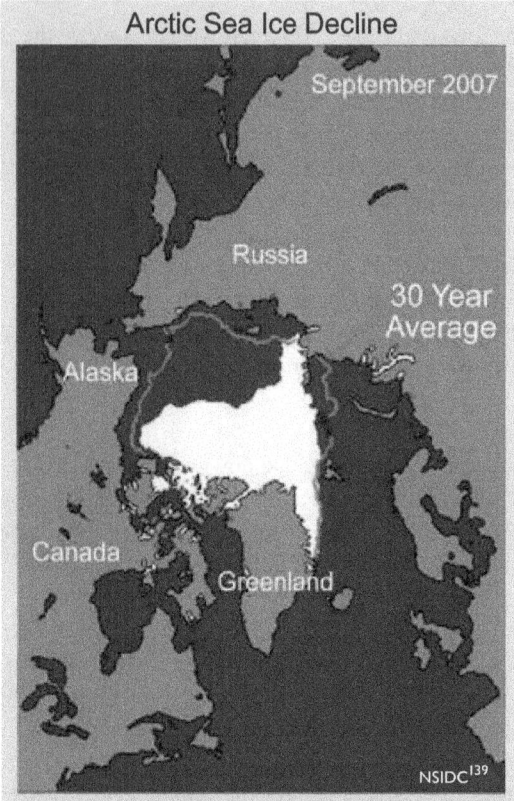

Arctic Sea Ice Decline

September 2007

Russia

30 Year Average

Alaska

Canada

Greenland

NSIDC[139]

The pink line shows the average September sea ice extent from 1979 through the present. The white area shows September 2007 sea ice extent. In 2008, the extent was slightly larger than 2007, but the ice was thinner, resulting in a lower total volume of sea ice. In addition, recent years have had less ice that persisted over numerous years and more first-year ice, which melts more quickly.[139]

frost heaves that adversely affect the integrity of the road structure and its load-carrying capacity. The majority of Alaska's highways are located in areas where permafrost is discontinuous, and dealing with thaw settlement problems already claims a significant portion of highway maintenance dollars.

Bridges and large culverts are particularly sensitive to movement caused by thawing permafrost and are often much more difficult than roads to repair and modify for changing site conditions. Thus, designing these facilities to take climate change into account is even more critical than is the case for roads.

Another impact of climate change on bridges is increased scouring. Hotter, drier summers in Alaska have led to increased glacial melting and longer periods of high streamflows, causing both increased

sediment in rivers and scouring of bridge supporting piers and abutments. Temporary ice roads and bridges are commonly used in many parts of Alaska to access northern communities and provide support for the mining and oil and gas industries. Rising temperatures have already shortened the season during which these critical facilities can be used. Like the highway system, the Alaska Railroad crosses permafrost terrain, and frost heave and settlement from thawing affect some portions of the track, increasing maintenance costs.[28,132,220]

A significant number of Alaska's airstrips in the southwest, northwest, and interior of the state are built on permafrost. These airstrips will require major repairs or relocation if their foundations are compromised by thawing.

The cost of maintaining Alaska's public infrastructure is projected to increase 10 to 20 percent by 2030 due to warming, costing the state an additional $4 billion to $6 billion, with roads and airports accounting for about half of this cost.[230] Private infrastructure impacts have not been evaluated.[217]

The Trans-Alaska Pipeline System, which stretches from Prudhoe Bay in the north to the ice-free port of Valdez in the south, crosses a wide range of permafrost types and varying temperature conditions. More than half of the 800-mile pipeline is elevated on vertical supports over potentially unstable permafrost. Because the system was designed in the early 1970s on the basis of permafrost and climate conditions of the 1950 to 1970 period, it requires continuous monitoring and some supports have had to be replaced.

Travel over the tundra for oil and gas exploration and extraction is limited to the period when the ground is sufficiently frozen to avoid damage to the fragile tundra. In recent decades, the number of days that exploration and extraction equipment could be used has dropped from 200 days to 100 days per year due to warming.[220] With continued warming, the number of exploration days is expected to decline even more.

Agriculture

Key Messages:

- Many crops show positive responses to elevated carbon dioxide and low levels of warming, but higher levels of warming often negatively affect growth and yields.
- Extreme events such as heavy downpours and droughts are likely to reduce crop yields because excesses or deficits of water have negative impacts on plant growth.
- Weeds, diseases, and insect pests benefit from warming, and weeds also benefit from a higher carbon dioxide concentration, increasing stress on crop plants and requiring more attention to pest and weed control.
- Forage quality in pastures and rangelands generally declines with increasing carbon dioxide concentration because of the effects on plant nitrogen and protein content, reducing the land's ability to supply adequate livestock feed.
- Increased heat, disease, and weather extremes are likely to reduce livestock productivity.

Key Sources

Agriculture in the United States is extremely diverse in the range of crops grown and animals raised, and produces over $200 billion a year in food commodities, with livestock accounting for more than half. Climate change will increase productivity in certain crops and regions and reduce productivity in others (see for example *Midwest* and *Great Plains* regions).[193]

While climate change clearly affects agriculture, climate is also affected by agriculture, which contributes 13.5 percent of all human-induced greenhouse gas emissions globally. In the United States, agriculture represents 8.6 percent of the nation's total greenhouse gas emissions, including 80 percent of its nitrous oxide emissions and 31 percent of its methane emissions.[231]

Increased agricultural productivity will be required in the future to supply the needs of an increasing population. Agricultural productivity is dependent upon the climate and land resources. Climate change can have both beneficial and detrimental impacts on plants. Throughout history, agricultural enterprises have coped with changes in climate through changes in management and in crop or animal selection. However, under higher heat-trapping gas emissions scenarios, the projected climate changes are likely to increasingly challenge U.S. capacity to as efficiently produce food, feed, fuel, and livestock products.

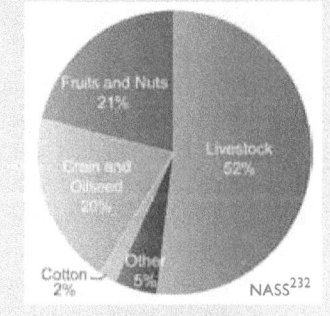

Relative Contributions to Agricultural Products, 2002

Fruits and Nuts 21%
Livestock 52%
Grain and Oilseed 20%
Cotton 2%
Other 5%

NASS[232]

Market Value of Agricultural Products Sold, 2002

1 Dot = $20 million
United States Total
$200.6 billion

NASS[232]

Many crops show positive responses to elevated carbon dioxide and low levels of warming, but higher levels of warming often negatively affect growth and yields.

Crop responses in a changing climate reflect the interplay among three factors: rising temperatures, changing water resources, and increasing carbon dioxide concentrations. Warming generally causes plants that are below their optimum temperature to grow faster, with obvious benefits. For some plants, such as cereal crops, however, faster growth means there is less time for the grain itself to grow and mature, reducing yields.[193] For some annual crops, this can be compensated for by adjusting the planting date to avoid late season heat stress.[164]

The grain-filling period (the time when the seed grows and matures) of wheat and other small grains shortens dramatically with rising temperatures. Analysis of crop responses suggests that even moderate increases in temperature will decrease yields of corn, wheat, sorghum, bean, rice, cotton, and peanut crops.[193]

Some crops are particularly sensitive to high nighttime temperatures, which have been rising even faster than daytime temperatures.[68] Nighttime temperatures are expected to continue to rise in the future. These changes in temperature are especially critical to the reproductive phase of growth because warm nights increase the respiration rate and reduce the amount of carbon that is captured during the day by photosynthesis to be retained in

the fruit or grain. Further, as temperatures continue to rise and drought periods increase, crops will be more frequently exposed to temperature thresholds at which pollination and grain-set processes begin to fail and quality of vegetable crops decreases. Grain, soybean, and canola crops have relatively low optimal temperatures, and thus will have reduced yields and will increasingly begin to experience failure as warming proceeds.[193] Common snap beans show substantial yield reduction when nighttime temperatures exceed 80°F.

Higher temperatures will mean a longer growing season for crops that do well in the heat, such as melon, okra, and sweet potato, but a shorter growing season for crops more suited to cooler conditions, such as potato, lettuce, broccoli, and spinach.[193] Higher temperatures also cause plants to use more water to keep cool. This is one example of how the interplay between rising temperatures and water availability is critical to how plants respond to climate change. But fruits, vegetables, and grains can suffer even under well-watered conditions if temperatures exceed the maximum level for pollen viability in a particular plant; if temperatures exceed the threshold for that plant, it won't produce seed and so it won't reproduce.[193]

Temperature increases will cause the optimum latitude for crops to move northward; decreases in temperature would cause shifts toward the equator. Where plants can be efficiently grown depends upon climate conditions, of which temperature is one of the major factors.

Plants need adequate water to maintain their temperature within an optimal range. Without water for cooling, plants will suffer heat stress. In many regions, irrigation water is used to maintain adequate temperature conditions for the growth of cool season plants (such as many vegetables), even in warm environments. With increasing demand and competition for freshwater supplies, the water needed for these crops might be increasingly limited. If water supply variability increases, it will affect plant growth and cause

Corn and Soybean Temperature Response

Corn

Soybean

Plant Growth Rate

Plant Growth Rate

Vegetative Response Curve ——
Optimum Range ▆▆▆
Reproductive Response Curve ——
Optimum Range ▆▆▆
Corn Failure at 95°F
Soybean Failure at 102°F

50 60 70 80 90 100
Air Temperature (°F)

50 60 70 80 90 100
Air Temperature (°F)

ARS USDA

For each plant variety, there is an optimal temperature for vegetative growth, with growth dropping off as temperatures increase or decrease. Similarly, there is a range of temperatures at which a plant will produce seed. Outside of this range, the plant will not reproduce. As the graphs show, corn will fail to reproduce at temperatures above 95°F and soybean above 102°F.

Increase in Percent of Very Warm Nights

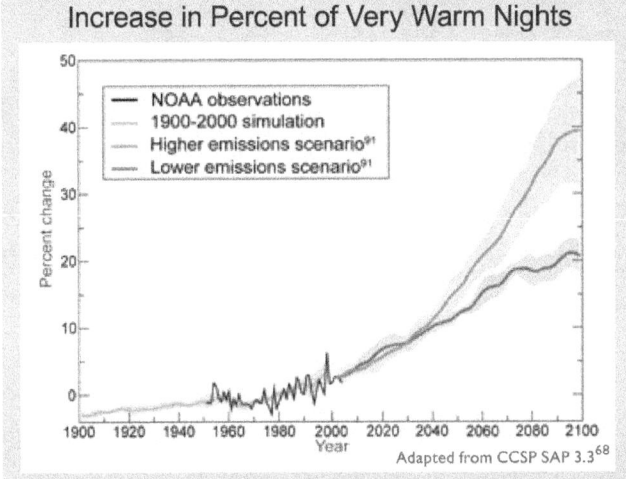

Adapted from CCSP SAP 3.3[68]

The graph shows the observed and projected change in percent of very warm nights from the 1950-1990 average in the United States. Under the lower emissions scenario,[91] the percentage of very warm nights is projected to increase about 20 percent by 2100. Under the higher emissions scenario,[91] it is projected to increase by about 40 percent.[68] The shaded areas show the likely ranges while the lines show the central projections from a set of climate models. The projections appear smooth because they show the calculated average of many models.

reduced yields. The amount and timing of precipitation during the growing season are also critical, and will be affected by climate change. Changes in season length are also important and affect crops differently.[193]

Higher carbon dioxide levels generally cause plants to grow larger. For some crops, this is not necessarily a benefit because they are often less nutritious, with reduced nitrogen and protein content. Carbon dioxide also makes some plants more water-use efficient, meaning they produce more plant material, such as grain, on less water.[193] This is a benefit in water-limited areas and in seasons with less than normal rainfall amounts.

In some cases, adapting to climate change could be as simple as changing planting dates, which can be an effective no- or low-cost option for taking advantage of a longer growing season or avoiding crop exposure to adverse climatic conditions such as high temperature stress or low rainfall periods. Effectiveness will depend on the region, crop, and the rate and amount of warming. It is unlikely to be effective if a farmer goes to market when the supply-demand balance drives prices down. Predicting the optimum planting date for maximum profits will be more challenging in a future with increased

uncertainty regarding climate effects on not only local productivity, but also on supply from competing regions.[193]

Another adaptation strategy involves changing to crop varieties with improved tolerance to heat or drought, or those that are adapted to take advantage of a longer growing season. This is less likely to be cost-effective for perennial crops, for which changing varieties is extremely expensive and new plantings take several years to reach maximum productivity. Even for annual crops, changing varieties is not always a low-cost option. Seed for new stress-tolerant varieties can be expensive, and new varieties often require investments in new planting equipment or require adjustments in a wide range of farming practices. In some cases, it is difficult to breed for genetic tolerance to elevated temperature or to identify an alternative variety that is adapted to the new climate and to local soils, practices, and market demands.

Fruits that require long winter chilling periods will experience declines. Many varieties of fruits (such as popular varieties of apples and berries) require between 400 and 1,800 cumulative hours below 45°F each winter to produce abundant yields the following summer and fall. By late this century, under higher emissions scenarios,[91] winter temperatures in many important fruit-producing regions such as the Northeast will be too consistently warm to meet these requirements. Cranberries have a particularly high chilling requirement, and there are no known low-chill varieties. Massachusetts and New Jersey supply nearly half the nation's cranberry crop. By the middle of this century, under higher emissions scenarios,[91] it is unlikely that these areas will support cranberry production due to a lack of the winter chilling they need.[233,234] Such impacts will vary by region. For example, though there will still be risks of early-season frosts and damaging winter thaws, warming is expected to improve the climate for fruit production in the Great Lakes region.[164]

A seemingly paradoxical impact of warming is that it appears to be increasing the risk of plant frost

Effects of Increased Air Pollution on Crop Yields

Ground-level ozone (a component of smog) is an air pollutant that is formed when nitrogen oxides emitted from fossil fuel burning interact with other compounds, such as unburned gasoline vapors, in the atmosphere,[237] in the presence of sunlight. Higher air temperatures result in greater concentrations of ozone. Ozone levels at the land surface have risen in rural areas of the United States over the past 50 years, and they are forecast to continue increasing with warming, especially under higher emissions scenarios.[91] Plants are sensitive to ozone, and crop yields are reduced as ozone levels increase. Some crops that are particularly sensitive to ozone pollution include soybeans, wheat, oats, green beans, peppers, and some types of cotton.[193]

damage. Mild winters and warm, early springs, which are beginning to occur more frequently as climate warms, induce premature plant development and blooming, resulting in exposure of vulnerable young plants and plant tissues to subsequent late-season frosts. For example, the 2007 spring freeze in the eastern United States caused widespread devastation of crops and natural vegetation because the frost occurred during the flowering period of many trees and during early grain development on wheat plants.[235] Another example is occurring in the Rocky Mountains where in addition to the process described above, reduced snow cover leaves young plants unprotected from spring frosts, with some plant species already beginning to suffer as a result[236] (see *Ecosystems* sector).

Extreme events such as heavy downpours and droughts are likely to reduce crop yields because excesses or deficits of water have negative impacts on plant growth.

One of the most pronounced effects of climate change is the increase in heavy downpours. Precipitation has become less frequent but more intense, and this pattern is projected to continue across the United States.[112] One consequence of excessive rainfall is delayed spring planting, which jeopardizes profits for farmers paid a premium for early season production of high-value crops such as melon, sweet corn, and tomatoes. Field flooding during the growing season causes crop losses due to low oxygen levels in the soil, increased susceptibility to root diseases, and increased soil compaction due to the use of heavy farm equipment on wet soils. In spring 2008, heavy rains caused the Mississippi River to rise to about 7 feet above flood

stage, inundating hundreds of thousands of acres of cropland. The flood hit just as farmers were preparing to harvest wheat and plant corn, soybeans, and cotton. Preliminary estimates of agricultural losses are around $8 billion.[213] Some farmers were put out of business and others will be recovering for years to come. The flooding caused severe erosion in some areas and also caused an increase in runoff and leaching of agricultural chemicals into surface water and groundwater.[233]

Another impact of heavy downpours is that wet conditions at harvest time result in reduced quality of many crops. Storms with heavy rainfall often are accompanied by wind gusts, and both strong winds and rain can flatten crops, causing significant damage. Vegetable and fruit crops are sensitive to even short-term, minor stresses, and as such are par-

U.S. Corn Yields 1960 to 2008

Yield: Bushels per Acre

Blight -18%

Wet spring, early frost -16%

Drought -17%

Drought -26%

Drought -29%

Flood

Unusual Climate Events

Year

Updated from NAST[219]

While technological improvements have resulted in a general increase in corn yields, extreme weather events have caused dramatic reductions in yields in particular years. Increased variation in yield is likely to occur as temperatures increase and rainfall becomes more variable during the growing season. Without dramatic technological breakthroughs, yields are unlikely to continue their historical upward trend as temperatures rise above the optimum level for vegetative and reproductive growth.

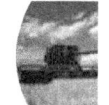

ticularly vulnerable to weather extremes.[193] More rainfall concentrated into heavy downpours also increases the likelihood of water deficiencies at other times because of reductions in rainfall frequency.

Drought frequency and severity are projected to increase in the future over much of the United States, particularly under higher emissions scenarios.[90,91] Increased drought will be occurring at a time when crop water requirements also are increasing due to rising temperatures. Water deficits are detrimental for all crops.[233]

Temperature extremes will also pose problems. Even crop species that are well-adapted to warmth, such as tomatoes, can have reduced yield and/or quality when daytime maximum temperatures exceed 90°F for even short periods during critical reproductive stages (see maps page 34).[112] For many high-value crops, just hours or days of moderate heat stress at critical growth stages can reduce grower profits by negatively affecting visual or flavor quality, even when total yield is not reduced.[238]

Weeds, diseases, and insect pests benefit from warming, and weeds also benefit from a higher carbon dioxide concentration, increasing stress on crop plants and requiring more attention to pest and weed control.

Weeds benefit more than cash crops from higher temperatures and carbon dioxide levels.[193] One concern with continued warming is the northward expansion of invasive weeds. Southern farmers currently lose more of their crops to weeds than do northern farmers. For example, southern farmers lose 64 percent of the soybean crop to weeds, while northern farmers lose 22 percent.[239] Some extremely aggressive weeds plaguing the South (such as kudzu) have historically been confined to areas where winter temperatures do not drop below specific thresholds. As temperatures continue to rise, these weeds will expand their ranges northward into important ag-

ricultural areas.[240] Kudzu currently has invaded 2.5 million acres of the Southeast and is a carrier of the fungal disease soybean rust, which represents a major and expanding threat to U.S. soybean production.[234]

Controlling weeds currently costs the United States more than $11 billion a year, with the majority spent on herbicides;[241] so both herbicide use and costs are likely to increase as temperatures and carbon dioxide levels rise. At the same time, the most widely used herbicide in the United States, glyphosate (RoundUp®), loses its efficacy on weeds grown at carbon dioxide levels that are projected to occur in the coming decades (see photos below). Higher concentrations of the chemical and more frequent spraying thus will be needed, increasing economic and environmental costs associated with chemical use.[233]

Many insect pests and crop diseases thrive due to warming, increasing losses and necessitating greater pesticide use. Warming aids insects and diseases in several ways. Rising temperatures allow both insects and pathogens to expand their ranges northward. In addition, rapidly rising winter temperatures allow more insects to survive over the winter, whereas cold winters once controlled their populations. Some of these insects, in addition to directly

Herbicide Loses Effectiveness at Higher CO₂

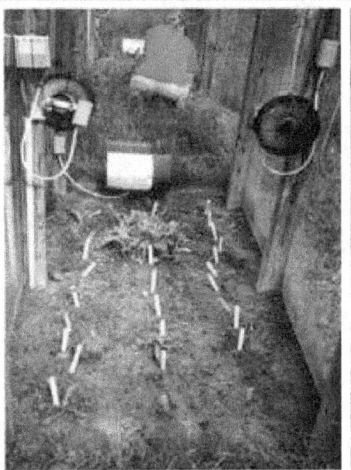 

Current CO₂ (380 ppm)   Potential Future CO₂ (680 ppm)

The left photo shows weeds in a plot grown at a carbon dioxide (CO₂) concentration of about 380 parts per million (ppm), which approximates the current level. The right photo shows a plot in which the CO₂ level has been raised to about 680 ppm. Both plots were equally treated with herbicide.[233]

damaging crops, also carry diseases that harm crops. Crop diseases in general are likely to increase as earlier springs and warmer winters allow proliferation and higher survival rates of disease pathogens and parasites.[193,234] The longer growing season will allow some insects to produce more generations in a single season, greatly increasing their populations. Finally, plants grown in higher carbon dioxide conditions tend to be less nutritious, so insects must eat more to meet their protein requirements, causing greater destruction to crops.[193]

Due to the increased presence of pests, spraying is already much more common in warmer areas than in cooler areas. For example, Florida sweet corn growers spray their fields 15 to 32 times a year to fight pests such as corn borer and corn earworm, while New York farmers average zero to five times.[193] In addition, higher temperatures are known to reduce the effectiveness of certain classes of pesticides (pyrethroids and spinosad).

A particularly unpleasant example of how carbon dioxide tends to favor undesirable plants is found in the response of poison ivy to rising carbon dioxide concentrations. Poison ivy thrives in air with extra carbon dioxide in it, growing bigger and producing a more toxic form of the oil, urushiol, which causes painful skin reactions in 80 percent of people. Contact with poison ivy is one of the most widely reported ailments at poison centers in the United States, causing more than 350,000 cases of contact dermatitis each year. The growth stimulation of poison ivy due to increasing carbon dioxide concentration exceeds that of most other woody species. Given continued increases in carbon dioxide emissions, poison ivy is expected to become more abundant and more toxic in the future, with implications for forests and human health.[234]

Higher temperatures, longer growing seasons, and increased drought will lead to increased agricultural water use in some areas. Obtaining the maxi-

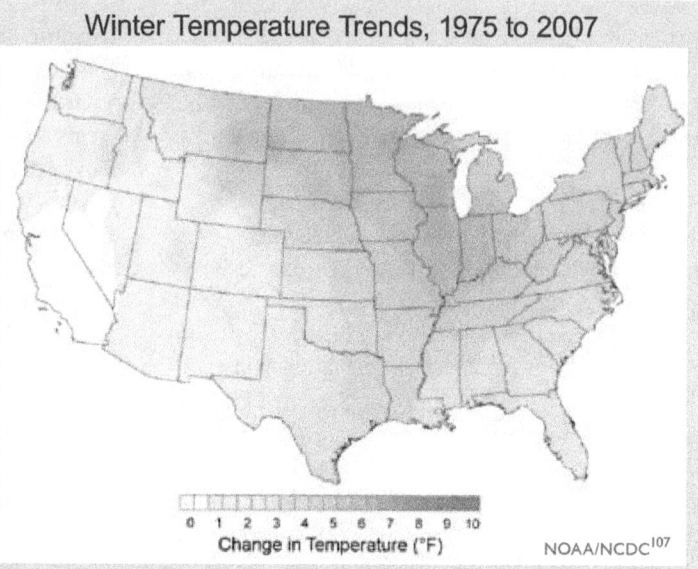

Winter Temperature Trends, 1975 to 2007

Change in Temperature (°F)

NOAA/NCDC[107]

Temperatures are rising faster in winter than in any other season, especially in many key agricultural regions. This allows many insect pests and crop diseases to expand and thrive, creating increasing challenges for agriculture. As indicated by the map, the Midwest and northern Great Plains have experienced increases of more than 7°F in average winter temperatures over the past 30 years.

mum "carbon dioxide fertilization" benefit often requires more efficient use of water and fertilizers that better synchronize plant demand with supply. Farmers are likely to respond to more aggressive and invasive weeds, insects, and pathogens with increased use of herbicides, insecticides, and fungicides. Where increases in water and chemical inputs become necessary, this will increase costs for the farmer, as well as having society-wide impacts by depleting water supply, increasing reactive nitrogen and pesticide loads to the environment, and increasing risks to food safety and human exposure to pesticides.

Forage quality in pastures and rangelands generally declines with increasing carbon dioxide concentration because of the effects on plant nitrogen and protein content, reducing the land's ability to supply adequate livestock feed.

Beef cattle production takes place in every state in the United States, with the greatest number raised in regions that have an abundance of native or planted pastures for grazing. Generally, eastern pasturelands are planted and managed, whereas western rangelands are native pastures, which are not seeded and receive much less rainfall. There are transformations now underway in many semi-arid rangelands as a result of increasing atmospheric carbon dioxide concentration and the associated climate change. These transformations include which species of grasses dominate, as well as the forage quality of the dominant grasses. Increases in carbon dioxide are generally reducing the quality of the forage, so that more acreage is needed to provide animals with the same nutritional value, resulting in an overall decline in livestock productivity. In addition, woody shrubs and invasive cheatgrass are encroaching into grasslands, further reducing their forage value.[193] The combination of these factors leads to an overall decline in livestock productivity.

While rising atmospheric carbon dioxide concentration increases forage quantity, it has negative impacts on forage quality because plant nitrogen and protein concentrations often decline with higher concentrations of carbon dioxide.[193] This reduction in protein reduces forage quality and counters

the positive effects of carbon dioxide enrichment on carbohydrates. Rising carbon dioxide concentration also has the potential to reduce the digestibility of forages that are already of poor quality. Reductions in forage quality could have pronounced detrimental effects on animal growth, reproduction, and survival, and could render livestock production unsustainable unless animal diets are supplemented with protein, adding more costs to production. On shortgrass prairie, for example, a carbon dioxide enrichment experiment reduced the protein concentration of autumn forage below critical maintenance levels for livestock in 3 out of 4 years and reduced the digestibility of forage by 14 percent in mid-summer and by 10 percent in autumn. Significantly, the grass type that thrived the most under excess carbon dioxide conditions also had the lowest protein concentration.[193]

At the scale of a region, the composition of forage plant species is determined mostly by climate and soils. The primary factor controlling the distribution and abundance of plants is water: both the

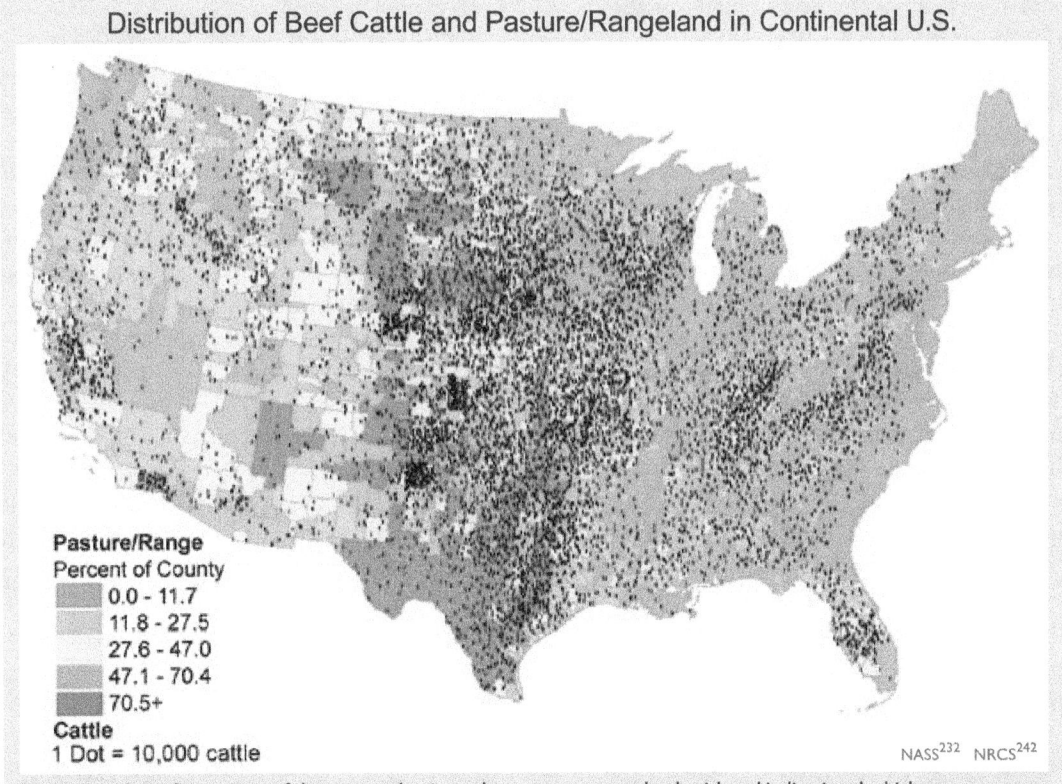

Distribution of Beef Cattle and Pasture/Rangeland in Continental U.S.

Pasture/Range
Percent of County
- 0.0 - 11.7
- 11.8 - 27.5
- 27.6 - 47.0
- 47.1 - 70.4
- 70.5+

Cattle
1 Dot = 10,000 cattle

NASS[232] NRCS[242]

The colors show the percent of the county that is cattle pasture or rangeland, with red indicating the highest percentage. Each dot represents 10,000 cattle. Livestock production occurs in every state. Increasing concentration of carbon dioxide reduces the quality of forage, necessitating more acreage and resulting in a decline in livestock productivity.

amount of water plants use and water availability over time and space. The ability to anticipate vegetation changes at local scales and over shorter periods is limited because at these scales the response of vegetation to global-scale changes depends on a variety of local processes including the rate of disturbances such as fire and grazing, and the rate at which plant species can move across sometimes-fragmented landscapes. Nevertheless, some general patterns of vegetation change are beginning to emerge. For example, experiments indicate that a higher carbon dioxide concentration favors weeds and invasive plants over native species because invasives have traits (such as rapid growth rate and prolific seed production) that allow a larger growth response to carbon dioxide. In addition, the effect of a higher carbon dioxide concentration on plant species composition appears to be greatest where the land has been disturbed (such as by fire or grazing) and nutrient and light availability are high.[193]

Increases in temperature lengthen the growing season, and thus are likely to extend forage production into the late fall and early spring. However, overall productivity remains dependent on precipitation during the growing season.[193]

Increased heat, disease, and weather extremes are likely to reduce livestock productivity.

Like human beings, cows, pigs, and poultry are warm-blooded animals that are sensitive to heat. In terms of production efficiency, studies show that the negative effects of hotter summers will outweigh the positive effects of warmer winters. The more the U.S. climate warms, the more production will fall. For example, an analysis projected that a warming in the range of 9 to 11°F (as in the higher emissions scenarios[91]) would cause a 10 percent decline in livestock yields in cow/calf and dairy operations in Appalachia, the Southeast (including the Mississippi Delta), and southern Plains regions, while a warming of 2.7°F would cause less than a 1 percent decline.

Temperature and humidity interact to cause stress in animals, just as in humans; the higher the heat and humidity, the greater the stress and discomfort,

and the larger the reduction in the animals' ability to produce milk, gain weight, and reproduce. Milk production declines in dairy operations, the number of days it takes for cows to reach their target weight grows longer in meat operations, conception rate in cattle falls, and swine growth rates decline due to heat. As a result, swine, beef, and milk production are all projected to decline in a warmer world.[193]

The projected increases in air temperatures will negatively affect confined animal operations (dairy, beef, and swine) located in the central United States, increasing production costs as a result of reductions in performance associated with lower feed intake and increased requirements for energy to maintain healthy livestock. These costs do not account for the increased death of livestock associated with extreme weather events such as heat waves. Nighttime recovery is an essential element of survival when livestock are stressed by extreme heat. A feature of recent heat waves is the lack of nighttime relief. Large numbers of deaths have occurred in recent heat waves, with individual states reporting losses of 5,000 head of cattle in a single heat wave in one summer.[193]

Warming also affects parasites and disease pathogens. The earlier arrival of spring and warmer winters allow greater proliferation and survival of parasites and disease pathogens.[193] In addition, changes in rainfall distributions are likely to lead to changes in diseases sensitive to moisture. Heat stress reduces animals' ability to cope with other stresses, such as diseases and parasites. Furthermore, changes in rainfall distributions could lead to changes in diseases sensitive to relative humidity.

Maintaining livestock production would require modifying facilities to reduce heat stress on animals, using the best understanding of the chronic and acute stresses that livestock will encounter to determine the optimal modification strategy.[193]

Changing livestock species as an adaptation strategy is a much more extreme, high-risk, and, in most cases, high-cost option than changing crop varieties. Accurate predictions of climate trends and development of the infrastructure and market for the new livestock products are essential to making this an effective response.

Ecosystems

Key Messages:

- Ecosystem processes, such as those that control growth and decomposition, have been affected by climate change.
- Large-scale shifts have occurred in the ranges of species and the timing of the seasons and animal migration, and are very likely to continue.
- Fires, insect pests, disease pathogens, and invasive weed species have increased, and these trends are likely to continue.
- Deserts and drylands are likely to become hotter and drier, feeding a self-reinforcing cycle of invasive plants, fire, and erosion.
- Coastal and near-shore ecosystems are already under multiple stresses. Climate change and ocean acidification will exacerbate these stresses.
- Arctic sea ice ecosystems are already being adversely affected by the loss of summer sea ice and further changes are expected.
- The habitats of some mountain species and coldwater fish, such as salmon and trout, are very likely to contract in response to warming.
- Some of the benefits ecosystems provide to society will be threatened by climate change, while others will be enhanced.

Key Sources

CCSP 1.2	CCSP 2.2	CCSP 4.1	CCSP 4.2
Past Climate	Carbon Cycle	Sea-Level Rise	Ecosystem Thresholds

CCSP 4.3	CCSP 4.4	IPCC	NAST	ACIA
Impacts	Ecosystem Adaptation	WG-2	U.S. Impacts	Arctic Impacts

The natural functioning of the environment provides both goods – such as food and other products that are bought and sold – and services, which our society depends upon. For example, ecosystems store large amounts of carbon in plants and soils; they regulate water flow and water quality; and they stabilize local climates. These services are not assigned a financial value, but society nonetheless depends on them. Ecosystem processes are the underpinning of these services: photosynthesis, the process by which plants capture carbon dioxide from the atmosphere and create new growth; the plant and soil processes that recycle nutrients from decomposing matter and maintain soil fertility; and the processes by which plants draw water from soils and return water to the atmosphere. These ecosystem processes are affected by climate and by the concentration of carbon dioxide in the atmosphere.[70]

The diversity of living things (biodiversity) in ecosystems is itself an important resource that maintains the ability of these systems to provide the services upon which society depends. Many factors affect biodiversity including: climatic conditions; the influences of competitors, predators, parasites, and diseases; disturbances such as fire; and other physical factors. Human-induced climate change,

in conjunction with other stresses, is exerting major influences on natural environments and biodiversity, and these influences are generally expected to grow with increased warming.[70]

Ecosystem processes, such as those that control growth and decomposition, have been affected by climate change.

Climate has a strong influence on the processes that control growth and development in ecosystems. Temperature increases generally speed up plant growth, rates of decomposition, and how rapidly the cycling of nutrients occurs, though other factors, such as whether sufficient water is available, also influence these rates. The growing season is lengthening as higher temperatures occur earlier in the spring. Forest growth has risen over the past several decades as a consequence of a number of factors – young forests reaching maturity, an increased concentration of carbon dioxide in the atmosphere, a longer growing season, and increased deposition of nitrogen from the atmosphere. Based on the current understanding of these processes, the individual effects are difficult to disentangle.[243]

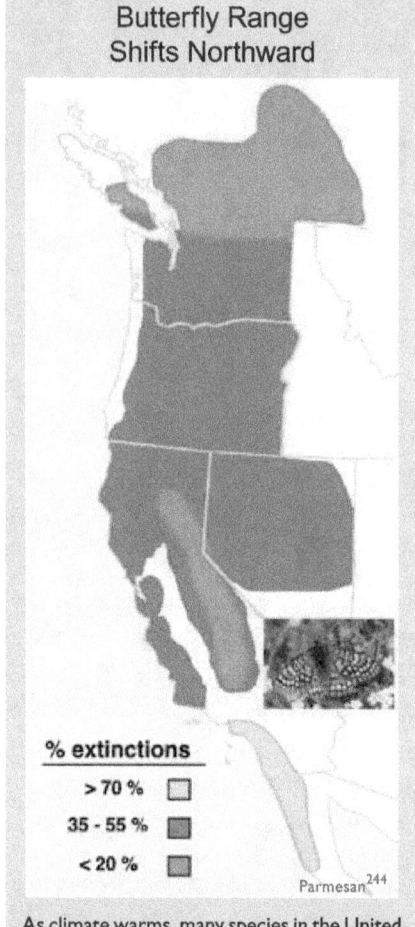

Butterfly Range Shifts Northward

% extinctions

> 70 % ☐
35 - 55 % ☐
< 20 % ☐

Parmesan[244]

As climate warms, many species in the United States are shifting their ranges northward and to higher elevations. The map shows the response of Edith's checkerspot butterfly populations to a warming climate over the past 136 years in the American West. Over 70 percent of the southernmost populations (shown in yellow) have gone extinct. The northernmost populations and those above 8,000 feet elevation in the cooler climate of California's Sierra Nevada (shown in green) are still thriving. These differences in numbers of population extinctions across the geographic range of the butterfly have resulted in the average location shifting northward and to higher elevations over the past century, illustrating how climate change is altering the ranges of many species. Because their change in range is slow, most species are not expected to be able to keep up with the rapid climate change projected in the coming decades.[244]

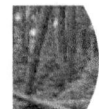

A higher atmospheric carbon dioxide concentration causes trees and other plants to capture more carbon from the atmosphere, but experiments show that trees put much of this extra carbon into producing fine roots and twigs, rather than new wood. The effect of carbon dioxide in increasing growth thus seems to be relatively modest, and generally is seen most strongly in young forests on fertile soils where there is also sufficient water to sustain this growth. In the future, as atmospheric carbon dioxide continues to rise, and as climate continues to change, forest growth in some regions is projected to increase, especially in relatively young forests on fertile soils.[243]

Forest productivity is thus projected to increase in much of the East, while it is projected to decrease in much of the West where water is scarce and projected to become more so. Wherever droughts increase, forest productivity will decrease and tree death will increase. In addition to occurring in much of the West, these conditions are projected to occur in parts of Alaska and in the eastern part of the Southeast.[243]

Large-scale shifts have occurred in the ranges of species and the timing of the seasons and animal migration, and are very likely to continue.

Climate change is already having impacts on animal and plant species throughout the United States. Some of the most obvious changes are related to the timing of the seasons: when plants bud in spring, when birds and other animals migrate, and so on. In the United States, spring now arrives an average of 10 days to two weeks earlier than it did 20 years ago. The growing season is lengthening over much of the continental United States. Many migratory bird species are arriving earlier. For example, a study of northeastern birds that migrate long distances found that birds wintering in the southern United States now arrive back in the Northeast an average of 13 days earlier than they did during the first half of the last century. Birds wintering in South America arrive back in the Northeast an average of four days earlier.[70]

Another major change is in the geographic distribution of species. The ranges of many species in the United States have shifted northward and upward in elevation. For example, the ranges of many butterfly species have expanded northward, contracted at the southern edge, and shifted to higher elevations as warming has continued. A study of Edith's checkerspot butterfly showed that 40 percent of the populations below 2,400 feet have gone extinct, despite the availability of otherwise suitable habitat and food supply. The checkerspot's most southern populations also have gone extinct, while new populations have been established north of the previous northern boundary for the species.[70]

For butterflies, birds, and other species, one of the concerns with such changes in geographic range and timing of migration is the potential for mismatches between species and the resources they need to survive. The rapidly changing landscape, such as new highways and expanding urban areas, can create barriers that limit habitat and increase species loss. Failure of synchronicity between butterflies and the resources they depend

upon has led to local population extinctions of the checkerspot butterfly during extreme drought and low-snowpack years in California.[70]

Tree species shifts

Forest tree species also are expected to shift their ranges northward and upslope in response to climate change, although specific quantitative predictions are very difficult to make because of the complexity of human land use and many other factors. This would result in major changes in the character of U.S. forests and the types of forests that will be most prevalent in different regions. In the United States, some common forests types are projected to expand, such as oak-hickory; others are projected to contract, such as maple-beech-birch. Still others, such as spruce-fir, are likely to disappear from the United States altogether.[243]

In Alaska, vegetation changes are already underway due to warming. Tree line is shifting northward into tundra, encroaching on the habitat for many migratory birds and land animals such as caribou that depend on the open tundra landscape.[245]

Marine species shifts and effects on fisheries

The distribution of marine fish and plankton are predominantly determined by climate, so it is not surprising that marine species in U.S. waters are moving northward and that the timing of plankton blooms is shifting. Extensive shifts in the ranges and distributions of both warmwater and coldwater species of fish have been documented.[70] For example, in the waters around Alaska, climate change already is causing significant alterations in marine ecosystems with important implications for fisheries and the people who depend on them (see *Alaska* region).

In the Pacific, climate change is expected to cause an eastward shift in the location of tuna stocks.[246] It is clear that such shifts are related to climate, including natural modes of climate variability such as the cycles of El Niño and La Niña. However, it is unclear how these modes of ocean variability will change as global climate continues to change, and therefore it is very difficult to predict quantitatively how

marine fish and plankton species' distributions might shift as a function of climate change.[70]

Breaking up of existing ecosystems

As warming drives changes in timing and geographic ranges for various species, it is important to note that entire communities of species do not shift intact. Rather, the range and timing of each species shifts in response to its sensitivity to climate change, its mobility, its lifespan, and the availability of the resources it needs (such as soil, moisture, food, and shelter). The speed with which species can shift their ranges is influenced by factors including their size, lifespan, and seed dispersal techniques in plants. In addition, migratory pathways must be available, such as northward flowing rivers which serve as conduits for fish. Some migratory pathways may be blocked by development and habitat fragmentation. All of these variations result in the breakup of existing ecosystems and formation of new ones, with unknown consequences.[220]

Extinctions and climate change

Interactions among impacts of climate change and other stressors can increase the risk of species extinction. Extinction rates of plants and animals have already risen considerably, with the vast majority of these extinctions attributed to loss of habitat or over-exploitation.[247] Climate change has been identified as a serious risk factor for the fu-

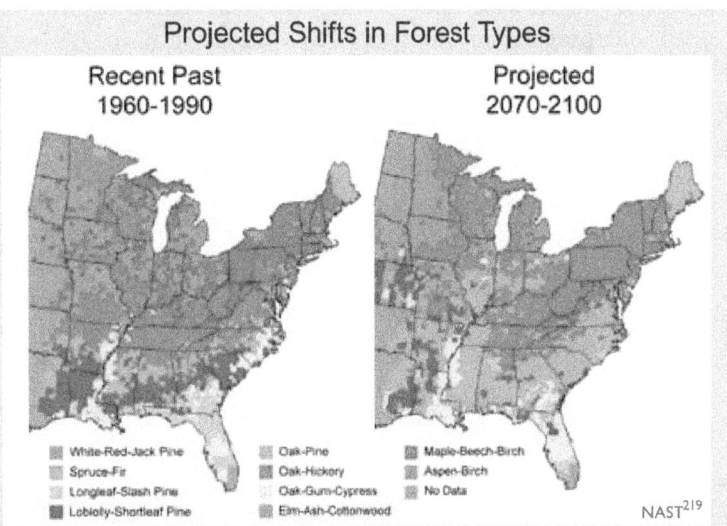

Projected Shifts in Forest Types

Recent Past 1960-1990

Projected 2070-2100

White-Red-Jack Pine
Spruce-Fir
Longleaf-Slash Pine
Loblolly-Shortleaf Pine

Oak-Pine
Oak-Hickory
Oak-Gum-Cypress
Elm-Ash-Cottonwood

Maple-Beech-Birch
Aspen-Birch
No Data

NAST[219]

The maps show current and projected forest types. Major changes are projected for many regions. For example, in the Northeast, under a mid-range warming scenario, the currently dominant maple-beech-birch forest type is projected to be completely displaced by other forest types in a warmer future.[243]

ture, however, since it is one of the environmental stresses on species and ecosystems that is continuing to increase.[247] The Intergovernmental Panel on Climate Change has estimated that if a warming of 3.5 to 5.5°F occurs, 20 to 30 percent of species that have been studied would be in climate zones that are far outside of their current ranges, and would therefore likely be at risk of extinction.[248] One reason this percentage is so high is that climate change would be superimposed on other stresses including habitat loss and continued overharvesting of some species, resulting in considerable stress on populations and species.

Fires, insect pests, disease pathogens, and invasive weed species have increased, and these trends are likely to continue.

Forest fires

In the western United States, both the frequency of large wildfires and the length of the fire season have increased substantially in recent decades, due primarily to earlier spring snowmelt and higher spring and summer temperatures.[294] These changes in climate have reduced the availability of moisture, drying out the vegetation that provides the fuel for fires. Alaska also has experienced large increases in fire, with the area burned more than doubling in recent decades. As in the western United States, higher air temperature is a key factor. In Alaska, for example, June air temperatures alone explained approximately 38 percent of the increase in the area burned annually from 1950 to 2003.[243]

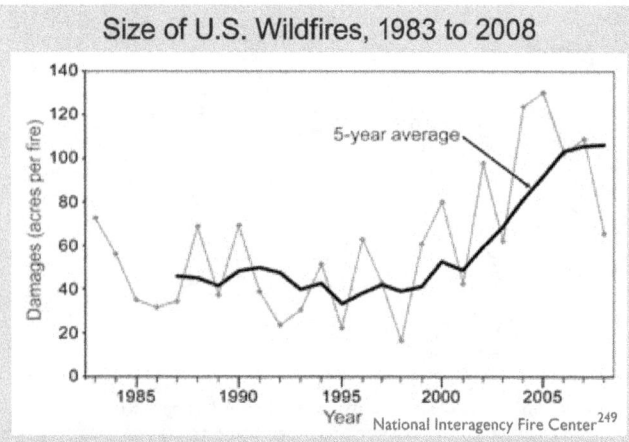

Size of U.S. Wildfires, 1983 to 2008

5-year average

National Interagency Fire Center[249]

Data on wildland fires in the United States show that the number of acres burned per fire has increased since the 1980s.

Insect pests

Insect pests are economically important stresses on forest ecosystems in the United States. Coupled with pathogens, they cost $1.5 billion in damage per year. Forest insect pests are sensitive to climatic variations in many stages of their lives. Changes in climate have contributed significantly to several major insect pest outbreaks in the United States and Canada over the past several decades. The mountain pine beetle has infested lodgepole pine in British Columbia. Over 33 million acres of forest have been affected, by far the largest such outbreak in recorded history. Another 1.5 million acres have been infested by pine beetle in Colorado. Spruce beetle has affected more than 2.5 million acres in Alaska (see *Alaska* region) and western Canada. The combination of drought and high temperatures also has led to serious insect infestations and death of piñon pine in the Southwest, and to various insect pest attacks throughout the forests of the eastern United States.[243]

Rising temperatures increase insect outbreaks in a number of ways. First, winter temperatures above a certain threshold allow more insects to survive the cold season that normally limits their numbers. Second, the longer warm season allows them to develop faster, sometimes completing two life cycles instead of one in a single growing season. Third, warmer conditions help expand their ranges northward. And fourth, drought stress reduces trees' ability to resist insect attack (for example, by pushing back against boring insects with the pressure of their sap). Spruce beetle, pine beetle, spruce budworm, and woolly adelgid (which attacks eastern hemlocks) are just some of the insects that are proliferating in the United States, devastating many forests. These outbreaks are projected to increase with ongoing warming. Trees killed by insects also provide more dry fuel for wildfires.[70,243,250]

Disease pathogens and their carriers

One consequence of a longer, warmer growing season and less extreme cold in winter is that opportunities are created for many insect pests and disease pathogens to flourish. Accumulating evidence links the spread of disease pathogens to a warming climate. For example, a recent study showed that widespread amphibian extinctions in the mountains of Costa Rica are linked to changes in climatic conditions which are thought to have enabled the proliferation of an amphibian disease.[70,251]

Diseases that affect wildlife and the living things that carry these diseases have been expanding their geographic ranges as climate heats up. Depending on their specific adaptations to current climate, many parasites, and the insects, spiders, and scorpions that carry and transmit diseases, die or fail to develop below threshold temperatures. Therefore, as temperatures rise, more of these disease-carrying creatures survive. For some species, rates of reproduction, population growth, and biting, tend to increase with increasing temperatures, up to a limit. Some parasites' development rates and infectivity periods also increase with temperature.[70] An analysis of diseases among marine species found that diseases were increasing for mammals, corals, turtles, and mollusks, while no trends were detected for sharks, rays, crabs, and shrimp.[70]

Invasive plants

Problems involving invasive plant species arise from a mix of human-induced changes, including disturbance of the land surface (such as through over grazing or clearing natural vegetation for development), deliberate or accidental transport of non-native species, the increase in available nitrogen through over-fertilization of crops, and the rising carbon dioxide concentration and the resulting climate change.[243] Human-induced climate change is not generally the initiating factor, nor the most important one, but it is becoming a more important part of the mix.

The increasing carbon dioxide concentration stimulates the growth of most plant species, and some invasive plants respond with greater growth rates than native plants. Beyond this, invasive plants appear to better tolerate a wider range of environmental conditions and may be more successful in a warming world because they can migrate and establish themselves in new sites more rapidly than native plants.[70] They are also not usually dependent on external pollinators or seed dispersers to reproduce. For all of these reasons, invasive plant species present a growing problem that is extremely difficult to control once unleashed.[70]

Deserts and drylands are likely to become hotter and drier, feeding a self-reinforcing cycle of invasive plants, fire, and erosion.

The arid Southwest is projected to become even drier in this century. There is emerging evidence that this is already underway.[34] Deserts in the United States are also projected to expand to the north, east, and upward in elevation in response to projected warming and associated changes in climate.

Increased drying in the region contributes to a variety of changes that exacerbate a cycle of desertification. Increased drought conditions cause perennial plants to die due to water stress and increased susceptibility to plant diseases. At the same time, non-native grasses have invaded the region. As these grasses increase in abundance, they provide more fuel for fires, causing fire frequency to increase in a self-reinforcing cycle that leads to further losses of vegetation. When it does rain, the rain tends to come in heavy downpours, and since there is less vegetation to protect the soil, water erosion increases. Higher air temperatures and decreased soil moisture reduce soil stability, further exacerbating erosion. And with a growing population needing water for urban uses, hydroelectric generation, and agriculture, there is increasing pressure on mountain water sources that would otherwise flow to desert river areas.[70,149]

Desertification of Arid Grassland
near Tucson, Arizona, 1902 to 2003

Photostation #233:
Huerfano Butte

Photomontage by *Rob Wu*
Original historic photos courtesy of
Santa Rita Experimental Range

1902   1951   2003

CCSP SAP 4.3[243]

The photo series shows the progression from arid grassland to desert (desertification) over a 100-year period. The change is the result of grazing management and reduced rainfall in the Southwest.[250,252,253]

The response of arid lands to climate change also depends on how other factors interact with climate at local scales. Large-scale, unregulated livestock grazing in the Southwest during the late 1800s and early 1900s is widely regarded as having contributed to widespread desertification. Grazing peaked around 1920 on public lands in the West. By the 1970s, grazing had been reduced by about 70 percent, but the arid lands have been very slow to recover from its impacts. Warmer and drier climate conditions are expected to slow recovery even more. In addition, the land resource in the Southwest is currently managed more for providing water for people than for protecting the productivity of the landscape. As a result, the land resource is likely to be further degraded and its recovery hampered.[243]

Coastal and near-shore ecosystems are already under multiple stresses. Climate change and ocean acidification will exacerbate these stresses.

Coastal and near-shore marine ecosystems are vulnerable to a host of climate change-related effects including increasing air and water temperatures, ocean acidification, changes in runoff from the land, sea-level rise, and altered currents. Some of these changes have already led to coral bleaching, shifts in species ranges, increased storm intensity in some regions, dramatic reductions in sea ice extent and thickness along the Alaskan coast,[137] and other significant changes to the nation's coastlines and marine ecosystems.[70]

The interface between land and sea is important, as many species, including many endangered species, depend on it at some point in their life cycle. In addition, coastal areas buffer inland areas from the effects of wave action and storms.[247] Coastal wetlands, intertidal areas, and other near-shore ecosystems are subject to a variety of environmental stresses.[254,255] Sea-level rise, increased coastal storm intensity, and rising temperatures contribute to increased vulnerability of coastal wetland ecosystems. It has been estimated that 3 feet of sea-level rise (within the range of projections for this century) would inundate about 65 percent of the coastal marshlands and swamps in the contiguous United States.[256] The combination of sea-level rise,

local land sinking, and related factors already have resulted in substantially higher relative sea-level rise along the Gulf of Mexico and the mid-Atlantic coast, more so than on the Pacific Coast.[43,254] In Louisiana alone, over one-third of the coastal plain that existed a century ago has since been lost,[254] which is mostly due to local land sinking.[70] Barrier islands are also losing land at an increasing rate[257] (see *Southeast* region), and they are particularly important in protecting the coastline in some regions vulnerable to sea-level rise and storm surge.

Coral reefs
Coral reefs are very diverse ecosystems that support many other species by providing food and habitat. In addition to their ecological value, coral reefs provide billions of dollars in services including tourism, fish breeding habitat, and protection of coastlines. Corals face a host of challenges associated with human activities such as poorly regulated tourism, destructive fishing, and pollution, in addition to climate change-related stresses.[70]

Corals are marine animals that host symbiotic algae which help nourish the animals and give the corals their color. When corals are stressed by increases in water temperatures or ultraviolet light, they lose their algae and turn white, a process called coral bleaching. If the stress persists, the corals die. Intensities and frequencies of bleaching events, clearly driven by warming in surface water, have increased substantially over the past 30 years, leading to the death or severe damage of about one-third of the world's corals.[70]

The United States has extensive coral reef ecosystems in the Caribbean, Atlantic, and Pacific oceans. In 2005, the Caribbean basin experienced unprecedented water temperatures that resulted in dramatic coral bleaching with some sites in the U.S. Virgin Islands seeing 90 percent of the coral bleached. Some corals began to recover when water temperatures decreased, but later that year disease appeared, striking the previously bleached and weakened coral. To date, 50 percent of the corals in Virgin Islands National Park have died from the bleaching and disease events. In the Florida Keys, summer bleaching in 2005 was also followed by disease in September.[70]

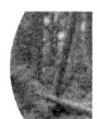

But rising temperature is not the only stress coral reefs face. As the carbon dioxide concentration in the air increases, more carbon dioxide is absorbed into the world's oceans, leading to their acidification. This makes less calcium carbonate available for corals and other sea life to build their skeletons and shells.[258] If carbon dioxide concentrations continue to rise and the resulting acidification proceeds, eventually, corals and other ocean life that rely on calcium carbonate will not be able to build these skeletons and shells at all. The implications of such extreme changes in ocean ecosystems are not clear, but there is now evidence that in some ocean areas, such as along the Northwest coast, acidification is already occurring[70,259] (see *Coasts* region for more discussion of ocean acidification).

Arctic sea ice ecosystems are already being adversely affected by the loss of summer sea ice and further changes are expected.

Perhaps most vulnerable of all to the impacts of warming are Arctic ecosystems that rely on sea ice, which is vanishing rapidly and is projected to disappear entirely in summertime within this century. Algae that bloom on the underside of the sea ice form the base of a food web linking microscopic animals and fish to seals, whales, polar bears, and people. As the sea ice disappears, so too do these algae. The ice also provides a vital platform for ice-dependent seals (such as the ringed seal) to give birth, nurse their pups, and rest. Polar bears use the ice as a platform from which to hunt their prey. The walrus rests on the ice near the continental shelf between its dives to eat clams and other shellfish. As the ice edge retreats away from the shelves to deeper areas, there will be no clams nearby.[70,132,220]

The Bering Sea, off the west coast of Alaska, produces our nation's largest commercial fish harvests as well as providing food for many Native Alaskan peoples. Ultimately, the fish populations (and animals including seabirds, seals, walruses, and whales) depend on plankton blooms regulated by the extent and location of the ice edge in spring. As the sea ice continues to decline, the location, timing, and species composition of the blooms is changing. The spring melt of sea ice in the

Bering Sea has long provided material that feeds the clams, shrimp, and other life forms on the ocean floor that, in turn, provide food for the walruses, gray whales, bearded seals, eider ducks, and many fish. The earlier ice melt resulting from warming, however, leads to later phytoplankton blooms that are largely consumed by microscopic animals near the sea surface, vastly decreasing the amount of food reaching the living things on the ocean floor. This will radically change the species composition of the fish and other creatures, with significant repercussions for both subsistence and commercial fishing.[70]

Ringed seals give birth in snow caves on the sea ice, which protect their pups from extreme cold and predators. Warming leads to earlier snow melt, which causes the snow caves to collapse before the pups are weaned. The small, exposed pups may die of hypothermia or be vulnerable to predation by arctic foxes, polar bears, gulls, and ravens. Gulls and ravens are arriving in the Arctic earlier as springs become warmer, increasing the birds' opportunity to prey on the seal pups.[70]

Polar bears are the top predators of the sea ice ecosystem. Because they prey primarily on ice-associated seals, they are especially vulnerable to the disappearance of sea ice. The bears' ability to catch seals depends on the presence of sea ice. In that habitat, polar bears take advantage of the fact that seals must surface to breathe in limited openings in the ice cover. In the open ocean, bears lack a hunting platform, seals are not restricted in where they can surface, and successful hunting is very rare. On shore, polar bears feed little, if at all.

About two-thirds of the world's polar bears are projected to be gone by the middle of this century. It is projected that there will be no wild polar bears in Alaska in 75 years.[70]

In addition, the rapid rate of warming in Alaska and the rest of the Arctic in recent decades is sharply reducing the snow cover in which polar bears build dens and the sea ice they use as foraging habitat. Female polar bears build snow dens in which they hibernate for four to five months each year and in which they give birth to their cubs. Born weighing only about 1 pound, the tiny cubs depend on the snow den for warmth.

About two-thirds of the world's polar bears are projected to be gone by the middle of this century. It is projected that there will be no wild polar bears left in Alaska in 75 years.[70]

Continued warming will inevitably entail major changes in the sea ice ecosystem, to the point that its viability is in jeopardy. Some species will become extinct, while others might adapt to new habitats. The chances of species surviving the current changes may depend critically on the rate of change. The current rates of change in the sea ice ecosystem are very rapid relative to the life spans of animals including seals, walruses, and polar bears, and as such, are a major threat to their survival.[70]

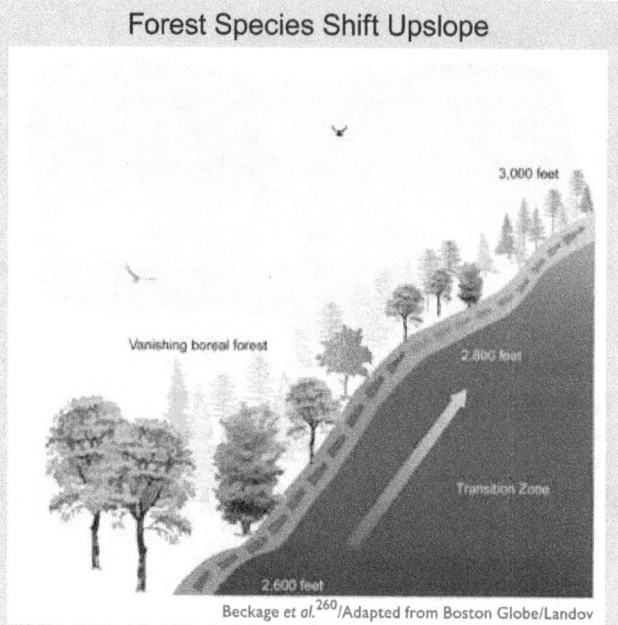

Forest Species Shift Upslope

Beckage et al.[260]/Adapted from Boston Globe/Landov

As climate warms, hardwood trees out-compete evergreen trees that are adapted to colder conditions.

vulnerable is that their suitable habitats are being compressed as climatic zones shift upward in elevation. Some species try to shift uphill with the changing climate, but may face constraints related to food, other species present, and so on. In addition, as species move up the mountains, those near the top simply run out of habitat.[70]

Fewer wildflowers are projected to grace the slopes of the Rocky Mountains as global warming causes earlier spring snowmelt. Larkspur, aspen fleabane, and aspen sunflower grow at an altitude of about 9,500 feet where the winter snows are deep. Once the snow melts, the flowers form buds and prepare to bloom. But warmer springs mean that the snow melts earlier, leaving the buds exposed to frost. (The percentage of buds that were frosted has doubled over the past decade.) Frost does not kill the plants, but it does make them unable to seed and reproduce, meaning there will be no next generation. Insects and other animal species depend on the flowers for food, and other species depend on those species, so the loss is likely to propagate through the food chain.[236]

Shifts in tree species on mountains in New England, where temperatures have risen 2 to 4°F in the last 40 years, offer another example. Some mountain tree species have shifted uphill by 350

The pika, pictured above, is a small mammal whose habitat is limited to cold areas near the tops of mountains. As climate warms, little suitable habitat is left. Of 25 pika populations studied in the Great Basin between the Rocky Mountains and the Sierra Nevada, more than one-third have gone extinct in recent decades.[261,262]

The habitats of some mountain species and coldwater fish, such as salmon and trout, are very likely to contract in response to warming.

Animal and plant species that live in the mountains are among those particularly sensitive to rapid climate change. They include animal species such as the grizzly bear, bighorn sheep, pika, mountain goat, and wolverine. Major changes have already been observed in the pika as previously reported populations have disappeared entirely as climate has warmed over recent decades.[70] One reason mountain species are so feet in the last 40 years. Tree communities were relatively unchanged at low and high elevations, but in the transition zone in between (at about 2,600 feet elevation) the changes have been dramatic. Cold-loving tree species declined from 43 to 18 percent, while warmer-loving trees increased from 57 to 82 percent. Overall, the transition zone has shifted about 350 feet uphill in just a few decades, a surprisingly rapid rate since these are trees that live for hundreds of years. One possibility is that as trees were damaged or killed by air pollution, it left an opportunity for the warming-induced transition to occur more quickly. These results indicate that the composition of high elevation forests is changing rapidly.[260]

Coldwater fish

Salmon and other coldwater fish species in the United States are at particular risk from warming. Salmon are under threat from a variety of human activities, but global warming is a growing source of stress. Rising temperatures affect salmon in several important ways. As precipitation increasingly falls as rain rather than snow, it feeds floods that wash away salmon eggs incubating in the streambed. Warmer water leads eggs to hatch earlier in the year, so the young are smaller and more vulnerable to predators. Warmer conditions increase the fish's metabolism, taking energy away from growth and forcing the fish to find more food, but earlier hatching of eggs could put them out of sync with the insects they eat. Earlier melting of snow leaves rivers and streams warmer and shallower in summer and fall. Diseases and parasites tend to flourish in warmer water. Studies suggest that up to 40 percent of Northwest salmon populations may be lost by 2050.[263]

Large declines in trout populations are also projected to occur around the United States. Over half of the wild trout populations are likely to disappear from the southern Appalachian Mountains because of the effects of rising stream temperatures. Losses of western trout populations may exceed 60 percent in certain regions. About 90 percent of bull trout, which live in western rivers in some of the country's most wild places, are projected to be lost due to warming. Pennsylvania is predicted to lose 50 percent of its trout habitat in the coming decades. Projected losses of trout habitat for some warmer

states, such as North Carolina and Virginia, are up to 90 percent.[264]

Some of the benefits ecosystems provide to society will be threatened by climate change, while others will be enhanced.

Human well-being depends on the Earth's ecosystems and the services that they provide to sustain and fulfill human life.[265] These services are important to human well-being because they contribute to basic material needs, physical and psychological health, security, and economic activity. A recent assessment reported that of 24 vital ecosystem services, 15 were being degraded by human activity.[247] Climate change is one of several human-induced stresses that threaten to intensify and extend these adverse impacts to biodiversity, ecosystems, and the services they provide. Two of many possible examples follow.

Forests and carbon storage

Forests provide many services important to the well-being of Americans: air and water quality maintenance, water flow regulation, and watershed protection; wildlife habitat and biodiversity conservation; recreational opportunities and aesthetic and spiritual fulfillment; raw materials for wood and paper products; and climate regulation and carbon storage. A changing climate will alter forests and the services they provide. Most of these changes are likely to be detrimental.

In the United States, forest growth and long-lived forest products currently offset about 20 percent of U.S. fossil fuel carbon emissions.[140,257] This carbon "sink" is an enormous service provided by forests and its persistence or growth will be important to limiting the atmospheric carbon dioxide concentration. The scale of the challenge of increasing this sink is very large. To offset an additional 10 percent of U.S. emissions through tree planting would require converting one-third of current croplands to forests.[243]

Recreational opportunities

Tourism is one of the largest economic sectors in the world, and it is also one of the fastest

growing;[266] the jobs created by recreational tourism provide economic benefits not only to individuals but also to communities. Slightly more than 90 percent of the U.S. population participates in some form of outdoor recreation, representing nearly 270 million participants,[267] and several billion days spent each year in a wide variety of outdoor recreation activities.

Since much recreation and tourism occurs outside, increased temperature and precipitation have a direct effect on the enjoyment of these activities, and on the desired number of visitor days and associated level of visitor spending as well as tourism employment. Weather conditions are an important factor influencing tourism visits. In addition, outdoor recreation and tourism often depends on the availability and quality of natural resources,[268] such as beaches, forests, wetlands, snow, and wildlife, all of which will be affected by climate change.

Thus, climate change can have direct effects on the natural resources that people enjoy. The length of the season for, and desirability of, several of the most popular activities – walking; visiting a beach, lakeshore, or river; sightseeing; swimming; and picnicking[267] – are likely to be enhanced by small near-term increases in temperature. Other activities are likely to be harmed by even small increases in warming, such as snow- and ice-dependent activities including skiing, snowmobiling, and ice fishing.

The net economic effect of near-term climate change on recreational activities is likely to be positive. In the longer term, however, as climate change effects on ecosystems and seasonality become more pronounced, the net economic effect on tourism and recreation is not known with certainty.[172]

Adaptation: Preserving Coastal Wetlands

Coastal wetlands are rich ecosystems that protect the shore from damage during storm surges and provide society with other services. One strategy designed to preserve coastal wetlands as sea level rises is the "rolling easement." Rolling easements allow some development near the shore, but prohibit construction of seawalls or other armoring to protect buildings; they recognize nature's right-of-way to advance inland as sea level rises. Massachusetts and Rhode Island prohibit shoreline armoring along the shores of some estuaries so that ecosystems can migrate inland, and several states limit armoring along ocean shores.[269,270]

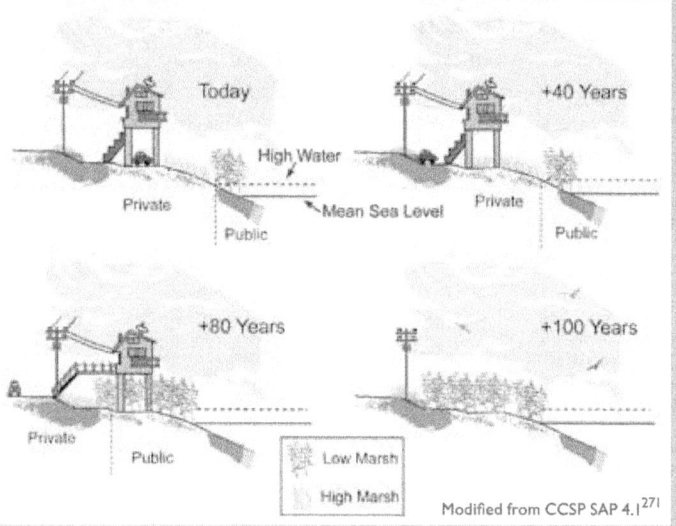

Modified from CCSP SAP 4.1[271]

In the case shown here, the coastal marsh would reach the footprint of the house 40 years in the future. Because the house is on pilings, it could still be occupied if it is connected to a community sewage treatment system; a septic system would probably fail due to proximity to the water table. After 80 years, the marsh would have taken over the yard, and the footprint of the house would extend onto public property. The house could still be occupied but reinvestment in the property would be unlikely. After 100 years, this house would be removed, although some other houses in the area could still be occupied. Eventually, the entire area would return to nature. A home with a rolling easement would depreciate in value rather than appreciate like other coastal real estate. But if the loss were expected to occur 100 years from now, it would only reduce the current property value by 1 to 5 percent, for which the owner could be compensated.[271]

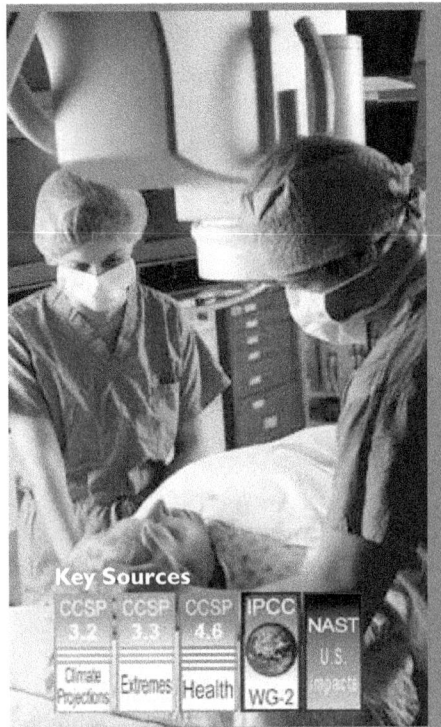

Human Health

Key Messages:

- Increases in the risk of illness and death related to extreme heat and heat waves are very likely. Some reduction in the risk of death related to extreme cold is expected.
- Warming is likely to make it more challenging to meet air quality standards necessary to protect public health.
- Extreme weather events cause physical and mental health problems. Some of these events are projected to increase.
- Some diseases transmitted by food, water, and insects are likely to increase.
- Rising temperature and carbon dioxide concentration increase pollen production and prolong the pollen season in a number of plants with highly allergenic pollen, presenting a health risk.
- Certain groups, including children, the elderly, and the poor, are most vulnerable to a range of climate-related health effects.

Key Sources

CCSP 3.2 | CCSP 3.3 | CCSP 4.6 | IPCC | NAST
Climate Projections | Extremes | Health | WG-2 | U.S. impacts

Climate change poses unique challenges to human health. Unlike health threats caused by a particular toxin or disease pathogen, there are many ways that climate change can lead to potentially harmful health effects. There are direct health impacts from heat waves and severe storms, ailments caused or exacerbated by air pollution and airborne allergens, and many climate-sensitive infectious diseases.[163]

Realistically assessing the potential health effects of climate change must include consideration of the capacity to manage new and changing climate conditions.[163] Whether or not increased health risks due to climate change are realized will depend largely on societal responses and underlying vulnerability. The probability of exacerbated health risks due to climate change points to a need to maintain a strong public health infrastructure to help limit future impacts.[163]

Increased risks associated with diseases originating outside the United States must also be considered because we live in an increasingly globalized world. Many poor nations are expected to suffer even greater health consequences from climate change.[272] With global trade and travel, disease flare-ups in any part of the world can potentially reach the United States. In addition, weather and climate extremes such as severe storms and drought can undermine public health infrastructure, further stress environmental resources, destabilize economies, and potentially create security risks both within the United States and internationally.[219]

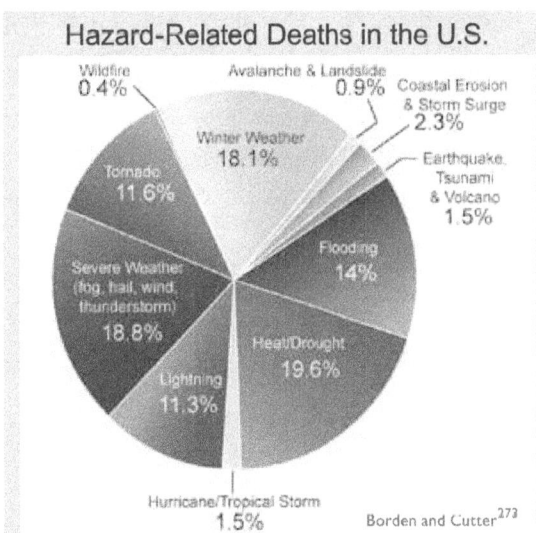

Hazard-Related Deaths in the U.S.

Wildfire 0.4%
Avalanche & Landslide 0.9%
Coastal Erosion & Storm Surge 2.3%
Winter Weather 18.1%
Earthquake, Tsunami & Volcano 1.5%
Tornado 11.6%
Flooding 14%
Severe Weather (fog, hail, wind, thunderstorm) 18.8%
Heat/Drought 19.6%
Lightning 11.3%
Hurricane/Tropical Storm 1.5%
Borden and Cutter[273]

The pie chart shows the distribution of deaths for 11 hazard categories as a percent of the total 19,958 deaths due to these hazards from 1970 to 2004. Heat/drought ranks highest, followed by severe weather, which includes events with multiple causes such as lightning, wind, and rain.[273] This analysis ended prior to the 2005 hurricane season which resulted in approximately 2,000 deaths.[229]

Increases in the risk of illness and death related to extreme heat and heat waves are very likely. Some reduction in the risk of death related to extreme cold is expected.

Temperatures are rising and the probability of severe heat waves is increasing. Analyses suggest that currently rare extreme heat waves will become much more common in the future (see *National Climate Change*).[68] At the same time, the U.S. population is aging, and older people are more vulnerable to hot weather and heat waves. The percentage of the U.S. population over age 65 is currently 12 percent and is projected to be 21 percent by 2050 (over 86 million people).[163,274] Diabetics are also at greater risk of heat-related death, and the prevalence of obesity and diabetes is increasing. Heat-related illnesses range from heat exhaustion to kidney stones.[275,276]

Heat is already the leading cause of weather-related deaths in the United States. More than 3,400 deaths between 1999 and 2003 were reported as resulting from exposure to excessive heat.[277] An analysis of nine U.S. cities shows that deaths due to heat increase with rising temperature and humidity.[278] From the 1970s to the 1990s, however, heat-related deaths declined.[279] This likely resulted from a rapid

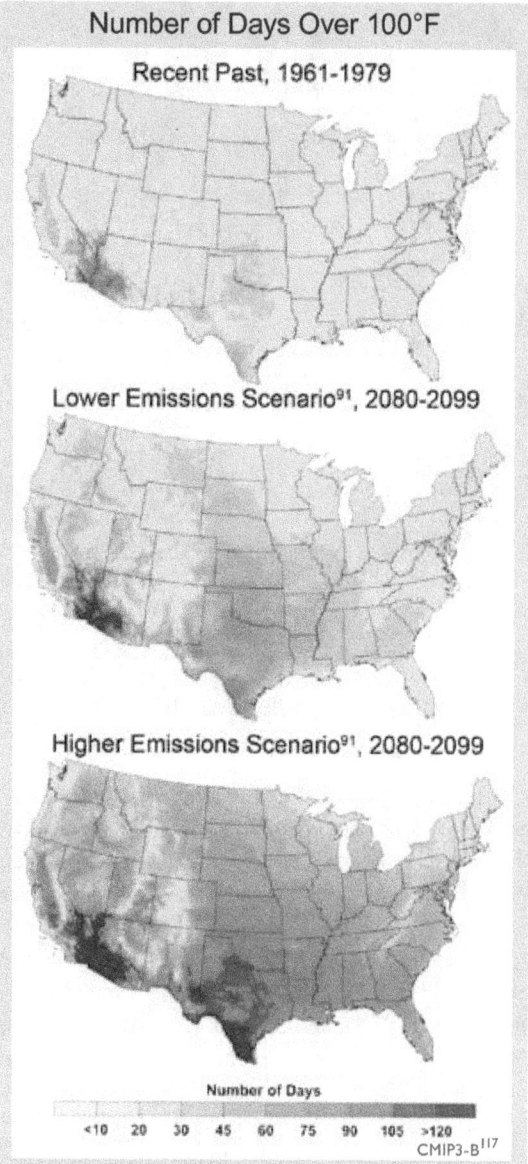

The number of days in which the temperature exceeds 100°F by late this century, compared to the 1960s and 1970s, is projected to increase strongly across the United States. For example, parts of Texas that recently experienced about 10 to 20 days per year over 100°F are expected to experience more than 100 days per year in which the temperature exceeds 100°F by the end of the century under the higher emissions scenario.[91]

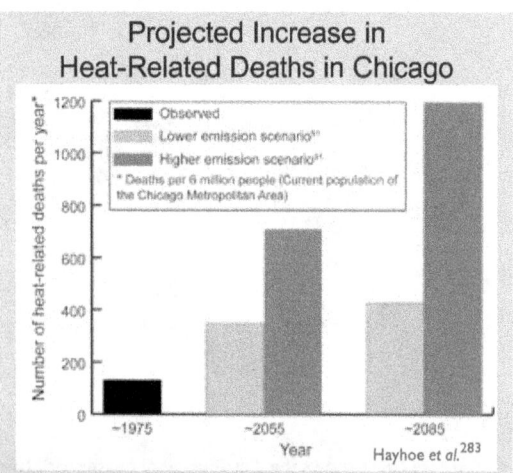

Increases in heat-related deaths are projected in cities around the nation, especially under higher emissions scenarios.[91] This analysis included some, but not all possible, adaptation measures. The graph shows the projected number of deaths per year, averaged over a three-decade period around 1975, 2055, and 2085 for the City of Chicago under lower and higher emissions.[91]

increase in the use of air conditioning. In 1978, 44 percent of households were without air conditioning, whereas in 2005, only 16 percent of the U.S. population lived without it (and only 3 percent did not have it in the South).[280,281] With air conditioning reaching near saturation, a recent study found that the general decline in heat-related deaths seems to have leveled off since the mid-1990s.[282]

As human-induced warming is projected to raise average temperatures by about 6 to 11°F in this century under a higher emissions scenario,[91] heat waves are expected to continue to increase in frequency, severity, and duration.[68,112] For example, by the end of this century, the number of heat-wave days in Los Angeles is projected to double,[284] and the number in Chicago to quadruple,[285] if emissions are not reduced.

Projections for Chicago suggest that the average number of deaths due to heat waves would more than double by 2050 under a lower emissions scenario[91] and quadruple under a high emissions scenario[91] (see figure page 90).[283]

A study of climate change impacts in California projects that, by the 2090s, annual heat-related deaths in Los Angeles would increase by two to three times under a lower emissions scenario and by five to seven times under a higher emissions scenario, compared to a 1990s baseline of about 165 deaths. These estimates assume that people will have become somewhat more accustomed to higher temperatures. Without such acclimatization, these estimates are projected to be about 20 to 25 percent higher.[284]

The full effect of global warming on heat-related illness and death involves a number of factors including actual changes in temperature (averages, highs, and lows); and human population characteristics, such as age, wealth, and fitness. In addition, adaptation at the scale of a city includes options such as heat wave early warning systems, urban

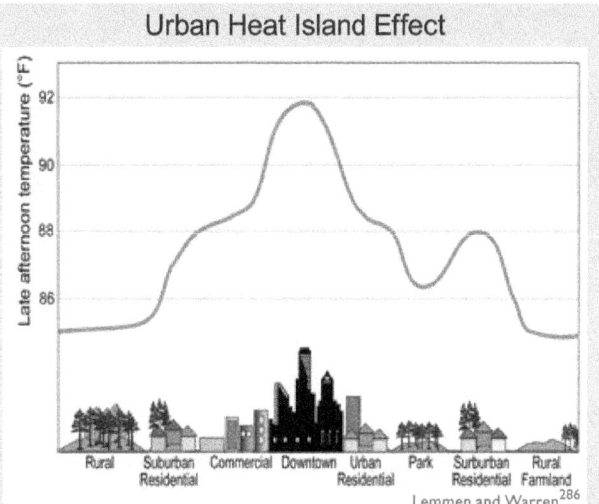

Urban Heat Island Effect

Late afternoon temperature (°F)

Rural | Suburban Residential | Commercial | Downtown | Urban Residential | Park | Surburban Residential | Rural Farmland

Lemmen and Warren[286]

Large amounts of concrete and asphalt in cities absorb and hold heat. Tall buildings prevent heat from dissipating and reduce air flow. At the same time, there is generally little vegetation to provide shade and evaporative cooling. As a result, parts of cities can be up to 10°F warmer than the surrounding rural areas, compounding the temperature increases that people experience as a result of human-induced warming.[313]

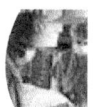

design to reduce heat loads, and enhanced services during heat waves.[163]

Reduced extreme cold

In a warmer world, the number of deaths caused by extremely low temperatures would be expected to drop, although in general, it is uncertain how climate change will affect net mortality.[163] Nevertheless, a recent study that analyzed daily mortality and weather data with regard to 6,513,330 deaths in 50 U.S. cities between 1989 and 2000 shows a marked difference between deaths resulting from hot and cold temperatures. The researchers found that, on average, cold snaps increased death rates

Adaptation: Reducing Deaths During Heat Waves

In the mid-1990s, Philadelphia became the first U.S. city to implement a system for reducing the risk of death during heat waves. The city focuses its efforts on the elderly, homeless, and poor. During a heat wave, a heat alert is issued and news organizations are provided with tips on how vulnerable people can protect themselves. The health department and thousands of block captains use a buddy system to check on elderly residents in their homes; electric utilities voluntarily refrain from shutting off services for non-payment; and public cooling places extend their hours. The city operates a "Heatline" where nurses are standing by to assist callers experiencing health problems; if callers are deemed "at risk," mobile units are dispatched to the residence. The city has also implemented a "Cool Homes Program" for elderly, low-income residents, which provides measures such as roof coatings and roof insulation that save energy and lower indoor temperatures. Philadelphia's system is estimated to have saved 117 lives over its first 3 years of operation.[287,288]

by 1.6 percent, while heat waves triggered a 5.7 percent increase in death rates.[289] The analysis found that the reduction in deaths as a result of relatively milder winters attributable to global warming will be substantially less than the increase in deaths due to summertime heat extremes.

Many factors contribute to winter deaths, including highly seasonal diseases such as influenza and pneumonia. It is unclear how these diseases are affected by temperature.[163]

Warming is likely to make it more challenging to meet air quality standards necessary to protect public health.

Poor air quality, especially in cities, is a serious concern across the United States. Half of all Americans, 158 million people, live in counties where air pollution exceeds national health standards.[290] While the Clean Air Act has improved air quality, higher temperatures and associated stagnant air masses are expected to make it more challenging to meet air quality standards, particularly for ground-level ozone (a component of smog).[13] It

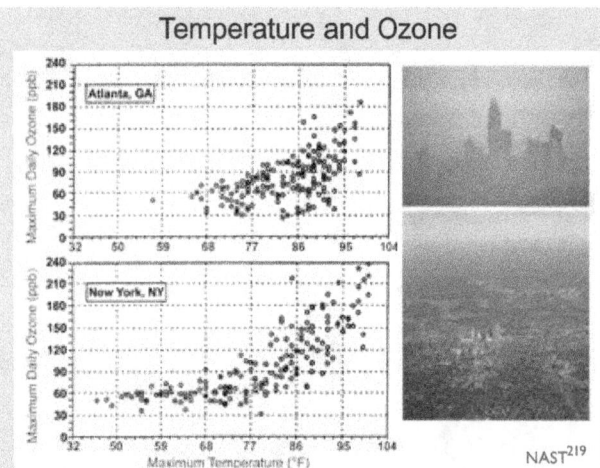

Temperature and Ozone

NAST[219]

The graphs illustrate the observed association between ground-level ozone (a component of smog) concentration in parts per billion (ppb) and temperature in Atlanta and New York City (May to October 1988 to 1990).[219] The projected higher temperatures across the United States in this century are likely to increase the occurrence of high ozone concentrations, although this will also depend on emissions of ozone precursors and meteorological factors. Ground-level ozone can exacerbate respiratory diseases and cause short-term reductions in lung function.

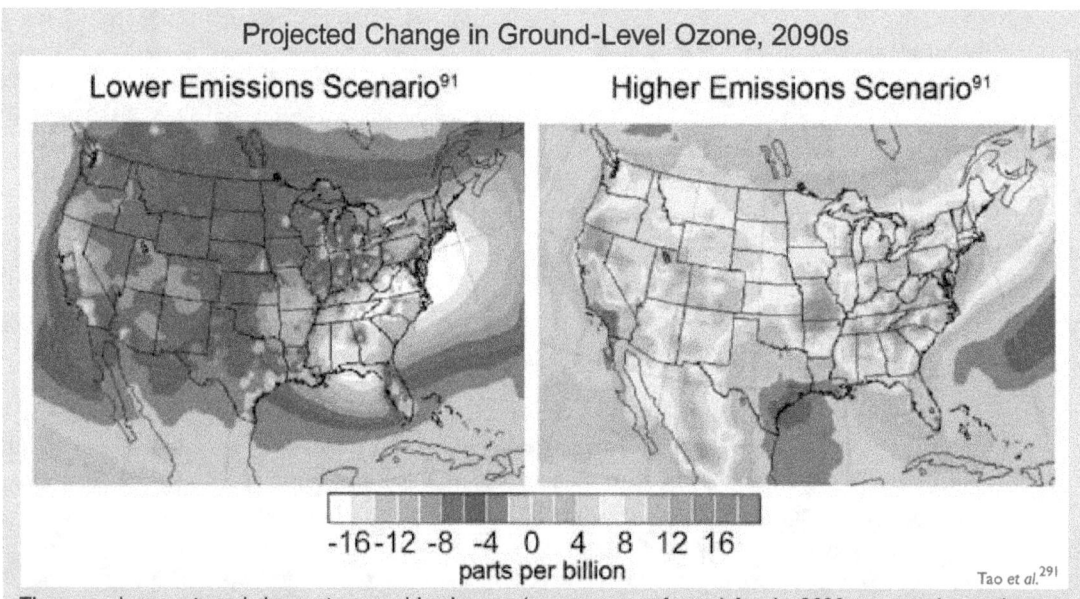

Projected Change in Ground-Level Ozone, 2090s

Lower Emissions Scenario[91]

Higher Emissions Scenario[91]

-16 -12 -8 -4 0 4 8 12 16
parts per billion

Tao et al.[291]

The maps show projected changes in ground-level ozone (a component of smog) for the 2090s, averaged over the summer months (June through August), relative to 1996-2000, under lower and higher emissions scenarios, which include both greenhouse gases and emissions that lead to ozone formation (some of which decrease under the lower emissions scenario).[91] By themselves, higher temperatures and other projected climate changes would increase ozone levels under both scenarios. However, the maps indicate that future projections of ozone depend heavily on emissions, with the higher emissions scenario[91] increasing ozone by large amounts, while the lower emissions scenario[91] results in an overall decrease in ground-level ozone by the end of the century.[291]

has been firmly established that breathing ozone results in short-term decreases in lung function and damages the cells lining the lungs. It also increases the incidence of asthma-related hospital visits and premature deaths.[272] Vulnerability to ozone effects is greater for those who spend time outdoors, especially with physical exertion, because this results

in a higher cumulative dose to their lungs. As a result, children, outdoor workers, and athletes are at higher risk for these ailments.[163]

Ground-level ozone concentrations are affected by many factors including weather conditions, emissions of gases from vehicles and industry that lead

Spotlight on Air Quality in California

Californians currently experience the worst air quality in the nation. More than 90 percent of the population lives in areas that violate state air quality standards for ground-level ozone or small particles. These pollutants cause an estimated 8,800 deaths and over a billion dollars in health care costs every year in California.[292] Higher temperatures are projected to increase the frequency, intensity, and duration of conditions conducive to air pollution formation, potentially increasing the number of days conducive to air pollution by 75 to 85 percent in Los Angeles and the San Joaquin Valley, toward the end of this century, under a higher emissions scenario, and by 25 to 35 percent under a lower emissions scenario.[293] Air quality could be further compromised by wildfires, which are already increasing as a result of warming.[252,294]

Adaptation: Improving Urban Air Quality

Because ground-level ozone is related to temperature (see figure at top of previous page), air quality is projected to become worse with human-induced climate change. Many areas in the country already have plans in place for responding to air quality problems. For example, the Air Quality Alert program in Rhode Island encourages residents to reduce air pollutant emissions by limiting car travel and the use of small engines, lawn mowers, and charcoal lighter fluids on days when ground-level ozone is high. Television weather reports include alerts when ground-level ozone is high, warning especially susceptible people to limit their time outdoors. To help cut down on the use of cars, all regular bus routes are free on Air Quality Alert days.[295]

Pennsylvania offers the following suggestions for high ozone days:
- Refuel vehicles after dark. Avoid spilling gasoline and stop fueling when the pump shuts off automatically.
- Conserve energy. Do not overcool homes. Turn off lights and appliances that are not in use. Wash clothes and dishes only in full loads.
- Limit daytime driving. Consider carpooling or taking public transportation. Properly maintain vehicles, which also helps to save fuel.
- Limit outdoor activities, such as mowing the lawn or playing sports, to the evening hours.
- Avoid burning leaves, trash, and other materials.

Traffic restrictions imposed during the 1996 summer Olympics in Atlanta quantified the direct respiratory health benefits of reducing the number of cars and the amount of their tailpipe emissions from an urban environment. Peak morning traffic decreased by 23 percent, and peak ozone levels dropped by 28 percent. As a result, childhood asthma-related emergency room visits fell by 42 percent.[296]

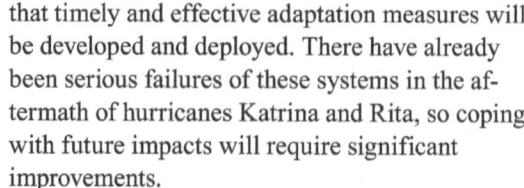

to ozone formation (especially nitrogen oxides and volatile organic compounds [VOCs]), natural emissions of VOCs from plants, and pollution blown in from other places.[290,297] A warmer climate is projected to increase the natural emissions of VOCs, accelerate ozone formation, and increase the frequency and duration of stagnant air masses that allow pollution to accumulate, which will exacerbate health symptoms.[298] Increased temperatures and water vapor due to human-induced carbon dioxide emissions have been found to increase ozone more in areas with already elevated concentrations, meaning that global warming tends to exacerbate ozone pollution most in already polluted areas. Under constant pollutant emissions, by the middle of this century, Red Ozone Alert Days (when the air is unhealthy for everyone) in the 50 largest cities in the eastern United States are projected to increase by 68 percent due to warming alone.[298] Such conditions would challenge the ability of communities to meet health-based air quality standards such as those in the Clean Air Act.

Health risks from heat waves and air pollution are not necessarily independent. The formation of ground-level ozone occurs under hot and stagnant conditions – essentially the same weather conditions accompanying heat waves (see box page 102). Such interactions among risk factors are likely to increase as climate change continues.

Extreme weather events cause physical and mental health problems. Some of these events are projected to increase.

Injury, illness, emotional trauma, and death are known to result from extreme weather events.[68] The number and intensity of some of these events are already increasing and are projected to increase further in the future.[68,112] Human health impacts in the United States are generally expected to be less severe than in poorer countries where the emergency preparedness and public health infrastructure is less developed. For example, early warning and evacuation systems and effective sanitation lessen the health impacts of extreme events.[68]

This assumes that medical and emergency relief systems in the United States will function well and that timely and effective adaptation measures will be developed and deployed. There have already been serious failures of these systems in the aftermath of hurricanes Katrina and Rita, so coping with future impacts will require significant improvements.

Extreme storms

Over 2,000 Americans were killed in the 2005 hurricane season, more than double the average number of lives lost to hurricanes in the United States over the previous 65 years.[163] But the human health impacts of extreme storms go beyond direct injury and death to indirect effects such as carbon monoxide poisoning from portable electric generators in use following hurricanes, an increase in stomach and intestinal illness among evacuees, and mental health impacts such as depression and post-traumatic stress disorder.[163] Failure to fully account for both direct and indirect health impacts might result in inadequate preparation for and response to future extreme weather events.[163]

Floods

Heavy downpours have increased in recent decades and are projected to increase further as the world continues to warm.[68,112] In the United States, the amount of precipitation falling in the heaviest 1 percent of rain events increased by 20 percent in the past century, while total precipitation increased by 7 percent. Over the last century, there was a 50 percent increase in the frequency of days with precipitation over 4 inches in the upper Midwest.[112] Other regions, notably the South, have also seen strong increases in heavy downpours, with most of these coming in the warm season and almost all of the increase coming in the last few decades.

Heavy rains can lead to flooding, which can cause health impacts including direct injuries as well as increased incidence of waterborne diseases due to pathogens such as *Cryptosporidium* and *Giardia*.[163] Downpours can trigger sewage overflows, contaminating drinking water and endangering beachgoers. The consequences will be particularly severe in the roughly 770 U.S. cities and towns, including New York, Chicago, Washington DC, Milwaukee, and Philadelphia, that have "combined sewer systems;" an older design that carries storm water and sewage in the same pipes.[299] During heavy rains, these systems often cannot handle the volume, and raw sewage spills into lakes or waterways, including drinking-water supplies and places where people swim.[252]

In 1994, the Environmental Protection Agency (EPA) established a policy that mandates that communities substantially reduce or eliminate their combined sewer overflow, but this mandate remains unfulfilled.[300] In 2004, the EPA estimated it would cost $55 billion to correct combined sewer overflow problems in publicly owned wastewater treatment systems.[301]

Using 2.5 inches of precipitation in one day as the threshold for initiating a combined sewer overflow event, the frequency of these events in Chicago is expected to rise by 50 percent to 120 percent by the end of this century,[302] posing further risks to drinking and recreational water quality.

Wildfires

Wildfires in the United States are already increasing due to warming. In the West, there has been a nearly fourfold increase in large wildfires in recent decades, with greater fire frequency, longer fire durations, and longer wildfire seasons. This increase is strongly associated with increased spring and summer temperatures and earlier spring snowmelt, which have caused drying of soils and vegetation.[163,252,294] In addition to direct injuries and deaths due to burns, wildfires can cause eye and respiratory illnesses due to fire-related air pollution.[163]

Some diseases transmitted by food, water, and insects are likely to increase.

A number of important disease-causing agents (pathogens) commonly transmitted by food, water,

Spotlight on West Nile Virus

The first outbreak of West Nile virus in the United States occurred in the summer of 1999, likely a result of international air transport. Within five years, the disease had spread across the continental United States, transmitted by mosquitoes that acquire the virus from infected birds. While bird migrations were the primary mode of disease spread, during the epidemic summers of 2002 to 2004, epicenters of West Nile virus were linked to locations with either drought or above average temperatures.

Since 1999, West Nile virus has caused over 28,000 reported cases, and over 1,100 Americans have died from it.[303] During 2002, a more virulent strain of West Nile virus emerged in the United States. Recent analyses indicate that this mutated strain responds strongly to higher temperatures, suggesting that greater risks from the disease may result from increases in the frequency of heatwaves,[304] though the risk will also depend on the effectiveness of mosquito control programs.

While West Nile virus causes mild flu-like symptoms in most people, about one in 150 infected people develop serious illness, including the brain inflammation diseases encephalitis and meningitis.

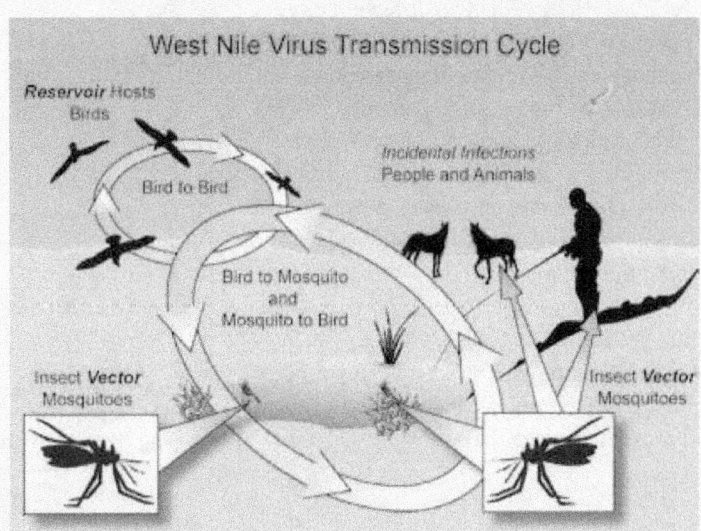

West Nile Virus Transmission Cycle

or animals are susceptible to changes in replication, survival, persistence, habitat range, and transmission as a result of changing climatic conditions such as increasing temperature, precipitation, and extreme weather events.[163]

- Cases of food poisoning due to *Salmonella* and other bacteria peak within one to six weeks of the highest reported ambient temperatures.[163]
- Cases of waterborne *Cryptosporidium* and *Giardia* increase following heavy downpours. These parasites can be transmitted in drinking water and through recreational water use.[163]
- Climate change affects the life cycle and distribution of the mosquitoes, ticks, and rodents that carry West Nile virus, equine encephalitis, Lyme disease, and hantavirus. However, moderating factors such as housing quality, land use patterns, pest control programs, and a robust public health infrastructure are likely to prevent the large-scale spread of these diseases in the United States.[163,305]
- Heavy rain and flooding can contaminate certain food crops with feces from nearby livestock or wild animals, increasing the likelihood of food-borne disease associated with fresh produce.[163]
- *Vibrio* sp. (shellfish poisoning) accounts for 20 percent of the illnesses and 95 percent of the deaths associated with eating infected shellfish, although the overall incidence of illness from *Vibrio* infection remains low. There is a close association between temperature, *Vibrio* sp. abundance, and clinical illness. The U.S. infection rate increased 41 percent from 1996 to 2006,[163] concurrent with rising temperatures.
- As temperatures rise, tick populations that carry Rocky Mountain spotted fever are projected to shift from south to north.[306]
- The introduction of disease-causing agents from other regions of the world is an additional threat.[163]

While the United States has programs such as the Safe Drinking Water Act that help protect against some of these problems, climate change will present new challenges.

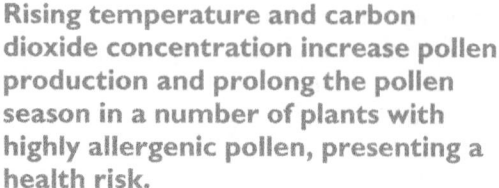

Rising temperature and carbon dioxide concentration increase pollen production and prolong the pollen season in a number of plants with highly allergenic pollen, presenting a health risk.

Rising carbon dioxide levels have been observed to increase the growth and toxicity of some plants that cause health problems. Climate change has caused an earlier onset of the spring pollen season in the United States.[272] It is reasonable to conclude that allergies caused by pollen have also experienced associated changes in seasonality.[272] Several laboratory studies suggest that increasing carbon dixoide concentrations and temperatures increase ragweed pollen production and prolong the ragweed pollen season.[163,272]

Poison ivy growth and toxicity is also greatly increased by carbon dioxide, with plants growing larger and more allergenic. These increases exceed those of most beneficial plants. For example, poison ivy vines grow twice as much per year in air with a doubled preindustrial carbon dioxide concentration as they do in unaltered air; this is nearly five times the increase reported for tree species in

Pollen Counts Rise with Increasing Carbon Dioxide

Pollen production (grams per plant)

| CO₂ | 280 ppm | 370 ppm | 720 ppm |
| Year | ~1900 | ~2000 | ~2075 |

Ziska and Caulfield[307]

Pollen production from ragweed grown in chambers at the carbon dioxide concentration of a century ago (about 280 parts per million [ppm]) was about 5 grams per plant; at today's approximate carbon dioxide level, it was about 10 grams; and at a level projected to occur about 2075 under the higher emissions scenario,[91] it was about 20 grams.[307]

Poison ivy

other analyses.[308] Recent and projected increases in carbon dioxide also have been shown to stimulate the growth of stinging nettle and leafy spurge, two weeds that cause rashes when they come into contact with human skin.[309,310]

Certain groups, including children, the elderly, and the poor, are most vulnerable to a range of climate-related health effects.

Infants and children, pregnant women, the elderly, people with chronic medical conditions, outdoor workers, and people living in poverty are especially at risk from a variety of climate related health effects. Examples of these effects include increasing heat stress, air pollution, extreme weather events, and diseases carried by food, water, and insects.[163]

Children's small ratio of body mass to surface area and other factors make them vulnerable to heat-related illness and death. Their increased breathing rate relative to body size, additional time spent outdoors, and developing respiratory tracts, heighten their sensitivity to air pollution. In addition, children's immature immune systems increase their risk of serious consequences from waterborne and food-borne diseases, while developmental factors make them more vulnerable to complications from severe infections such as *E. coli* or *Salmonella*.[163]

The greatest health burdens related to climate change are likely to fall on the poor, especially

those lacking adequate shelter and access to other resources such as air conditioning.[163]

Elderly people are more likely to have debilitating chronic diseases or limited mobility. The elderly are also generally more sensitive to extreme heat for several reasons. They have a reduced ability to regulate their own body temperature or sense when they are too hot. They are at greater risk of heart failure, which is further exacerbated when cardiac demand increases in order to cool the body during a heat wave.[318] Also, people taking medications, such as diuretics for high blood pressure, have a higher risk of dehydration.[163]

The multiple health risks associated with diabetes will increase the vulnerability of the U.S. population to increasing temperatures. The number of Americans with diabetes has grown to about 24 million people, or roughly 8 percent of the U.S. population. Almost 25 percent of the population 60 years and older had diabetes in 2007.[311] Fluid imbalance and dehydration create higher risks for diabetics during heat waves. People with diabetes-related heart disease are at especially increased risk of dying in heat waves.[318]

High obesity rates in the United States are a contributing factor in currently high levels of diabetes. Similarly, a factor in rising obesity rates is a sedentary lifestyle and automobile dependence; 60 percent of Americans do not meet minimum daily exercise requirements. Making cities more walkable and bikeable would thus have multiple benefits: improved personal fitness and weight loss;

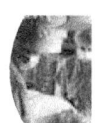

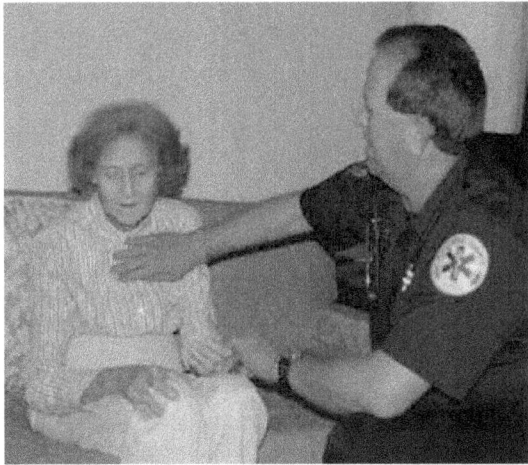

reduced local air pollution and associated respiratory illness; and reduced greenhouse gas emissions.[312]

The United States has considerable capacity to adapt to climate change, but during recent extreme weather and climate events, actual practices have not always protected people and property. Vulnerability to extreme events is highly variable, with disadvantaged groups and communities (such as the poor, infirm, and elderly) experiencing consider-

able damage and disruptions to their lives. Adaptation tends to be reactive, unevenly distributed, and focused on coping rather than preventing problems. Future reduction in vulnerability will require consideration of how best to incorporate planned adaptation into long-term municipal and public service planning, including energy, water, and health services, in the face of changing climate-related risks combined with ongoing changes in population and development patterns.[163,164]

Geographic Vulnerability of U.S. Residents to Selected Climate-Related Health Impacts

a) **Location of Hurricane Landfalls, 1995 to 2000**

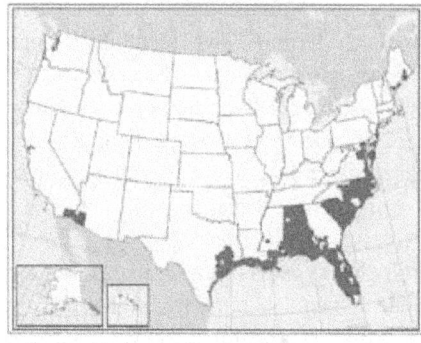

c) **Percentage of Population Aged 65 or Older, 2000**

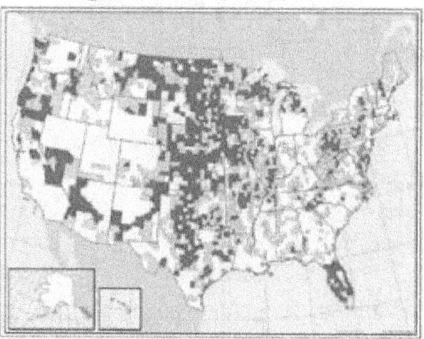

b) **Location of Extreme Heat Events, 1995 to 2000**

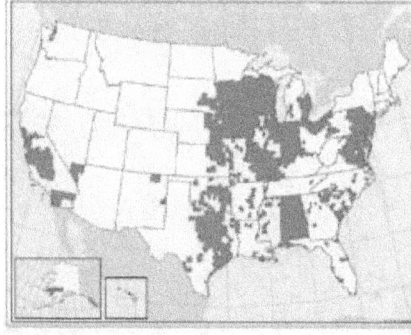

d) **West Nile Virus Cases 2004**

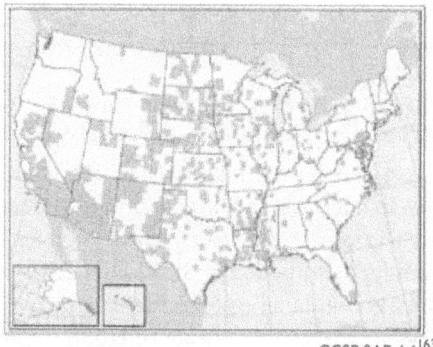

CCSP SAP 4.6[163]

Maps indicating U.S. counties, or in some cases states, with existing vulnerability to climate-sensitive health outcomes: a) location of hurricane landfalls; b) extreme heat events (defined by the Centers for Disease Control as temperatures 10 or more degrees F above the average high temperature for the region and lasting for several weeks); c) percentage of population over age 65 (dark blue indicates that percentage is over 17.6 percent, light blue 14.4 to 17.5 percent); d) locations of West Nile virus cases reported in 2004. These examples demonstrate both the diversity of climate-sensitive health outcomes and the geographic variability of where they occur. Events over short time spans, in particular West Nile virus cases, are not necessarily predictive of future vulnerability.

Society

Key Messages:

- Population shifts and development choices are making more Americans vulnerable to the expected impacts of climate change.
- Vulnerability is greater for those who have few resources and few choices.
- City residents and city infrastructure have unique vulnerabilities to climate change.
- Climate change affects communities through changes in climate-sensitive resources that occur both locally and at great distances.
- Insurance is one of the industries particularly vulnerable to increasing extreme weather events such as severe storms, but it can also help society manage the risks.
- The United States is connected to a world that is unevenly vulnerable to climate change and thus will be affected by impacts in other parts of the world.

Climate change will affect society through impacts on the necessities and comforts of life: water, energy, housing, transportation, food, natural ecosystems, and health. This section focuses on some characteristics of society that make it vulnerable to the potential impacts of climate change and how the risks and costs may be distributed. Many impacts of climate change on society, for example, sea-level rise and increased water scarcity, are covered in other sections of this report. This section is not a comprehensive analysis of societal vulnerabilities, but rather highlights key examples.

Because societies and their built environments have developed under a climate that has fluctuated within a relatively confined range of conditions, most impacts of a rapidly changing climate will present challenges. Society is especially vulnerable to extremes, such as heat waves and floods, many of which are increasing as climate changes.[313] And while there are likely to be some benefits and opportunities in the early stages of warming, as climate continues to change, negative impacts are projected to dominate.[164]

Climate change will affect different segments of society differently because of their varying exposures and adaptive capacities. The impacts of climate change also do not affect society in isolation. Rather, impacts can be exacerbated when climate change occurs in combination with the effects of an aging and growing population, pollution, poverty, and natural environmental fluctuations.[164,172,274] Unequal adaptive capacity in the world as a whole also will pose challenges to the United States. Poorer countries are projected to be disproportionately affected by the impacts of climate change and the United States is strongly connected to the world beyond its borders through markets, trade, investments, shared resources, migrating species, health, travel and tourism, environmental refugees (those fleeing deteriorating environmental conditions), and security.

Cedar Rapids, Iowa, June 12, 2008

Population shifts and development choices are making more Americans vulnerable to the expected impacts of climate change.

Climate is one of the key factors in Americans' choices of where to live. As the U.S. population grows, ages, and becomes further concentrated in cities and coastal areas, society is faced with additional challenges. Climate change is likely to exacerbate these challenges as changes in temperature, precipitation, sea levels, and extreme weather events increasingly affect homes, communities, water supplies, land resources, transportation, urban infrastructure, and regional characteristics that people have come to value and depend on.

Population growth in the United States over the past century has been most rapid in the South, near the coasts, and in large urban areas (see figure on page 55 in the *Energy* sector). The four most populous states in 2000 – California, Texas, Florida, and New York – accounted for 38 percent of the total growth in U.S. population during that time, and share significant vulnerability to coastal storms, severe drought, sea-level rise, air pollution, and urban heat island effects.[313] But migration patterns are now shifting: the population of the Mountain West (Montana, Idaho, Wyoming, Nevada, Utah, Colorado, Arizona, and New Mexico) is projected to increase by 65 percent from 2000 to 2030, representing one-third of all U.S. population growth.[274,314] Southern coastal areas on both the Atlantic and the Gulf of Mexico are projected to continue to see population growth.[313]

Overlaying projections of future climate change and its impacts on expected changes in U.S. population and development patterns reveals a critical insight: more Americans will be living in the areas that are most vulnerable to the effects of climate change.[274]

America's coastlines have seen pronounced population growth in regions most at risk of hurricane activity, sea-level rise, and storm surge – putting more people and property in harm's way as the probability of harm increases.[274] On the Atlantic and Gulf coasts where hurricane activity is prevalent, the coastal land in many areas is sinking while sea level is rising. Human activities are exacerbat-

ing the loss of coastal wetlands that once helped buffer the coastline from erosion due to storms. The devastation caused by recent hurricanes highlights the vulnerability of these areas.[224]

The most rapidly growing area of the country is the Mountain West, a region projected to face more frequent and severe wildfires and have less water available, particularly during the high-demand period of summer. Continued population growth in these arid and semi-arid regions would stress water supplies. Because of high demand for irrigating agriculture, overuse of rivers and streams is common in the arid West, particularly along the Front Range of the Rocky Mountains in Colorado, in Southern California, and in the Central Valley of California. Rapid population and economic growth in these arid and semi-arid regions has dramatically increased vulnerability to water shortages (see *Water Resources* sector and *Southwest* region).[274]

Many questions are raised by ongoing development patterns in the face of climate change. Will growth continue as projected in vulnerable areas, despite the risks? Will there be a retreat from the coastline as it becomes more difficult to insure vulnerable properties? Will there be pressure for the government to insure properties that private insurers have rejected? How can the vulnerability of new development be minimized? How can we ensure that communities adopt measures to manage the significant changes that are projected in sea level, temperature, rainfall, and extreme weather events?

Development choices are based on people's needs and desires for places to live, economies that provide employment, ecosystems that provide services, and community-based social activities. Thus, the future vulnerability of society will be influenced by how and where people choose to live. Some choices, such as expanded development in coastal regions, can increase vulnerabilities to climate-related events, even without any change in climate.

Vulnerability is greater for those who have few resources and few choices.

Vulnerabilities to climate change depend not only on where people are but also on their circumstances. In general, groups that are especially vulnerable include the very young, the very old, the sick, and the poor. These groups represent a more significant portion of the total population in some regions and localities than others. For example, the elderly more often cite a warm climate as motivating their choice of where to live and thus make up a larger share of the population in warmer areas.[305]

In the future (as in the past), the impacts of climate change are likely to fall disproportionately on the disadvantaged.[313] People with few resources often live in conditions that increase their vulnerability to the effects of climate change.[172] For example, the experience with Hurricane Katrina showed that the poor and elderly were the most vulnerable because of where they lived and their limited ability to get out of harm's way. Thus, those who had the least proportionately lost the most. And it is clear that people with access to financial resources, including insurance, have a greater capacity to adapt to, recover, or escape from adverse impacts of climate change than those who do not have such access.[305, 316] The fate of the poor can be permanent dislocation, leading to the loss of social relationships and community support networks provided by schools, churches, and neighborhoods.

Native American communities have unique vulnerabilities. Native Americans who live on established reservations are restricted to reservation boundaries and therefore have limited relocation options.[219] In Alaska, over 100 villages on the coast and in low-lying areas along rivers are subject to increased flooding and erosion due to warming.[315] Warming also reduces the availability and accessibility of many traditional food sources for Native Alaskans, such as seals that live on ice and caribou whose migration patterns depend on being able to cross frozen rivers and wetlands. These vulnerable people face losing their current livelihoods, their communities, and in some cases, their culture, which depends on traditional ways of collecting and sharing food.[132,220] Native cultures in the Southwest are particularly vulnerable to impacts of climate change on water quality and availability.

Chalmette, Louisiana after Hurricane Katrina

City residents and city infrastructure have unique vulnerabilities to climate change.

Over 80 percent of the U.S. population resides in urban areas, which are among the most rapidly changing environments on Earth. In recent decades, cities have become increasingly spread out, complex, and interconnected with regional and national economies and infrastructure.[319] Cities also experience a host of social problems, including neighborhood degradation, traffic congestion, crime, unemployment, poverty, and inequities in health and well-being.[320] Climate-related changes such as increased heat, water shortages, and extreme weather events will add further stress to existing problems. The impacts of climate change on cities are compounded by aging infrastructure, buildings, and populations, as well as air pollution and population growth. Further, infrastructure designed to handle past variations in climate can instill a false confidence in its ability to handle future changes. However, urban areas also present opportunities for adaptation through technology, infrastructure, planning, and design.[313]

As cities grow, they alter local climates through the urban heat island effect. This effect occurs because cities absorb, produce, and retain more heat than the surrounding countryside. The urban heat island effect has raised average urban air temperatures by 2 to 5°F more than surrounding areas over the past 100 years, and by up to 20°F more at night.[321] Such temperature increases, on top of the general increase caused by human-induced warming, affect urban dwellers in many ways, influencing health, comfort, energy costs, air quality, water quality and availability, and even violent crime (which increases at high temperatures) (see *Human Health, Energy,* and *Water Resources* sectors).[172,313,322,323]

More frequent heavy downpours and floods in urban areas will cause greater property damage, a heavier burden on emergency management, increased clean-up and rebuilding costs, and a growing financial toll on businesses and homeowners. The Midwest floods of 2008 provide a recent vivid example of such tolls. Heavy downpours and urban floods can also overwhelm combined sewer and storm-water systems and release pollutants to waterways.[313] Unfortunately, for many cities, current

planning and existing infrastructure are designed for the historical one-in-100 year event, whereas cities are likely to experience this same flood level much more frequently as a result of the climate change projected over this century.[146,164,324]

Cities are also likely to be affected by climate change in unforeseen ways, necessitating diversion of city funds for emergency responses to extreme weather.[313] There is the potential for increased summer electricity blackouts owing to greater demand for air conditioning.[325] For example, there were widespread power outages in Chicago during the 1995 heat wave and in some parts of New York City during the 1999 heat wave. In southern California's cities, additional summer electricity demand will intensify conflicts between hydropower and flood-control objectives.[164] Increased costs of repairs and maintenance are projected for transportation systems, including roads, railways, and airports, as they are negatively affected by heavy downpours

Heat, Drought, and Stagnant Air Degrade Air Quality and Quality of Life

Heat waves and poor air quality already threaten the lives of thousands of people each year.[292] Experience and research have shown that these events are interrelated as the atmospheric conditions that produce heat waves are often accompanied by stagnant air and poor air quality.[326] The simultaneous occurrence of heat waves, drought, and stagnant air negatively affects quality of life, especially in cities.

One such event occurred in the United States during the summer of 1988, causing 5,000 to 10,000 deaths and economic losses of more than $70 billion (in 2002 dollars).[229,327] Half of the nation was affected by drought, and 5,994 all-time daily high temperature records were set around the country in July alone (more than three times the most recent 10-year average).[328,329] Poor air quality resulting from the lack of rainfall, high temperatures, and stagnant conditions led to an unprecedented number of unhealthy air quality days throughout large parts of the country.[327,329] Continued climate change is projected to increase the likelihood of such episodes.[68,330]

Interactions such as those between heat wave and drought will affect adaptation planning. For example, electricity use increases during heat waves due to increased air conditioning demand.[330,331] During droughts, cooling water availability is at its lowest. Thus, during a simultaneous heat wave and drought, electricity demand for cooling will be high when power plant cooling water availability is at its lowest.[340]

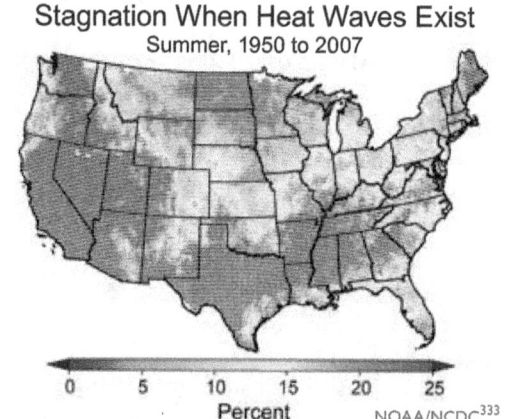

Stagnation When Heat Waves Exist
Summer, 1950 to 2007

0   5   10   15   20   25
Percent   NOAA/NCDC[333]

The map shows the frequency of occurrence of stagnant air conditions when heat wave conditions were also present. Since 1950, across the Southeast, southern Great Plains, and most of the West, the air was stagnant more than 25 percent of the time during heat waves.

and extreme heat[190] (see *Transportation* sector). Coping with increased flooding will require replacement or improvements in storm drains, flood channels, levees, and dams.

In addition, coastal cities are also vulnerable to sea-level rise, storm surge, and increased hurricane intensity. Cities such as New Orleans, Miami, and New York are particularly at risk, and would have difficulty coping with the sea-level rise projected by the end of the century under a higher emissions scenario.[91,164] Remnants of hurricanes moving inland also threaten cities of the Appalachian Mountains, which are vulnerable if hurricane frequency or intensity increases. Since most large U.S. cities are on coasts, rivers, or both, climate change will lead to increased potential flood damage. The largest impacts are expected when sea-level rise, heavy runoff, high tides, and storms coincide.[313] Analyses of New York and Boston indicate that the potential impacts of climate change are likely to be negative, but that vulnerability can be reduced by behavioral and policy changes.[313,334-336]

Urban areas concentrate the human activities that are largely responsible for heat-trapping emissions. The demands of urban residents are also associated with a much larger footprint on areas far removed from these population centers.[337] On the other hand, concentrating activities such as transportation can make them more efficient. Cities have a large role to play in reducing heat-trapping emissions, and many are pursuing such actions. For example, over 900 cities have committed to the U.S. Mayors' Climate Protection Agreement to advance emissions reduction goals.[317]

Cities also have considerable potential to adapt to climate change through technological, institutional, structural, and behavioral changes. For example, a number of cities have warning programs in place to reduce heat-related illness and death (see *Human Health* sector). Relocating development away from low-lying areas, building new infrastructure with future sea-level rise in mind, and promoting water conservation are examples of structural and institutional strategies. Choosing road materials that can handle higher temperatures is an adaptation option that relies on new technology (see *Transportation* sector). Cities can reduce heat loads by increasing

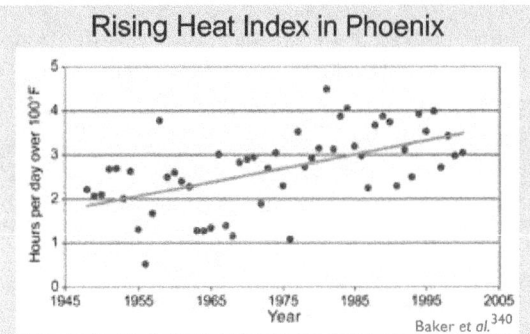

Rising Heat Index in Phoenix

Baker et al.[340]

The average number of hours per summer day in Phoenix that the temperature was over 100°F has doubled over the past 50 years, in part as a result of the urban heat island effect. Hot days take a toll on both quality of life and loss of life. Arizona's heat-related deaths are the highest of any state, at three to seven times the national average.[340,341]

reflective surfaces and green spaces. Some actions have multiple benefits. For example, increased planting of trees and other vegetation in cities has been shown to be associated with a reduction in crime,[338] in addition to reducing local temperatures, and thus energy demand for air conditioning.

Human well-being is influenced by economic conditions, natural resources and amenities, public health and safety, infrastructure, government, and social and cultural resources. Climate change will influence all of these, but an understanding of the many interacting impacts, as well as the ways society can adapt to them, remains in its infancy.[305,339]

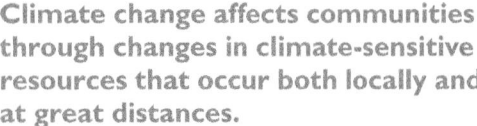

Climate change affects communities through changes in climate-sensitive resources that occur both locally and at great distances.

Human communities are intimately connected to resources beyond their geographical boundaries. Thus, communities will be vulnerable to the potential impacts of climate change on sometimes-distant resources. For example, communities that have developed near areas of agricultural production, such as the Midwest corn belt or the wine-producing regions of California and the Northwest, depend on the continued productivity of those regions, which would be compromised by increased temperature or severe weather.[313] Some agricultural production that is linked to cold climates is likely to disappear entirely: recent warming has altered the required temperature patterns for maple syrup production, shifting production northward from New England into Canada. Similarly, cranberries require a long winter chill period, which is shrinking as climate warms[234] (see *Northeast* region). Most cities depend on water supplies from distant watersheds, and those depending on diminishing supplies (such as the Sierra Nevada snowpack) are vulnerable. Northwest communities also depend upon forest resources for their economic base, and many island, coastal, and "sunbelt" communities depend on tourism.

Recreation and tourism play important roles in the economy and quality of life of many Americans. In some regions tourism and recreation are major job creators, bringing billions of dollars to regional economies. Across the nation, fishing, hunting, skiing, snowmobiling, diving, beach-going, and other outdoor activities make important economic contributions and are a part of family traditions that have value that goes beyond financial returns. A changing climate will mean reduced opportunities for some activities and locations and expanded opportunities for others.[305,342] Hunting and fishing will change as animals' habitats shift and as relationships among species in natural communities are disrupted by their different responses to rapid climate change. Water-dependent recreation in areas projected to get drier, such as the Southwest, and beach recreation in areas that are expected to see rising sea levels, will suffer. Some regions will see an expansion of the season for warm weather recreation such as hiking and bicycle riding.

Insurance is one of the industries particularly vulnerable to increasing extreme weather events such as severe storms, but it can also help society manage the risks.

Insurance – the world's largest industry – is one of the primary mechanisms through which the costs of climate change are distributed across society.[344,351]

Most of the climate change impacts described in this report have economic consequences. A significant portion of these flow through public and private insurance markets, which essentially aggregate and distribute society's risk. Insurance thus provides a window into the myriad ways in which the costs of climate change will manifest, and serves as a form of economic adaptation and a messenger of these impacts through the terms and price signals it sends its customers.[344]

In an average year, about 90 percent of insured catastrophe losses worldwide are weather-related. In the United States, about half of all these losses are insured, which amounted to $320 billion between 1980 and 2005 (inflation-adjusted to 2005 dollars). While major events such as hurricanes grab headlines, the aggregate effect of smaller events accounts for at least 60 percent of total insured losses on average.[344] Many of the smallest scale property losses and weather-related life/health losses are unquantified.[345]

Escalating exposures to catastrophic weather events, coupled with private insurers' withdrawal from various markets, are placing the federal government at increased financial risk as insurer of last resort. The National Flood Insurance Program would have gone bankrupt after the storms of 2005 had they not been given the ability to borrow about $20 billion from the U.S. Treasury.[172] For public and private insurance programs alike, rising losses require a combination of risk-based premiums and improved loss prevention.

Examples of Impacts On Recreation

Recreational Activity	Potential Impacts of Climate Change	Estimated Economic Impacts
Skiing, Northeast	20 percent reduction in ski season length	$800 million loss per year, potential resort closures[234]
Snowmobiling, Northeast	Reduction of season length under higher emissions scenario[91]	Complete loss of opportunities in New York and Pennsylvania within a few decades, 80 percent reduction in season length for region by end of century[234,342]
Beaches, North Carolina	Many beaches are eroded, and some lost by 2080[343]	Reduced opportunities for beach and fishing trips,[343] without additional costs for adaptation measures

While economic and demographic factors have no doubt contributed to observed increases in losses,[346] these factors do not fully explain the upward trend in costs or numbers of events.[344,347] For example, during the time period covered in the figure to the right, population increased by a factor of 1.3 while losses increased by a factor of 15 to 20 in inflation-corrected dollars. Analyses asserting little or no role of climate change in increasing the risk of losses tend to focus on a highly limited set of hazards and locations. They also often fail to account for the vagaries of natural cycles and inflation adjustments, or to normalize for countervailing factors such as improved pre- and post-event loss prevention (such as dikes, building codes, and early warning systems).[348]

What is known with far greater certainty is that future increases in losses will be attributable to climate change as it increases the frequency and intensity of many types of extreme weather, such as severe thunderstorms and heat waves.[131,350]

Insurance is emblematic of the increasing globalization of climate risks. Because large U.S.-based companies operate around the world, their customers and assets are exposed to climate impacts wherever they occur. Most of the growth in the insurance industry is in emerging markets, which will structurally increase U.S. insurers' exposure to climate risk because those regions are more vulnerable and are experiencing particularly high rates of population growth and development.[351]

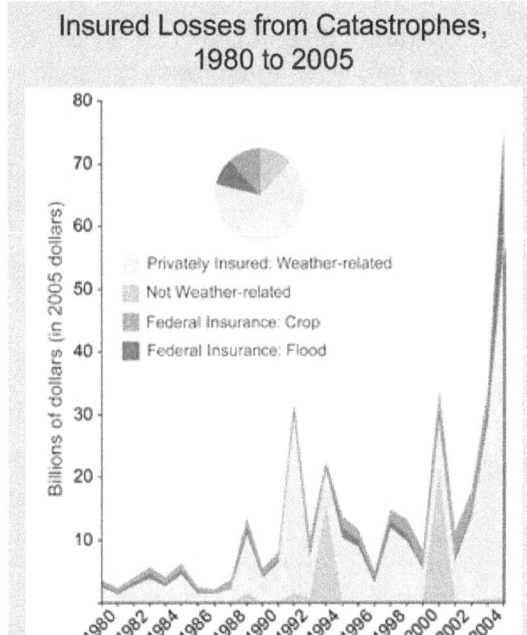

Insured Losses from Catastrophes, 1980 to 2005

US GAO[352]

Weather-related insurance losses in the United States are increasing. Typical weather-related losses today are similar to those that resulted from the 9/11 attack (shown in gray at 2001 in the graph). About half of all economic losses are insured, so actual losses are roughly twice those shown on the graph. Data on smaller-scale losses (many of which are weather-related) are significant but are not included in this graph as they are not comprehensively reported by the U.S. insurance industry.

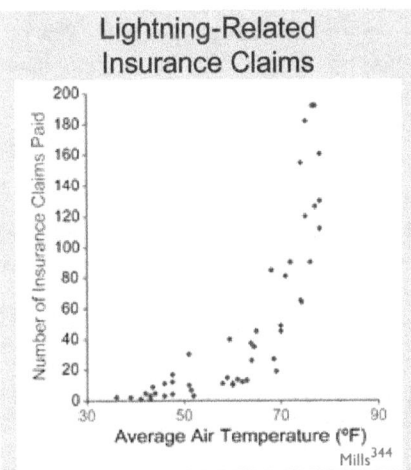

Lightning-Related Insurance Claims

Mills[344]

There is a strong observed correlation between higher temperatures and the frequency of lightning-induced insured losses in the United States. Each marker represents aggregate monthly U.S. lightning-related insurance claims paid by one large national insurer over a five-year period, 1991-1995. All else being equal, these claims are expected to increase with temperature.[344,353,354]

The movement of populations into harm's way creates a rising baseline of insured losses upon which the consequences of climate change will be superimposed. These observations reinforce a recurring theme in this report: the past can no longer be used as the basis for planning for the future.

It is a challenge to design insurance systems that properly price risks, reward loss prevention, and do not foster risk taking (for example by repeatedly rebuilding flooded homes). This challenge is particularly acute in light of insurance market distortions such as prices that inhibit insurers' ability to recover rising losses, combined with information gaps on the impacts of climate change and adaptation strategies. Rising losses[252] are already affecting the availability and affordability of insurance. Several million customers in the United States, no longer able to purchase private insurance coverage, are taking refuge in state-mandated insurance pools, or going without insurance altogether. Offsetting rising insurance costs is one benefit of mitigation and adaptation investments to reduce the impacts of climate change.

Virtually all segments of the insurance industry are vulnerable to the impacts of climate change. Examples include damage to property, crops, forest products, livestock, and transportation infrastructure; business and supply-chain interruptions caused by weather extremes, water shortages, and electricity outages; legal consequences;[355] and compromised health or loss of life. Increasing risks to insurers and their customers are driven by many factors including reduced periods of time between loss events, increasing variability, shifting types and location of events, and widespread simultaneous losses.

In light of these challenges, insurers are emerging as partners in climate science and the formulation of public policy and adaptation strategies.[356] Some have promoted adaptation by providing premium incentives for customers who fortify their properties, engaging in the process of determining building codes and land-use plans, and participating in the development and financing of new technologies and practices. For example, the Federal Emergency Management Agency (FEMA) Community Rating System is a point system that rewards communities that undertake floodplain management activities to reduce flood risk beyond the minimum requirement set by the National Flood Insurance Program. Everyone in these communities is rewarded with lower flood insurance premiums (−5 to −45 percent).[357] Others have recognized that mitigation and adaptation can work hand in hand in a coordinated climate risk-management strategy and are offering "green" insurance products designed to capture these dual benefits.[351,349]

The United States is connected to a world that is unevenly vulnerable to climate change and thus will be affected by impacts in other parts of the world.

American society will not experience the potential impacts of climate change in isolation. In an increasingly connected world, impacts elsewhere will have political, social, economic, and environmental ramifications for the United States. As in the United States, vulnerability to the potential impacts of climate change worldwide varies by location, population characteristics, and economic status.

The rising concentration of people in cities is occurring globally, but is most prevalent in lower-income countries. Many large cities are located in vulnerable areas such as floodplains and coasts. In most of these cities, the poor often live in the most marginal of these environments, in areas that are susceptible to extreme events, and their ability to adapt is limited by their lack of financial resources.[172]

In addition, over half of the world's population – including most of the world's major cities – depends on glacier melt or snowmelt to supply water for drinking and municipal uses. Today, some locations are experiencing abundant water supplies and even frequent floods due to increases in glacier melt rates due to increased temperatures worldwide. Soon, however, this trend is projected to reverse as even greater temperature increases reduce glacier mass and cause more winter precipitation to fall as rain and less as snow.[90]

As conditions worsen elsewhere, the number of people wanting to immigrate to the United States will increase. The direct cause of potential increased migration, such as extreme climatic events, will be difficult to separate from other forces that drive people to migrate. Climate change also has the potential to alter trade relationships by changing the comparative trade advantages of regions or nations. As with migration, shifts in trade can have multiple causes.

Accelerating emissions in economies that are rapidly expanding, such as China and India, pose future threats to the climate system and already are associated with air pollution episodes that reach the United States.[297]

Meeting the challenge of improving conditions for the world's poor has economic implications for the United States, as does intervention and resolution of intra- and intergroup conflicts. Where climate change exacerbates such challenges, for example by limiting access to scarce resources or increasing incidence of damaging weather events, consequences are likely for the U.S. economy and security.[358]

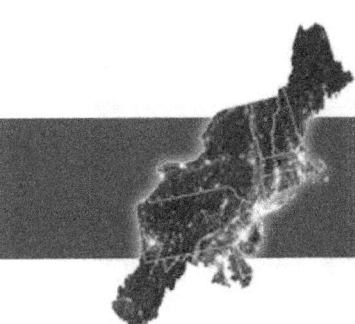

Northeast

The Northeast has significant geographic and climatic diversity within its relatively small area. The character and economy of the Northeast have been shaped by many aspects of its climate including its snowy winters, colorful autumns, and variety of extreme events such as nor'easters, ice storms, and heat waves. This familiar climate has already begun changing in noticeable ways.

Since 1970, the annual average temperature in the Northeast has increased by 2°F, with winter temperatures rising twice this much.[150] Warming has resulted in many other climate-related changes, including:

- More frequent days with temperatures above 90°F
- A longer growing season
- Increased heavy precipitation
- Less winter precipitation falling as snow and more as rain
- Reduced snowpack
- Earlier breakup of winter ice on lakes and rivers
- Earlier spring snowmelt resulting in earlier peak river flows
- Rising sea surface temperatures and sea level

Each of these observed changes is consistent with the changes expected in this region from global warming. The Northeast is projected to face continued warming and more extensive climate-related changes, some of which could dramatically alter the region's economy, landscape, character, and quality of life.

Over the next several decades, temperatures in the Northeast are projected to rise an additional 2.5 to 4°F in winter and 1.5 to 3.5°F in summer. By mid-century and beyond, however, today's emissions choices would generate starkly different climate futures; the lower the emissions, the smaller the climatic changes and resulting impacts.[150,359] By late this century, under a higher emissions scenario[91]:

- Winters in the Northeast are projected to be much shorter with fewer cold days and more precipitation.
- The length of the winter snow season would be cut in half across northern New York, Vermont, New Hampshire, and Maine, and reduced to a week or two in southern parts of the region.
- Cities that today experience few days above 100°F each summer would average 20 such days per summer, while certain cities, such as Hartford and Philadelphia, would average nearly 30 days over 100°F.
- Short-term (one- to three-month) droughts are projected to occur as frequently as once each summer in the Catskill and Adirondack Mountains, and across the New England states.
- Hot summer conditions would arrive three weeks earlier and last three weeks longer into the fall.
- Sea level in this region is projected to rise more than the global average, see *Global* and *National Climate Change* and *Coasts* sections for more information on sea-level rise (pages 25, 37, 150).

Climate on the Move:
Changing Summers in New Hampshire

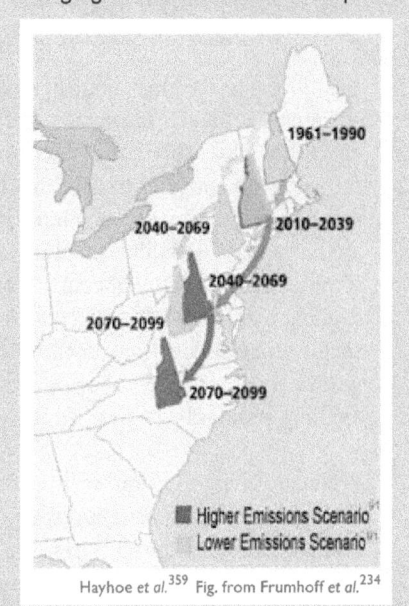

Hayhoe et al.[359] Fig. from Frumhoff et al.[234]

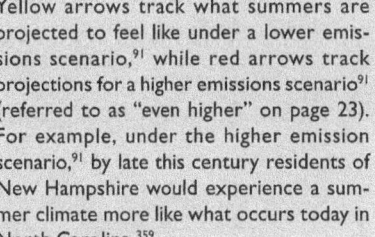

Yellow arrows track what summers are projected to feel like under a lower emissions scenario,[91] while red arrows track projections for a higher emissions scenario[91] (referred to as "even higher" on page 23). For example, under the higher emission scenario,[91] by late this century residents of New Hampshire would experience a summer climate more like what occurs today in North Carolina.[359]

Extreme heat and declining air quality are likely to pose increasing problems for human health, especially in urban areas.

Heat waves, which are currently rare in the region, are projected to become much more commonplace in a warmer future, with major implications for human health (see *Human Health* sector).[163,68]

In addition to the physiological stresses associated with hotter days and nights,[360] for cities that now experience ozone pollution problems, the number of days that fail to meet federal air quality standards is projected to increase with rising temperatures if there are no additional controls on ozone-causing pollutants[163,361] (see *Human Health* sector). Sharp reductions in emissions will be needed to keep ozone within existing standards.

Projected changes in summer heat (see figure below) provide a clear sense of how different the climate of the Northeast is projected to be under lower versus higher emissions scenarios. Changes of this kind will require greater use of air conditioning (see *Energy* sector).

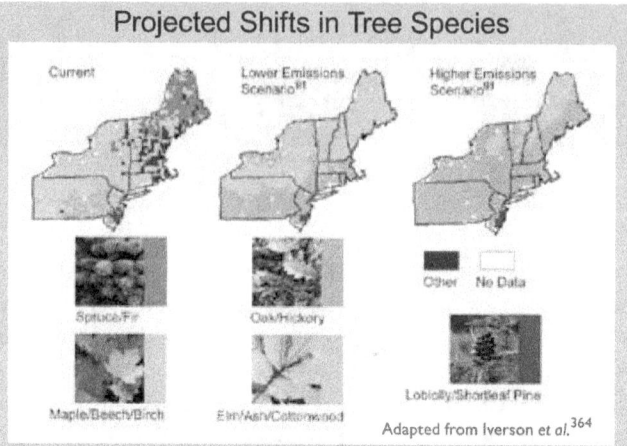

Projected Shifts in Tree Species

Adapted from Iverson et al.[364]

Much of the Northeast's forest is composed of the hardwoods maple, beech, and birch, while mountain areas and more northern parts of the region are dominated by spruce/fir forests. As climate changes over this century, suitable habitat for spruce and fir is expected to contract dramatically. Suitable maple/beech/birch habitat is projected to shift significantly northward under a higher emissions scenario (referred to as "even higher" on page 23),[91] but to shift far less under a lower emissions scenario.[91,363] Other studies of tree species shifts suggest even more dramatic changes than those shown here (see page 81).

Agricultural production, including dairy, fruit, and maple syrup, are likely to be adversely affected as favorable climates shift.

Large portions of the Northeast are likely to become unsuitable for growing popular varieties of apples, blueberries, and cranberries under a higher emissions scenario.[91,362,363] Climate conditions suitable for maple/beech/birch forests are projected to shift dramatically northward (see figure above), eventually leaving only a small portion of the Northeast with a maple sugar business.[364]

The dairy industry is the most important agricultural sector in this region, with annual production worth $3.6 billion.[365] Heat stress in dairy cows depresses both milk production and birth rates for periods of weeks to months.[193,366] By late this century, all but the northern parts of Maine, New Hampshire, New York, and Vermont are projected to suffer declines in July milk production under the higher emissions scenario. In parts of Connecticut, Massachusetts, New Jersey, New York, and Pennsylvania, a large decline in milk production, up to 20 percent or greater, is projected. Under the lower emissions scenario, however, reductions in milk production of up to 10 percent remain confined primarily to the southern parts of the region.

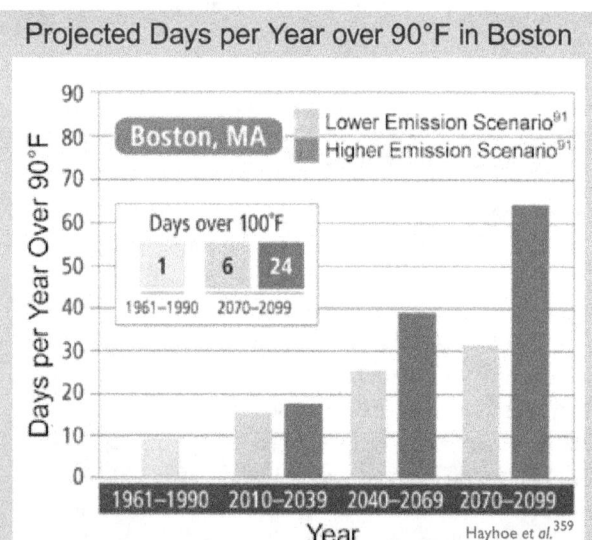

Projected Days per Year over 90°F in Boston

The graph shows model projections of the number of summer days with temperatures over 90°F in Boston, Massachusetts, under lower and higher (referred to as "even higher" on page 23) emissions scenarios.[91] The inset shows projected days over 100°F.[359]

This analysis used average monthly temperature and humidity data that do not capture daily variations in heat stress and projected increases in extreme heat. Nor did the analysis directly consider farmer responses, such as installation of potentially costly cooling systems. On balance, these projections are likely to underestimate impacts on the dairy industry.[150]

Severe flooding due to sea-level rise and heavy downpours is likely to occur more frequently.

The densely populated coasts of the Northeast face substantial increases in the extent and frequency of storm surge, coastal flooding, erosion, property damage, and loss of

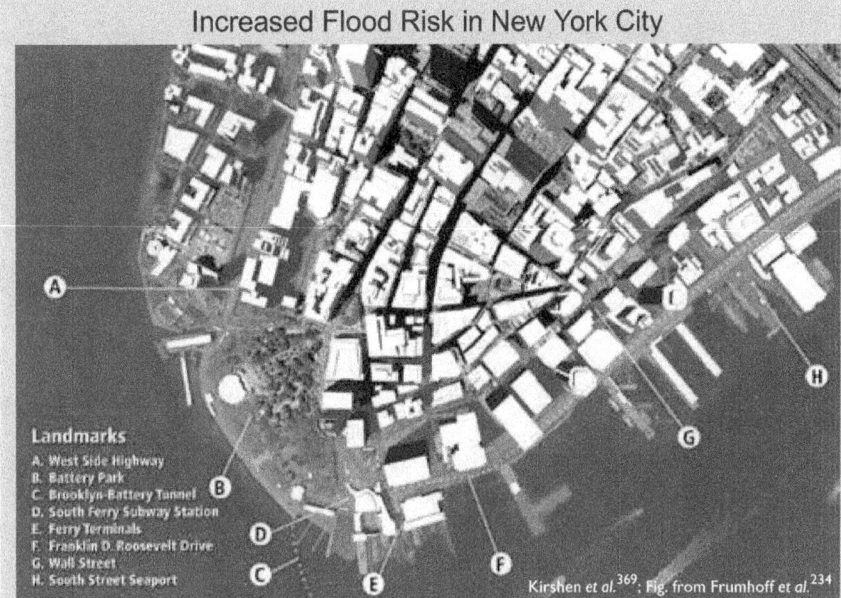

Increased Flood Risk in New York City

Landmarks
A. West Side Highway
B. Battery Park
C. Brooklyn-Battery Tunnel
D. South Ferry Subway Station
E. Ferry Terminals
F. Franklin D. Roosevelt Drive
G. Wall Street
H. South Street Seaport

Kirshen et al.[369]; Fig. from Frumhoff et al.[234]

The light blue area above depicts today's FEMA 100-year flood zone for the city (the area of the city that is expected to be flooded once every 100 years). With rising sea levels, a 100-year flood at the end of this century (not mapped here) is projected to inundate a far larger area of New York City, especially under the higher emissions scenario.[91] Critical transportation infrastructure located in the Battery area of lower Manhattan could be flooded far more frequently unless protected. The increased likelihood of flooding is causing planners to look into building storm-surge barriers in New York Harbor to protect downtown New York City. [234,370,371]

wetlands.[367,369] New York state alone has more than $2.3 trillion in insured coastal property.[368] Much of this coastline is exceptionally vulnerable to sea-level rise and related impacts. Some major insurers have withdrawn coverage from thousands of homeowners in coastal areas of the Northeast, including New York City.

Rising sea level is projected to increase the frequency and severity of damaging storm surges and flooding. Under a higher emissions scenario,[91] what is now considered a once-in-a-century coastal flood in New York City is projected to occur at least twice as often by mid-century, and 10 times as often (or once per decade

Adaptation: **Raising a Sewage Treatment Plant in Boston**

Boston's Deer Island sewage treatment plant was designed and built taking future sea-level rise into consideration. Because the level of the plant relative to the level of the ocean at the outfall is critical to the amount of rainwater and sewage that can be treated, the plant was built 1.9 feet higher than it would otherwise have been to accommodate the amount of sea-level rise projected to occur by 2050, the planned life of the facility.

The planners recognized that the future would be different from the past and they decided to plan for the future based on the best available information. They assessed what could be easily and inexpensively changed at a later date versus those things that would be more difficult and expensive to change later. For example, increasing the plant's height would be less costly to incorporate in the original design, while protective barriers could be added at a later date, as needed, at a relatively small cost.

on average) by late this century. With a lower emissions scenario,[91] today's 100-year flood is projected to occur once every 22 years on average by late this century.[369]

The projected reduction in snow cover will adversely affect winter recreation and the industries that rely upon it.

Winter snow and ice sports, which contribute some $7.6 billion annually to the regional economy, will be particularly affected by warming.[342] Of this total, alpine skiing and other snow sports (not including snowmobiling) account for $4.6 billion annually. Snowmobiling, which now rivals skiing as the largest winter recreation industry in the nation, accounts for the remaining $3 billion.[372] Other winter traditions, ranging from skating and ice fishing on frozen ponds and lakes, to cross-country (Nordic) skiing, snowshoeing, and dog sledding, are integral to the character of the Northeast, and for many residents and visitors, its desirable quality of life.

Warmer winters will shorten the average ski and snowboard seasons, increase artificial snowmaking requirements, and drive up operating costs. While snowmaking can enhance the prospects for ski resort success, it requires a great deal of water and energy, as well as very cold nights, which are becoming less frequent. Without the opportunity

to benefit from snowmaking, the prospects for the snowmobiling industry are even worse. Most of the region is likely to have a marginal or non-existent snowmobile season by mid-century.

The center of lobster fisheries is projected to continue its northward shift and the cod fishery on Georges Bank is likely to be diminished.

Lobster catch has increased dramatically in the Northeast as a whole over the past three decades, though not uniformly.[374,375] Catches in the southern part of the region peaked in the mid-1990s, and have since declined sharply, beginning with a 1997 die-off in Rhode Island and Buzzards Bay (Massachusetts) associated with the onset of a temperature-sensitive bacterial shell disease, and accelerated by a 1999 lobster die-off in Long Island Sound. Currently, the southern extent of the commercial lobster harvest appears to be limited by this temperature-sensitive shell disease, and these effects are expected to increase as near-shore water temperatures rise above the threshold for this disease. Analyses also suggest that lobster survival and settlement in northern regions of the Gulf of Maine could be increased by warming water, a longer growing season, more rapid growth, an earlier hatching season, an increase in nursery grounds suitable for larvae, and faster development of plankton.[376]

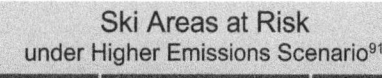

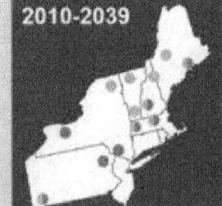 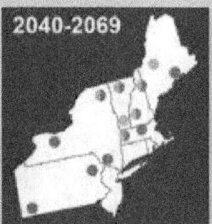 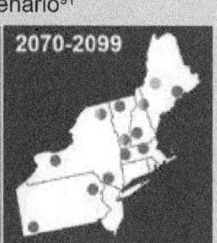

● highly vulnerable　● vulnerable　● viable

Scott et al.[342]; Fig. from Frumhoff et al.[234]

The ski resorts in the Northeast have three climate-related criteria that need to be met for them to remain viable: the average length of the ski season must be at least 100 days; there must be a good probability of being open during the lucrative winter holiday week between Christmas and the New Year; and there must be enough nights that are sufficiently cold to enable snowmaking operations. By these standards, only one area in the region (not surprisingly, the one located farthest north) is projected to be able to support viable ski resorts by the end of this century under a higher emissions scenario (referred to as "even higher" on page 23).[91,373]

Cod populations throughout the North Atlantic are adapted to a wide range of seasonal ocean temperatures, including average annual temperatures near the seafloor ranging from 36 to 54°F. Large populations of cod are generally not found above the 54°F threshold.[377] Temperature also influences both the location and timing of spawning, which in turn affects the subsequent growth and survival of young cod. Increases in average annual bottom temperatures above 47°F lead to a decline in growth and survival.[378,379] Projections of warming indicate that both the 47°F and the 54°F thresholds will be met or exceeded in this century under a higher emissions scenario.[234] Climate change will thus introduce an additional stress to an already-stressed fishery.[377]

Southeast

The climate of the Southeast is uniquely warm and wet, with mild winters and high humidity, compared with the rest of the continental United States. The average annual temperature of the Southeast did not change significantly over the past century as a whole. Since 1970, however, annual average temperature has risen about 2°F, with the greatest seasonal increase in temperature occurring during the winter months. The number of freezing days in the Southeast has declined by four to seven days per year for most of the region since the mid-1970s.

Average autumn precipitation has increased by 30 percent for the region since 1901. The decline in fall precipitation in South Florida contrasts strongly with the regional average. There has been an increase in heavy downpours in many parts of the region,[380,381] while the percentage of the region experiencing moderate to severe drought increased over the past three decades. The area of moderate to severe spring and summer drought has increased by 12 percent and 14 percent, respectively, since the mid-1970s. Even in the fall months, when precipitation tended to increase in most of the region, the extent of drought increased by 9 percent.

Climate models project continued warming in all seasons across the Southeast and an increase in the rate of warming through the end of this century. The projected rates of warming are more than double those experienced in the Southeast since 1975, with the greatest temperature increases projected to occur in the summer months. The number of very hot days is projected to rise at a greater rate than the average temperature. Under a lower emissions scenario,[91]

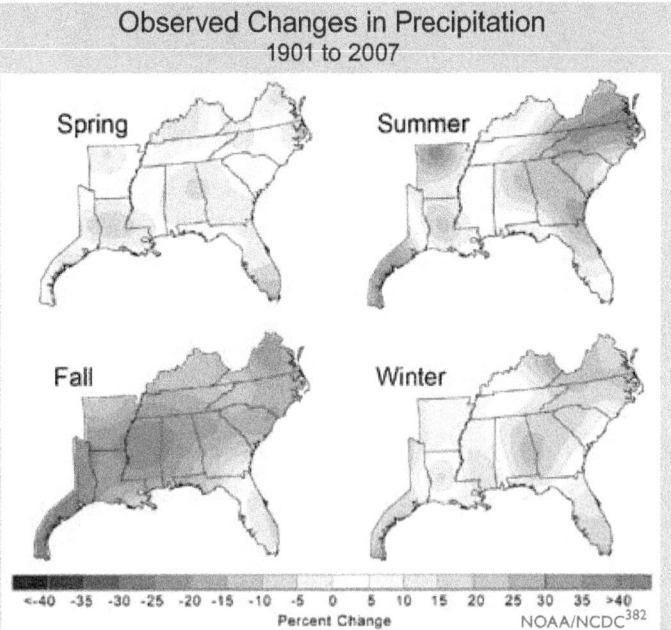

Observed Changes in Precipitation 1901 to 2007

Spring   Summer

Fall   Winter

<-40 -35 -30 -25 -20 -15 -10 -5 0 5 10 15 20 25 30 35 >40
Percent Change   NOAA/NCDC[382]

While average fall precipitation in the Southeast increased by 30 percent since the early 1900s, summer and winter precipitation declined by nearly 10 percent in the eastern part of the region. Southern Florida has experienced a nearly 10 percent drop in precipitation in spring, summer, and fall. The percentage of the Southeast region in drought has increased over recent decades.

average temperatures in the region are projected to rise by about 4.5°F by the 2080s, while a higher emissions scenario[91] yields about 9°F of average warming (with about a 10.5°F increase in summer, and a much higher heat index). Spring and summer rainfall is projected to decline in South Florida during this century. Except for indications that the amount of rainfall from individual hurricanes will increase,[68] climate models provide divergent

Average Change in Temperature and Precipitation in the Southeast					
	Temperature Change in °F			**Precipitation change in %**	
	1901-2008	1970-2008		1901-2008	1970-2008
Annual	0.3	1.6	Annual	6.0	-7.7
Winter	0.2	2.7	Winter	1.2	-9.6
Spring	0.4	1.2	Spring	1.7	-29.2
Summer	0.4	1.6	Summer	-4.0	3.6
Fall	0.2	1.1	Fall	27.4	0.1

Observed temperature and precipitation changes in the Southeast are summarized above for two different periods.[383] Southeast average temperature declined from 1901 to 1970 and then increased strongly since 1970.

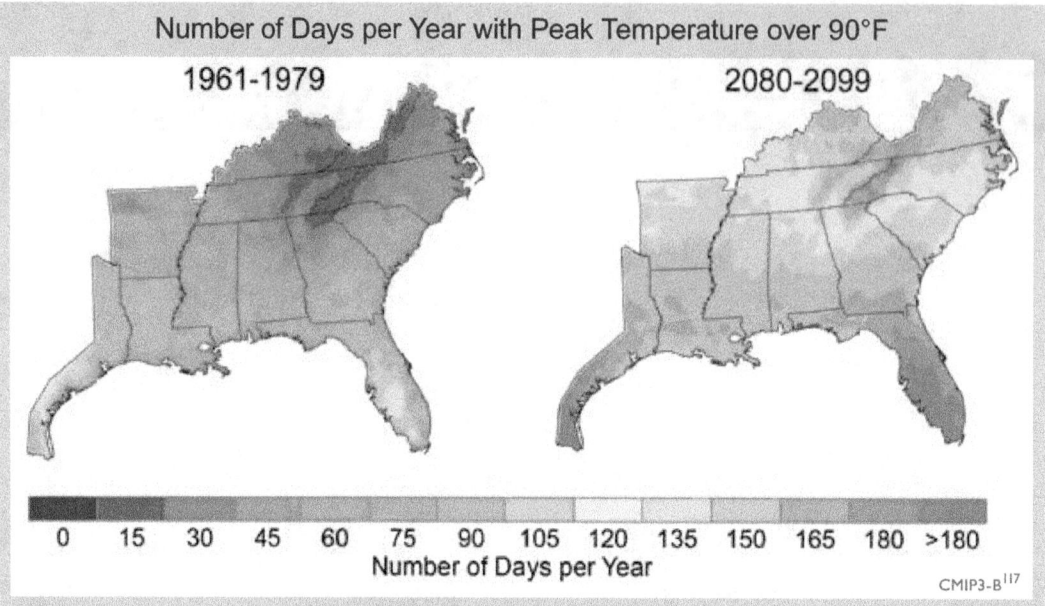

Number of Days per Year with Peak Temperature over 90°F

1961-1979

2080-2099

0   15   30   45   60   75   90   105   120   135   150   165   180   >180
Number of Days per Year

CMIP3-B[117]

The number of days per year with peak temperature over 90°F is expected to rise significantly, especially under a higher emissions scenario[91] as shown in the map above. By the end of the century, projections indicate that North Florida will have more than 165 days (nearly six months) per year over 90°F, up from roughly 60 days in the 1960s and 1970s. The increase in very hot days will have consequences for human health, drought, and wildfires.

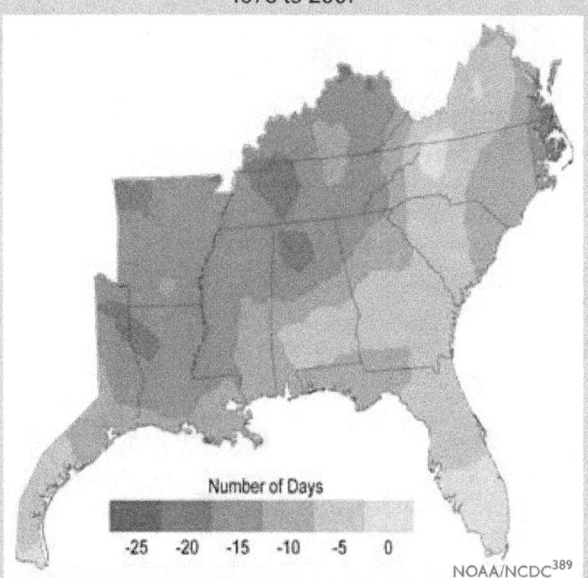

Change In Freezing Days per Year
1976 to 2007

Number of Days

-25   -20   -15   -10   -5   0

NOAA/NCDC[389]

Since the mid-1970s, the number of days per year in which the temperature falls below freezing has declined by four to seven days over much of the Southeast. Some areas, such as western Louisiana, have experienced more than 20 fewer freezing days. Climate models project continued warming across the region, with the greatest increases in temperature expected in summer, and the number of very hot days increasing at a greater rate than the average temperature.

results for future precipitation for the remainder of the Southeast. Models project that Gulf Coast states will tend to have less rainfall in winter and spring, compared with the more northern states in the region (see map on page 31 in the *National Climate Change* section). Because higher temperatures lead to more evaporation of moisture from soils and water loss from plants, the frequency, duration, and intensity of droughts are likely to continue to increase.

The destructive potential of Atlantic hurricanes has increased since 1970, correlated with an increase in sea surface temperature. A similar relationship with the frequency of landfalling hurricanes has not been established[98,384-387] (see *National Climate Change* section for a discussion of past trends and future projections). An increase in average summer wave heights along the U.S. Atlantic coastline since 1975 has been attributed to a progressive increase in hurricane power.[112,388] The intensity of Atlantic hurricanes is likely to increase during this century with higher peak wind speeds, rainfall intensity, and storm surge height and strength.[90,112] Even with no increase in hurricane intensity, coastal inundation and shoreline retreat would increase as sea-level rise accelerates, which is one of the most certain and most costly consequences of a warming climate.[164]

Projected increases in air and water temperatures will cause heat-related stresses for people, plants, and animals.

The warming projected for the Southeast during the next 50 to 100 years will create heat-related stress for people, agricultural crops, livestock, trees, transportation and other infrastructure, fish, and wildlife. The average temperature change is not as important for all of these sectors and natural systems as the projected increase in maximum and minimum temperatures. Examples of potential impacts include:

- Increased illness and death due to greater summer heat stress, unless effective adaptation measures are implemented.[164]
- Decline in forest growth and agricultural crop production due to the combined effects of thermal stress and declining soil moisture.[390]
- Increased buckling of pavement and railways.[217,222]
- Decline in dissolved oxygen in stream, lakes, and shallow aquatic habitats leading to fish kills and loss of aquatic species diversity.
- Decline in production of cattle and other rangeland livestock.[391] Significant impacts on beef cattle occur at continuous temperatures in the 90 to 100°F range, increasing in danger as the humidity level increases (see *Agriculture* sector).[391] Poultry and swine are primarily raised in indoor operations, so warming would increase energy requirements.[193]

A reduction in very cold days is likely to reduce the loss of human life due to cold-related stress, while heat stress and related deaths in the summer months are likely to increase. The reduction in cold-related deaths is not expected to offset the increase in heat-related deaths (see *Human Health* sector). Other effects of the projected increases in temperature include more frequent outbreaks of shellfish-borne diseases in coastal waters, altered distribution of native plants and animals, local loss of many threatened and endangered species, displacement of native species by invasive species, and more frequent and intense wildfires.

Decreased water availability is very likely to affect the region's economy as well as its natural systems.

Decreased water availability due to increased temperature and longer periods of time between rainfall events, coupled with an increase in societal demand is very likely to affect many sectors of the Southeast's economy. The amount and timing of water available to natural systems is also affected by climate change, as well as by human response strategies such as increasing storage capacity (dams)[142] and increasing acreage of irrigated cropland.[392] The 2007 water shortage in the Atlanta region created serious conflicts between three states, the U.S. Army Corps of Engineers (which operates the dam at Lake Lanier), and the U.S. Fish and Wildlife Service, which is charged with protecting endangered species. As humans seek to adapt to climate change by manipulating water resources, streamflow and biological diversity are likely to be reduced.[142] During droughts, recharge of groundwater will decline as the temperature and spacing between rainfall events increase. Responding by increasing groundwater pumping will further stress or deplete aquifers and place increasing strain on surface water resources. Increasing evaporation and plant water loss rates alter the balance of runoff and groundwater recharge, which is likely to lead to saltwater intrusion into shallow aquifers in many parts of the Southeast.[142]

In Atlanta and Athens, Georgia, 2007 was the second driest year on record. Among the numerous effects of the rainfall shortage were restrictions on water use in some cities and low water levels in area lakes. In the photo, a dock lies on dry land near Aqualand Marina on Lake Lanier (located northeast of Atlanta) in December 2007.

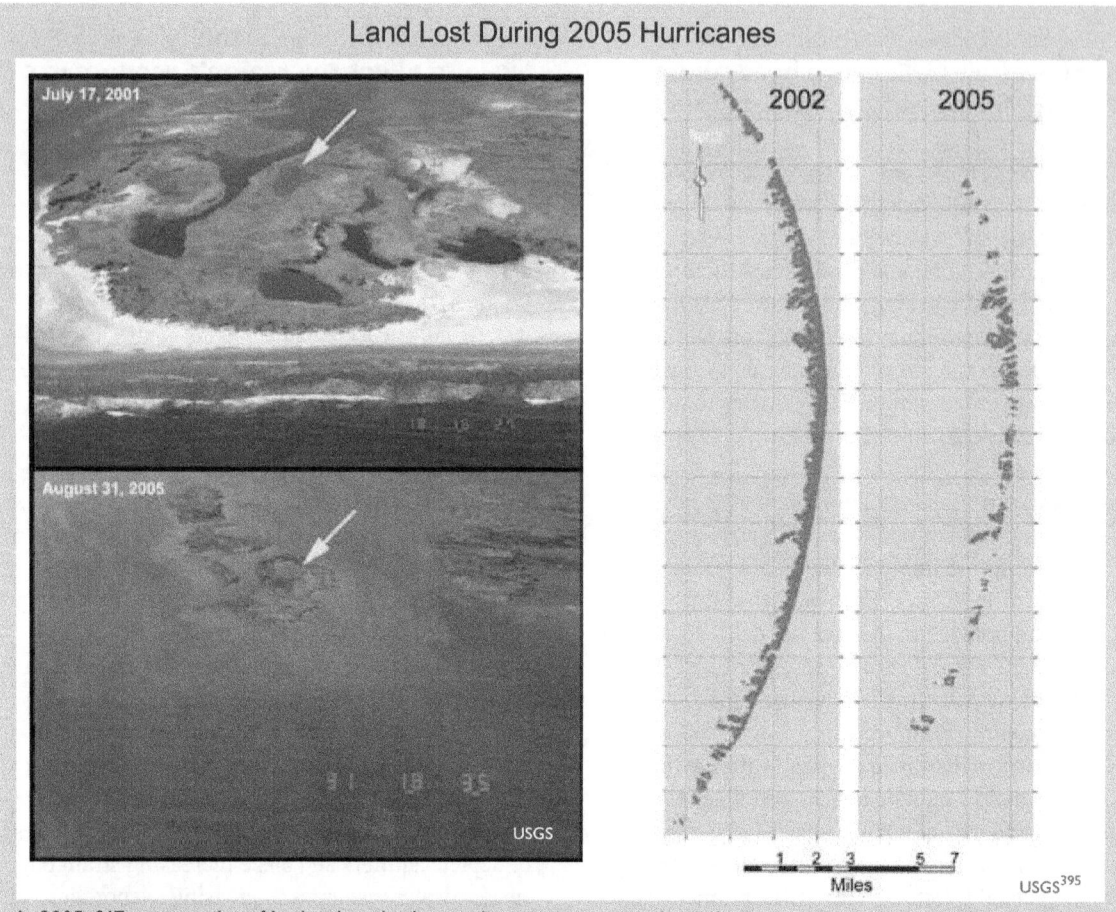

Land Lost During 2005 Hurricanes

In 2005, 217 square miles of land and wetlands were lost to open water during hurricanes Rita and Katrina. The photos and maps show the Chandeleur Islands, east of New Orleans, before and after the 2005 hurricanes; 85 percent of the islands' above-water land mass was eliminated.

Sea-level rise and the likely increase in hurricane intensity and associated storm surge will be among the most serious consequences of climate change.

An increase in average sea level of up to 2 feet or more and the likelihood of increased hurricane intensity and associated storm surge are likely to be among the most costly consequences of climate change for this region (see *National Climate Change* section). As sea level rises, coastal shorelines will retreat. Wetlands will be inundated and eroded away, and low-lying areas including some communities will be inundated more frequently – some permanently – by the advancing sea. Current buildings and infrastructure were not designed to withstand the intensity of the projected storm surge, which would cause catastrophic damage. As temperature increases and rainfall patterns change,

soil moisture and runoff to the coast are likely to be more variable. The salinity of estuaries, coastal wetlands, and tidal rivers is likely to increase in the southeastern coastal zone, thereby altering coastal ecosystems and displacing them farther inland if no barriers exist. More frequent storm surge flooding and permanent inundation of coastal ecosystems and communities is likely in some low-lying areas, particularly along the central Gulf Coast where the land surface is sinking.[393,394] Rapid acceleration in the rate of increase in sea-level rise could threaten a large portion of the Southeast coastal zone. The likelihood of a catastrophic increase in the rate of sea-level rise is dependent upon ice sheet response to warming, which is the subject of much scientific uncertainty (see *Global Climate Change* section).[90] Such rapid rise in sea level is likely to result in the destruction of barrier islands and wetlands.[257,390]

Compared to the present coastal situation, for which vulnerability is quite high, an increase in hurricane intensity will further affect low-lying coastal ecosystems and coastal communities along the Gulf and South Atlantic coastal margin. An increase in intensity is very likely to increase inland and coastal flooding, coastal erosion rates, wind damage to coastal forests, and wetland loss. Major hurricanes also pose a severe risk to people, personal property, and public infrastructure in the Southeast, and this risk is likely to be exacerbated.[393,394] Hurricanes have their greatest impact at the coastal margin where they make landfall, causing storm surge, severe beach erosion, inland flooding, and wind-related casualties for both cultural and natural resources. Some of these impacts extend farther inland, affecting larger areas. Recent examples of societal vulnerability to severe hurricanes include Katrina and Rita in 2005, which were responsible for the loss of more than 1,800 lives and the net loss of 217 square miles of low-lying coastal marshes and barrier islands in southern Louisiana.[390,396]

Sea Surface Temperature
Atlantic Hurricane Main Development Region
August through October, 1900 to 2008

NOAA/NCDC[397]

Ocean surface temperature during the peak hurricane season, August through October, in the main development region for Atlantic hurricanes.[397] Higher sea surface temperatures in this region of the ocean have been associated with more intense hurricanes. As ocean temperatures continue to increase in the future, it is likely that hurricane rainfall and wind speeds will increase in response to human-caused warming (see *National Climate Change* section).[68]

Ecological thresholds are expected to be crossed throughout the region, causing major disruptions to ecosystems and to the benefits they provide to people.

Ecological systems provide numerous important services that have high economic and cultural value in the Southeast. Ecological effects cascade among both living and physical systems, as illustrated in the following examples of ecological disturbances that result in abrupt responses, as opposed to gradual and proportional responses to warming:

- The sudden loss of coastal landforms that serve as a storm-surge barrier for natural resources and as a homeland for coastal communities (such as in a major hurricane).[254,390]
- An increase in sea level can have no apparent effect until an elevation is reached that allows widespread, rapid salt-water intrusion into coastal forests and freshwater aquifers.[398]
- Lower soil moisture and higher temperatures leading to intense wildfires or pest outbreaks (such as the southern pine beetle) in southeastern forests;[399] intense droughts leading to the drying of lakes, ponds, and wetlands; and the local or global extinction of riparian and aquatic species.[142]

Flooding damage in Louisiana due to Hurricane Katrina

- A precipitous decline of wetland-dependent coastal fish and shellfish populations due to the rapid loss of coastal marsh.[400]

Quality of life will be affected by increasing heat stress, water scarcity, severe weather events, and reduced availability of insurance for at-risk properties.

Over the past century, the southeastern "sunbelt" has attracted people, industry, and investment. The population of Florida more than doubled during the past three decades, and growth rates in most other southeastern states were in the range of 45 to 75 percent (see population map, page 55). Future population growth and the quality of life for existing residents is likely to be affected by the many challenges associated with climate change, such as reduced insurance availability, increased insurance cost, and increases in water scarcity, sea-level rise, extreme weather events, and heat stress. Some of these problems, such as increasing heat and declining air quality, will be especially acute in cities.

Adaptation: Reducing Exposure to Flooding and Storm Surge

Three different types of adaptation to sea-level rise are available for low-lying coastal areas.[173,269] One is to move buildings and infrastructure farther inland to get out of the way of the rising sea. Another is to accommodate rising water through changes in building design and construction, such as elevating buildings on stilts. Flood insurance programs even require this in some areas with high probabilities of floods. The third adaptation option is to try to protect existing development by building levees and river flood control structures. This option is being pursued in some highly vulnerable areas of the Gulf and South Atlantic coasts. Flood control structures can be designed to be effective in the face of higher sea level and storm surge.

Some hurricane levees and floodwalls were not just replaced after Hurricane Katrina, they were redesigned to withstand higher storm surge and wave action.[401]

The costs and environmental impacts of building such structures can be significant. Furthermore, building levees can actually increase future risks.[269] This is sometimes referred to as the levee effect or the safe-development paradox. Levees that provide protection from, for example, the storm surge from a Category 3 hurricane, increase real and perceived safety and thereby lead to increased development. This increased

Recent upgrades that raised the height of this earthen levee increased protection against storm surge in the New Orleans area.

development means there will be greater damage if and when the storm surge from a Category 5 hurricane tops the levee than there would have been if no levee had been constructed.[252]

In addition to levees, enhancement of key highways used as hurricane evacuation routes and improved hurricane evacuation planning is a common adaptation underway in all Gulf Coast states.[217] Other protection options that are being practiced along low-lying coasts include the enhancement and protection of natural features such as forested wetlands, saltmarshes, and barrier islands.[390]

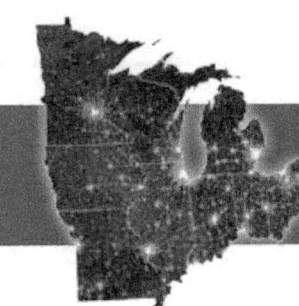

Midwest

The Midwest's climate is shaped by the presence of the Great Lakes and the region's location in the middle of the North American continent. This location, far from the temperature-moderating effects of the oceans, contributes to large seasonal swings in air temperature from hot, humid summers to cold winters. In recent decades, a noticeable increase in average temperatures in the Midwest has been observed, despite the strong year-to-year variations. The largest increase has been measured in winter, extending the length of the frost-free or growing season by more than one week, mainly due to earlier dates for the last spring frost. Heavy downpours are now twice as frequent as they were a century ago. Both summer and winter precipitation have been above average for the last three decades, the wettest period in a century. The Midwest has experienced two record-breaking floods in the past 15 years.[213] There has also been a decrease in lake ice, including on the Great Lakes. Since the 1980s, large heat waves have been more frequent in the Midwest than any time in the last century, other than the Dust Bowl years of the 1930s.[112,283,402-404]

During the summer, public health and quality of life, especially in cities, will be negatively affected by increasing heat waves, reduced air quality, and insect and waterborne diseases. In the winter, warming will have mixed impacts.

Heat waves that are more frequent, more severe, and longer lasting are projected. The frequency of hot days and the length of the heat-wave season both will be more than twice as great under the higher emissions scenario[91] compared to the lower emissions scenario.[91,283, 402,403,405] Events such as the Chicago heat wave of 1995, which resulted in over 700 deaths, will become more common. Under the lower emissions scenario,[91] such a heat wave is projected to occur every other year in Chicago by the end of the century, while under the higher emissions scenario,[91] there would be about three such heat waves per year. Even more severe heat waves, such as the one that claimed tens of thousands of lives in Europe in 2003, are projected to become more frequent in a warmer world, occurring as often as every other year in the Midwest by the end of this century under the higher emissions scenario.[91,283,403,406] Some health impacts can be reduced by better preparation for such events.[288]

Climate on the Move:
Changing Summers in the Midwest

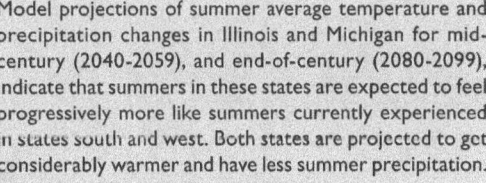

Lower Emissions Scenario[91] Higher Emissions Scenario[91]

Hayhoe et al.[283]

Model projections of summer average temperature and precipitation changes in Illinois and Michigan for mid-century (2040-2059), and end-of-century (2080-2099), indicate that summers in these states are expected to feel progressively more like summers currently experienced in states south and west. Both states are projected to get considerably warmer and have less summer precipitation.

During heat waves, high electricity demand combines with climate-related limitations on energy production capabilities (see *Energy Supply and Use* sector), increasing the likelihood of electricity shortages and resulting in brownouts or even blackouts. This combination can leave people without air conditioning and ventilation when they need it most, as occurred during the 1995 Chicago/Milwaukee heat wave. In general, electricity demand for air conditioning is projected to significantly increase in summer. Improved energy planning could reduce electricity disruptions.

The urban heat island effect can further add to high local daytime and nighttime temperatures (see *Human Health* sector). Heat waves take a greater toll in illness and death when there is little relief from the heat at night.

Another health-related issue arises from the fact that climate change can affect air quality. A warmer climate generally means more ground-level ozone (a component of smog), which can cause respiratory problems, especially for those who are young, old, or have asthma or allergies. Unless the emissions of pollutants that lead to ozone formation are reduced significantly, there will be more ground-level ozone as a result of the projected climate changes in the Midwest due to increased air temperatures, more stagnant air, and increased emissions from vegetation.[283,291,402,403,408-410]

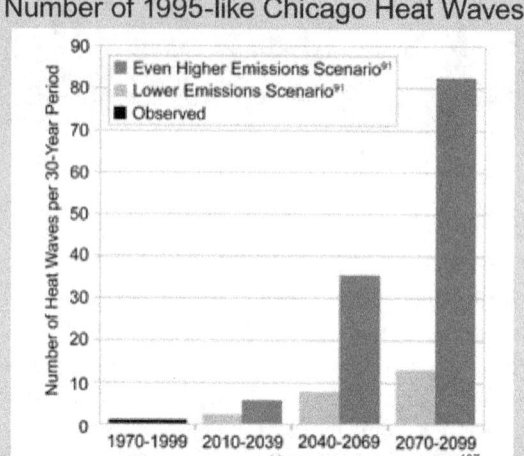

Number of 1995-like Chicago Heat Waves

Hayhoe et al.[407]

Over the last three decades of this century, heat waves like the one that occurred in Chicago in 1995 are projected to occur about once every three years under the lower emissions scenario.[91] Under the even higher emissions scenario, such events are projected to occur an average of nearly three times a year. In this analysis, heat waves were defined as at least one week of daily maximum temperatures greater than 90°F and nighttime minimum temperatures greater than 70°F, with at least two consecutive days with daily temperatures greater than 100°F and nighttime temperatures greater than 80°F.

Insects such as ticks and mosquitoes that carry diseases will survive winters more easily and produce larger populations in a warmer Midwest.[283,402,403] One potential risk is an increasing incidence of diseases such as West Nile

Adaptation: Chicago Tries to Cool the Urban Heat Island

Efforts to reduce urban heat island effects become even more important in a warming climate. The City of Chicago has produced a map of urban hotspots to use as a planning tool to target areas that could most benefit from heat-island reduction initiatives such as reflective or green roofing, and tree planting. Created using satellite images of daytime and nighttime temperatures, the map shows the hottest 10 percent of both day and night temperatures in red, and the hottest 10 percent of either day or night in orange.

Chicago's urban hot spots

"Green roofs" are cooler than the surrounding conventional roofs.

The City is working to reduce urban heat buildup and the need for air conditioning by using reflective roofing materials. This thermal image shows that the radiating temperature of the City Hall's "green roof" – covered with soil and vegetation – is up to 77°F cooler than the nearby conventional roofs.[411]

virus. Waterborne diseases will present an increasing risk to public health because many pathogens thrive in warmer conditions.[163]

In winter, oil and gas demand for heating will decline. Warming will also decrease the number of days with snow on the ground, which is expected to improve traffic safety.[222] On the other hand, warming will decrease outdoor winter recreational opportunities such as skiing, snowmobiling, ice skating, and ice fishing.

Significant reductions in Great Lakes water levels, which are projected under higher emissions scenarios, lead to impacts on shipping, infrastructure, beaches, and ecosystems.

The Great Lakes are a natural resource of tremendous significance, containing 20 percent of the planet's fresh surface water and serving as the dominant feature of the industrial heartland of the nation. Higher temperatures will mean more evaporation and hence a likely reduction in the Great Lakes water levels. Reduced lake ice increases

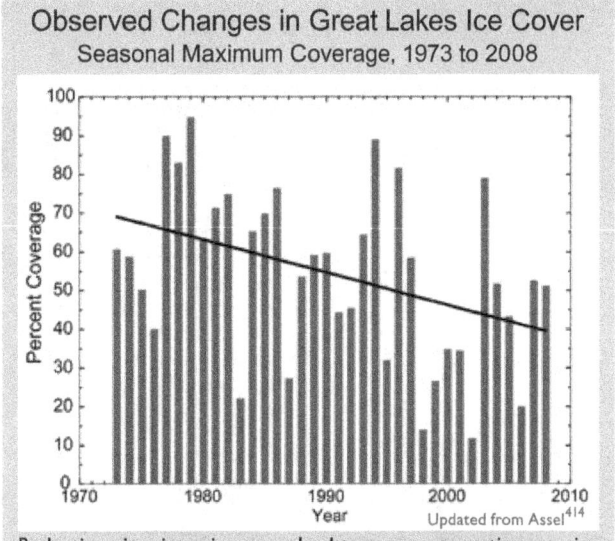

Observed Changes in Great Lakes Ice Cover
Seasonal Maximum Coverage, 1973 to 2008

Updated from Assel[414]

Reductions in winter ice cover lead to more evaporation, causing lake levels to drop even farther. While the graph indicates large year-to-year variations, there is a clear decrease in the extent of Great Lakes ice coverage, as shown by the black trend line.

evaporation in winter, contributing to the decline. Under a lower emissions scenario,[91] water levels in the Great Lakes are projected to fall no more than 1 foot by the end of the century, but under a higher emissions scenario,[91] they are projected to fall between 1 and 2 feet.[283] The greater the temperature rise, the higher the likelihood of a larger decrease in lake levels.[412] Even a decrease of 1 foot, combined with normal fluctuations, can result in significant lengthening of the distance to the lakeshore in many places. There are also potential impacts on beaches, coastal ecosystems, dredging requirements, infrastructure, and shipping. For example, lower lake levels reduce "draft," or the distance between the waterline and the bottom of a ship, which lessens a ship's ability to carry freight. Large vessels, sized for passage through the St. Lawrence Seaway, lose up to 240 tons of capacity for each inch of draft lost.[283,402,403,413] These impacts will have costs, including increased shipping, repair and maintenance costs, and lost recreation and tourism dollars.

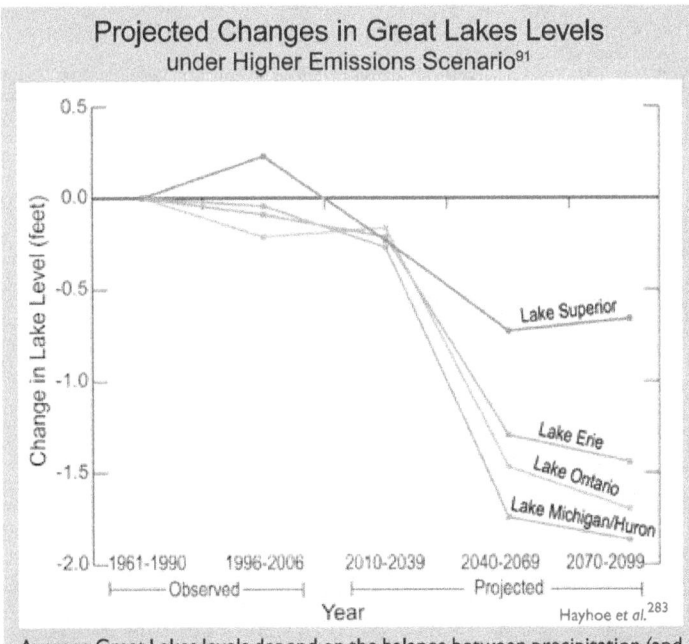

Projected Changes in Great Lakes Levels
under Higher Emissions Scenario[91]

Hayhoe et al.[283]

Average Great Lakes levels depend on the balance between precipitation (and corresponding runoff) in the Great Lakes Basin on one hand, and evaporation and outflow on the other. As a result, lower emissions scenarios[91] with less warming show less reduction in lake levels than higher emissions scenarios.[91] Projected changes in lake levels are based on simulations by the NOAA Great Lakes model for projected climate changes under a higher emissions scenario.[91]

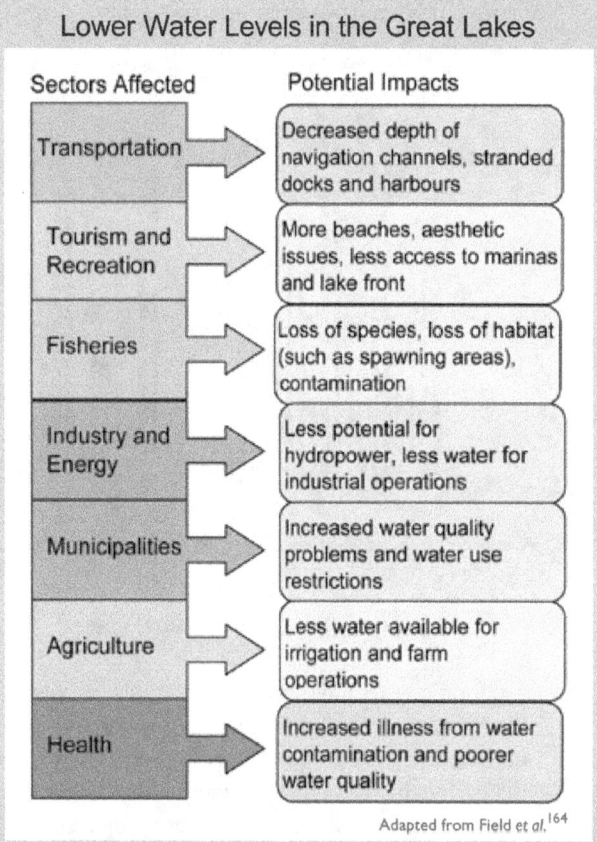

Lower Water Levels in the Great Lakes

Sectors Affected	Potential Impacts
Transportation	Decreased depth of navigation channels, stranded docks and harbours
Tourism and Recreation	More beaches, aesthetic issues, less access to marinas and lake front
Fisheries	Loss of species, loss of habitat (such as spawning areas), contamination
Industry and Energy	Less potential for hydropower, less water for industrial operations
Municipalities	Increased water quality problems and water use restrictions
Agriculture	Less water available for irrigation and farm operations
Health	Increased illness from water contamination and poorer water quality

Adapted from Field et al.[164]

Reduced water levels in the Great Lakes will have interconnected impacts across many sectors, creating mismatches between water supply and demand, and necessitating trade-offs. Regions outside the Midwest will also be affected. For example, a reduction in hydropower potential would affect the Northeast, and a reduction in irrigation water would affect regions that depend on agricultural produce from the Midwest.

The likely increase in precipitation in winter and spring, more heavy downpours, and greater evaporation in summer would lead to more periods of both floods and water deficits.

Precipitation is projected to increase in winter and spring, and to become more intense throughout the year. This pattern is expected to lead to more frequent flooding, increasing infrastructure damage, and impacts on human health. Such heavy downpours can overload drainage systems and water treatment facilities, increasing the risk of waterborne diseases. Such an incident occurred in Milwaukee in 1993 when the water supply was contaminated with the parasite *Cryptosporidium*, causing 403,000 reported cases of gastrointestinal illness and 54 deaths.[219]

In Chicago, rainfall of more than 2.5 inches per day is an approximate threshold beyond which combined water and sewer systems overflow into Lake Michigan (such events occurred 2.5 times per decade from 1961 to 1990). This generally results in beach closures to reduce the risk of disease transmission. Rainfall above this threshold is projected to occur twice as often by the end of this century under the lower emissions scenario[91] and three times as often under the higher emissions scenario.[91,283,403] Similar increases are expected across the Midwest.

More intense rainfall can lead to floods that cause significant impacts regionally and even nationally. For example, the Great Flood of 1993 caused catastrophic flooding along 500 miles of the Mississippi and Missouri river systems, affecting one-quarter of all U.S. freight (see *Transportation* sector).[222,415-417] Another example was a record-breaking 24-hour rainstorm in July 1996, which resulted in flash flooding in Chicago and its suburbs, causing extensive damage and disruptions, with some commuters not being able to reach Chicago for

The Great Flood of 1993 caused flooding along 500 miles of the Mississippi and Missouri river systems. The photo shows the flood's effects on U.S. Highway 54, just north of Jefferson City, Missouri.

three days (see *Transportation* sector).[222] There was also a record-breaking storm in August 2007. Increases in such events are likely to cause greater property damage, higher insurance rates, a heavier burden on emergency management, increased clean-up and rebuilding costs, and a growing financial toll on businesses, homeowners, and insurers.

In the summer, with increasing evaporation rates and longer periods between rainfalls, the likelihood of drought will increase and water levels in rivers, streams, and wetlands are likely to decline. Lower water levels also could create problems for river traffic, reminiscent of the stranding of more than 4,000 barges on the Mississippi River during the 1988 drought. Reduced summer water levels are also likely to reduce the recharge of groundwater, cause small streams to dry up (reducing native fish populations), and reduce the area of wetlands in the Midwest.

While the longer growing season provides the potential for increased crop yields, increases in heat waves, floods, droughts, insects, and weeds will present increasing challenges to managing crops, livestock, and forests.

The projected increase in winter and spring precipitation and flooding is likely to delay planting and crop establishment. Longer growing seasons and increased carbon dioxide have positive effects on some crop yields, but this is likely to be counterbalanced in part by the negative effects of additional disease-causing pathogens, insect pests, and weeds (including invasive weeds).[193] Livestock production is expected to become more costly as higher temperatures stress livestock, decreasing productivity and increasing costs associated with the needed ventilation and cooling equipment.[193]

Plant winter hardiness zones (each zone represents a 10°F change in minimum temperature) in the Midwest are likely to shift one-half to one full zone

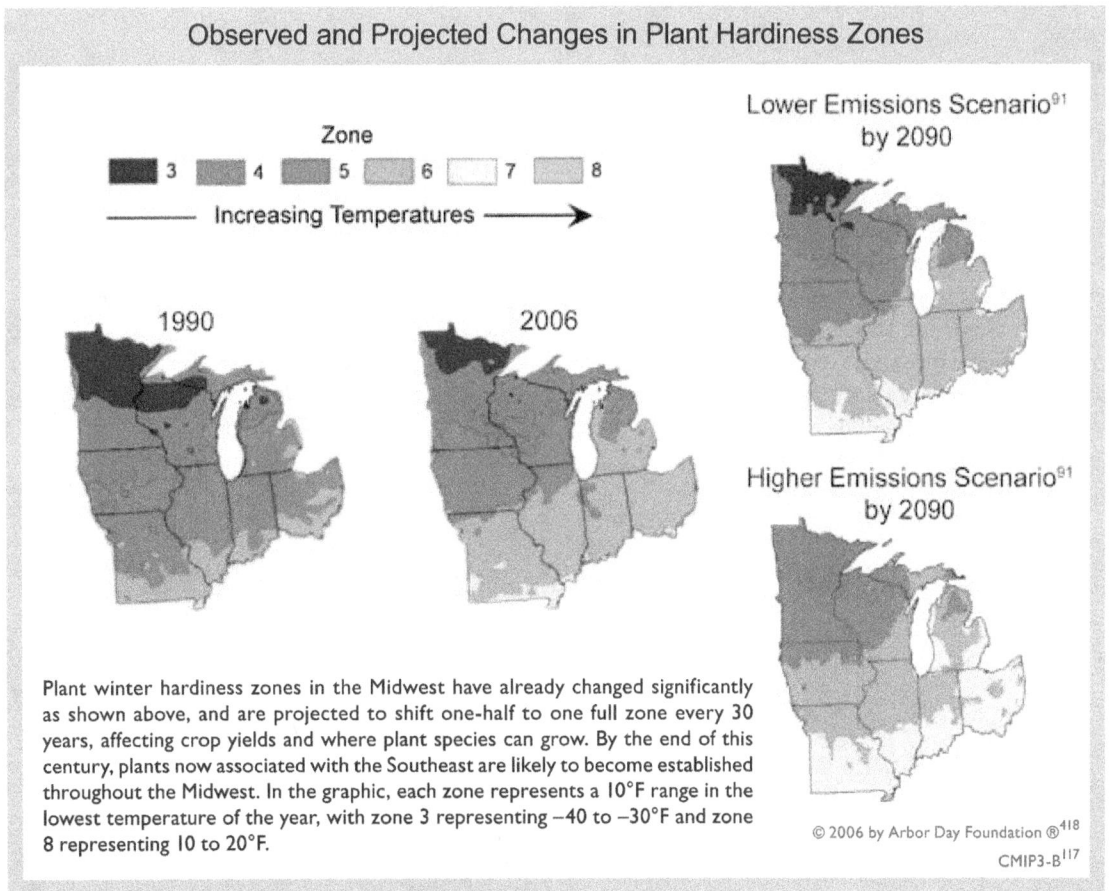

Observed and Projected Changes in Plant Hardiness Zones

Zone

3 4 5 6 7 8

⟶ Increasing Temperatures ⟶

1990   2006

Lower Emissions Scenario[91] by 2090

Higher Emissions Scenario[91] by 2090

Plant winter hardiness zones in the Midwest have already changed significantly as shown above, and are projected to shift one-half to one full zone every 30 years, affecting crop yields and where plant species can grow. By the end of this century, plants now associated with the Southeast are likely to become established throughout the Midwest. In the graphic, each zone represents a 10°F range in the lowest temperature of the year, with zone 3 representing −40 to −30°F and zone 8 representing 10 to 20°F.

© 2006 by Arbor Day Foundation ®[418]

CMIP3-B[117]

about every 30 years. By the end of the century, plants now associated with the Southeast are likely to become established throughout the Midwest.

Impacts on forests are likely to be mixed, with the positive effects of higher carbon dioxide and nitrogen levels acting as fertilizers potentially negated by the negative effects of decreasing air quality.[243] In addition, more frequent droughts, and hence fire hazards, and an increase in destructive insect pests, such as gypsy moths, hinder plant growth. Insects, historically controlled by cold winters, more easily survive milder winters and produce larger populations in a warmer climate (see *Agriculture* and *Ecosystems* sectors).

Native species are very likely to face increasing threats from rapidly changing climate conditions, pests, diseases, and invasive species moving in from warmer regions.

As air temperatures increase, so will water temperatures. In some lakes, this will lead to an earlier and longer period in summer during which mixing of the relatively warm surface lake water with the colder water below is reduced.[564] In such cases, this stratification can cut off oxygen from bottom layers, increasing the risk of oxygen-poor or oxygen-free "dead zones" that kill fish and other living things. In lakes with contaminated sediment, warmer water and low-oxygen conditions can more readily mobilize mercury and other persistent pollutants.[565] In such cases, where these increasing quantities of contaminants are taken up in the aquatic food chain, there will be additional potential for health hazards for species that eat fish from the lakes, including people.[566]

Populations of coldwater fish, such as brook trout, lake trout, and whitefish, are expected to decline dramatically, while populations of coolwater fish such as muskie, and warmwater species such as smallmouth bass and bluegill, will take their place. Aquatic ecosystem disruptions are likely to be compounded by invasions by non-native species, which tend to thrive under a wide range of environmental conditions. Native species, adapted to a narrower range of conditions, are expected to decline.

All major groups of animals, including birds, mammals, amphibians, reptiles, and insects, will be affected by impacts on local populations, and by competition from other species moving into the Midwest region.[70] The potential for animals to shift their ranges to keep pace with the changing climate will be inhibited by major urban areas and the presence of the Great Lakes.

Great Plains

The Great Plains is characterized by strong seasonal climate variations. Over thousands of years, records preserved in tree rings, sediments, and sand deposits provide evidence of recurring periods of extended drought (such as the Dust Bowl of the 1930s) alternating with wetter conditions.[97,419]

Today, semi-arid conditions in the western Great Plains gradually transition to a moister climate in the eastern parts of the region. To the north, winter days in North Dakota average 25°F, while it is not unusual to have a West Texas winter day over 75°F. In West Texas, there are between 70 and 100 days per year over 90°F, whereas North Dakota has only 10 to 20 such days on average.

Significant trends in regional climate are apparent over the last few decades. Average temperatures have increased throughout the region, with the largest changes occurring in winter months and over the northern states. Relatively cold days are becoming less frequent and relatively hot days more frequent.[420] Precipitation has also increased over most of the area.[149,421]

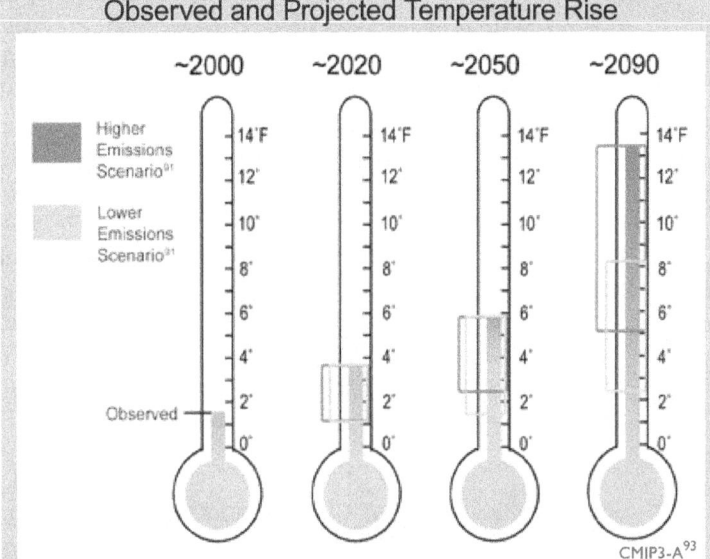

Observed and Projected Temperature Rise

The average temperature in the Great Plains already has increased roughly 1.5°F relative to a 1960s and 1970s baseline. By the end of the century, temperatures are projected to continue to increase by 2.5°F to more than 13°F compared with the 1960 to 1979 baseline, depending on future emissions of heat-trapping gases. The brackets on the thermometers represent the likely range of model projections, though lower or higher outcomes are possible.

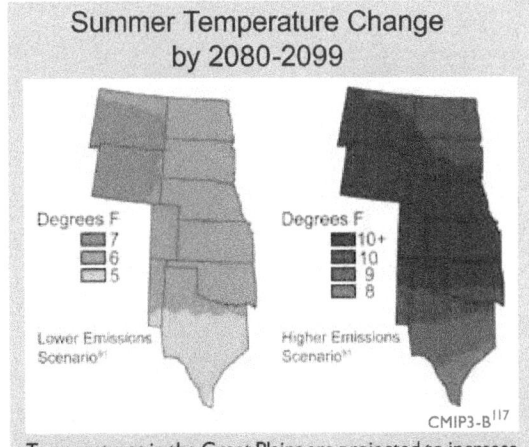

Summer Temperature Change by 2080-2099

Temperatures in the Great Plains are projected to increase significantly by the end of this century, with the northern part of the region experiencing the greatest projected increase in temperature.

Temperatures are projected to continue to increase over this century, with larger changes expected under scenarios of higher heat-trapping emissions as compared to lower heat-trapping emissions. Summer changes are projected to be larger than those in winter in the southern and central Great Plains.[108] Precipitation is also projected to change, particularly in winter and spring. Conditions are anticipated to become wetter in the north and drier in the south.

Projected changes in long-term climate and more frequent extreme events such as heat waves, droughts, and heavy rainfall will affect many aspects of life in the Great Plains. These include the region's already threatened water resources, essential agricultural and ranching activities, unique natural and protected areas, and the health and prosperity of its inhabitants.

Projected increases in temperature, evaporation, and drought frequency add to concerns about the region's declining water resources.

Water is the most important factor affecting activities on the Great Plains. Most of the water used in the Great Plains comes from the High Plains aquifer (sometimes referred to by the name of its largest formation, the Ogallala aquifer), which stretches from South Dakota to Texas. The aquifer holds both current recharge from precipitation and so-called "ancient" water, water trapped by silt and soil washed down from the Rocky Mountains during the last ice age.

As population increased in the Great Plains and irrigation became widespread, annual water withdrawals began to outpace natural recharge.[422]

Today, an average of 19 billion gallons of groundwater are pumped from the aquifer each day. This water irrigates 13 million acres of land and provides drinking water to over 80 percent of the region's population.[423] Since 1950, aquifer water levels have dropped an average of 13 feet, equivalent to a 9 percent decrease in aquifer storage. In heavily irrigated parts of Texas, Oklahoma, and Kansas, reductions are much larger, from 100 feet to over 250 feet.

Projections of increasing temperatures, faster evaporation rates, and more sustained droughts brought on by climate change will only add more stress to overtaxed water sources.[149,253,424,425] Current water use on the Great Plains is unsustainable, as the High Plains aquifer continues to be tapped faster than the rate of recharge.

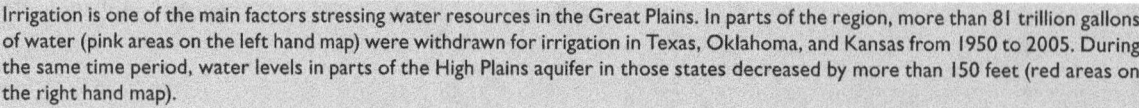

Irrigation is one of the main factors stressing water resources in the Great Plains. In parts of the region, more than 81 trillion gallons of water (pink areas on the left hand map) were withdrawn for irrigation in Texas, Oklahoma, and Kansas from 1950 to 2005. During the same time period, water levels in parts of the High Plains aquifer in those states decreased by more than 150 feet (red areas on the right hand map).

Average Annual Observed Precipitation 1971-2000

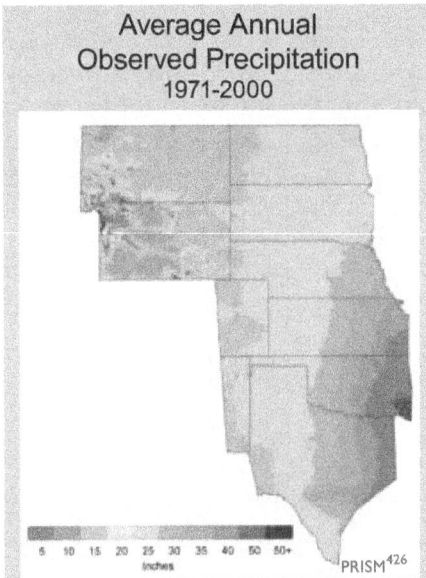

PRISM[426]

The Great Plains currently experiences a sharp precipitation gradient from east to west, from more than 50 inches of precipitation per year in eastern Oklahoma and Texas to less than 10 inches in some of the western parts of the region.

Projected Spring Precipitation Change by 2080s-2090s

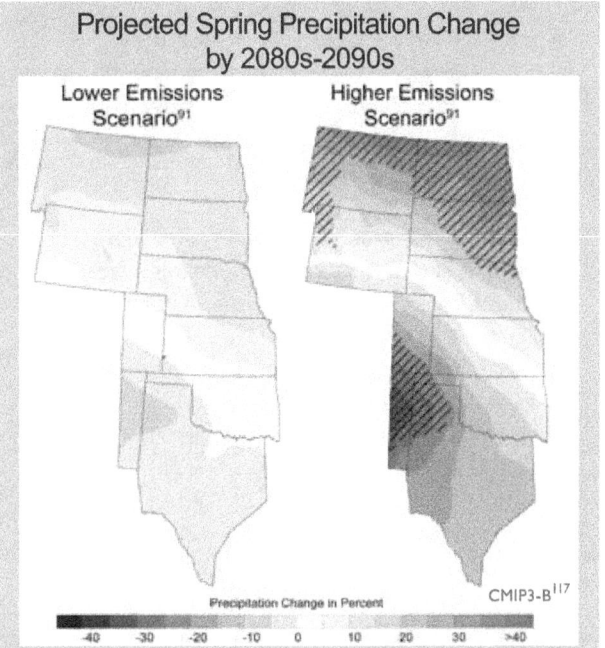

CMIP3-B[117]

Northern areas of the Great Plains are projected to experience a wetter climate by the end of this century, while southern areas are projected to experience a drier climate. The change in precipitation is compared with a 1960-1979 baseline. Confidence in the projected changes is highest in the hatched areas.

The Dust Bowl: Combined Effects of Land Use and Climate

Over the past century, large-scale conversion of grasslands to crops and ranchland has altered the natural environment of the Great Plains.[149] Irrigated fields have increased evaporation rates, reducing summer temperatures, and increasing local precipitation.[427,428]

The Dust Bowl of the 1930s epitomizes what can happen as a result of interactions between climate and human activity. In the 1920s, increasing demand for food encouraged poor agricultural practices. Small-scale producers ploughed under native grasses to plant wheat, removing the protective cover the land required to retain its moisture.

Dust Bowl of 1935 in Stratford, Texas

Variations in ocean temperature contributed to a slight increase in air temperatures, just enough to disrupt the winds that typically draw moisture from the south into the Great Plains. As the intensively tilled soils dried up, topsoil from an estimated 100 million acres of the Great Plains blew across the continent.

The Dust Bowl dramatically demonstrated the potentially devastating effects of poor land-use practices combined with climate variability and change.[429] Today, climate change is interacting with a different set of poor land-use practices. Water is being pumped from the Ogallala aquifer faster than it can recharge. In many areas, playa lakes are poorly managed (see page 127). Existing stresses on water resources in the Great Plains due to unsustainable water usage are likely to be exacerbated by future changes in temperature and precipitation, this time largely due to human-induced climate change.

Agriculture, ranching, and natural lands, already under pressure due to an increasingly limited water supply, are very likely to also be stressed by rising temperatures.

Agricultural, range, and croplands cover more than 70 percent of the Great Plains, producing wheat, hay, corn, barley, cattle, and cotton. Agriculture is fundamentally sensitive to climate. Heat and water stress from droughts and heat waves can decrease yields and wither crops.[430,431] The influence of long-term trends in temperature and precipitation can be just as great.[431]

As temperatures increase over this century, optimal zones for growing particular crops will shift. Pests that were historically unable to survive in the Great Plains' cooler areas are expected to spread northward. Milder winters and earlier springs also will encourage greater numbers and earlier emergence of insects.[149] Rising carbon dioxide levels in the atmosphere can increase crop growth, but also make some types of weeds grow even faster (see *Agriculture* sector).[432]

Projected increases in precipitation are unlikely to be sufficient to offset decreasing soil moisture and water availability in the Great Plains due to rising temperatures and aquifer depletion. In some areas, there is not expected to be enough water for agriculture to sustain even current usage.

With limited water supply comes increased vulnerability of agriculture to climate change. Further stresses on water supply for agriculture and ranching are likely as the region's cities continue to grow, increasing competition between urban and rural users.[433] The largest impacts are expected in heavily irrigated areas in the southern Great Plains, already plagued by unsustainable water use and greater frequency of extreme heat.[149]

Successful adaptation will require diversification of crops and livestock, as well as transitions from irrigated to rain-fed agriculture.[434-436] Producers who can adapt to changing climate conditions are likely to see their businesses survive; some might even thrive. Others, without resources or ability to adapt effectively, will lose out.

Climate change is likely to affect native plant and animal species by altering key habitats such as the wetland ecosystems known as prairie potholes or playa lakes.

Ten percent of the Great Plains is protected lands, home to unique ecosystems and wildlife. The region is a haven for hunters and anglers, with its ample supplies of wild game such as moose, elk, and deer; birds such as goose, quail, and duck; and fish such as walleye and bass.

Climate-driven changes are likely to combine with other human-induced stresses to further increase the vulnerability of natural ecosystems to pests, invasive species, and loss of native species. Changes in temperature and precipitation affect the composition and diversity of native animals and plants through altering their breeding patterns, water and food supply, and habitat availability.[149] In a changing climate, populations of some pests such as red fire ants and rodents, better adapted to a warmer climate, are projected to increase.[437,438] Grassland and plains birds, already besieged by habitat fragmentation, could experience significant shifts and reductions in their ranges.[439]

Urban sprawl, agriculture, and ranching practices already threaten the Great Plains' distinctive wetlands. Many of these are home to endangered and iconic species. In particular, prairie wetland ecosystems provide crucial habitat for migratory waterfowl and shorebirds.

Mallard ducks are one of the many species that inhabit the playa lakes, also known as prairie potholes.

Ongoing shifts in the region's population from rural areas to urban centers will interact with a changing climate, resulting in a variety of consequences.

Inhabitants of the Great Plains include a rising number of urban dwellers, a long tradition of rural communities, and extensive Native American

Playa Lakes and Prairie Potholes

Shallow ephemeral lakes dot the Great Plains, anomalies of water in the arid landscape. In the north they are known as prairie potholes; in the south, playa lakes. These lakes create unique microclimates that support diverse wildlife and plant communities. A playa can lie with little or no water for long periods, or have several wet/dry cycles each year. When it rains, what appeared to be only a few clumps of short, dry grasses just a few days earlier suddenly teems with frogs, toads, clam shrimp, and aquatic plants.

Playa lakes in west Texas fill up after a heavy spring rain.

The playas provide a perfect home for migrating birds to feed, mate, and raise their young. Millions of shorebirds and waterfowl, including Canada geese, mallard ducks, and Sandhill cranes, depend on the playas for their breeding grounds. From the prairie potholes of North Dakota to the playa lakes of West Texas, the abundance and diversity of native bird species directly depends on these lakes.[440,441]

Despite their small size, playa lakes and prairie potholes also play a critical role in supplying water to the Great Plains. The contribution of the playa lakes to this sensitively balanced ecosystem needs to be monitored and maintained in order to avoid unforeseen impacts on our natural resources. Before cultivation, water from these lakes was the primary source of recharge to the High Plains aquifer.[442] But many playas are disappearing and others are threatened by growing urban populations, extensive agriculture, and other filling and tilling practices.[443] In

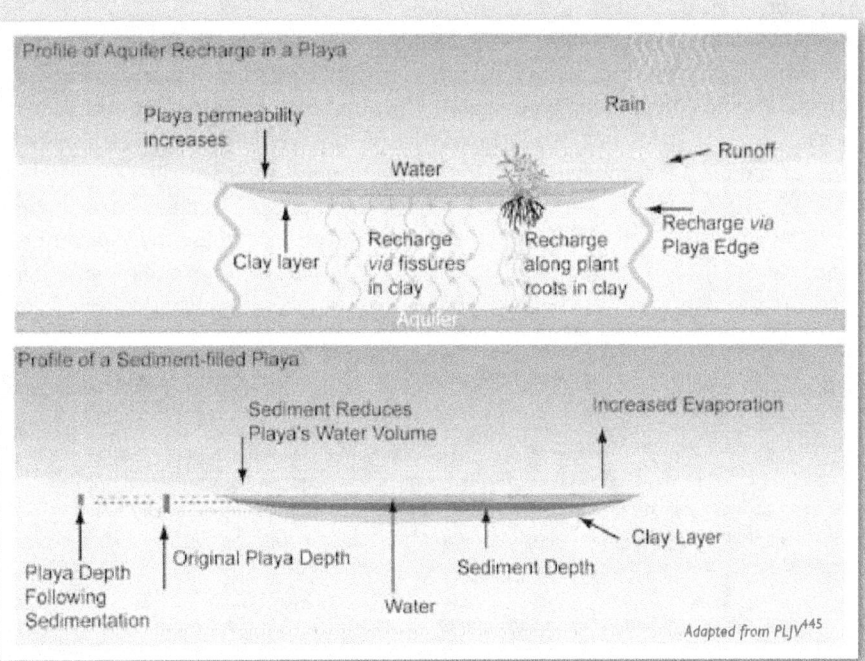

Profile of Aquifer Recharge in a Playa

Playa permeability increases

Rain

Runoff

Water

Recharge via Playa Edge

Clay layer

Recharge via fissures in clay

Recharge along plant roots in clay

Aquifer

Profile of a Sediment-filled Playa

Sediment Reduces Playa's Water Volume

Increased Evaporation

Playa Depth Following Sedimentation

Original Playa Depth

Sediment Depth

Clay Layer

Water

Adapted from PLJV[445]

recent years, agricultural demands have drawn down the playas to irrigate crops. Agricultural waste and fertilizer residues drain into playas, decreasing the quality of the water, or clogging them so the water cannot trickle down to refill the aquifer. Climate change is expected to add to these stresses, with increasing temperatures and changing rainfall patterns altering rates of evaporation, recharge, and runoff to the playa lake systems.[444]

populations. Although farming and ranching remain primary uses of the land – taking up much of the region's geographical area – growing cities provide housing and jobs for more than two-thirds of the population. For everyone on the Great Plains, though, a changing climate and a limited water supply are likely to challenge their ability to thrive, leading to conflicting interests in the allocation of increasingly scarce water resources.[313,433]

Native American communities

The Great Plains region is home to 65 Native American tribes. Native populations on rural tribal lands have limited capacities to respond to climate change.[313] Many reservations already face severe problems with water quantity and quality – problems likely to be exacerbated by climate change and other human-induced stresses.

Rural communities

As young adults move out of small, rural communities, the towns are increasingly populated by a vulnerable demographic of very old and very young people, placing them more at risk for health issues than urban communities. Combined effects of changing demographics and climate are likely to make it more difficult to supply adequate and efficient public health services and educational opportunities to rural areas. Climate-driven shifts in optimal crop types and increased risk of drought, pests, and extreme events will add more economic stress and tension to traditional communities.[430,433]

Urban populations

Although the Great Plains is not yet known for large cities, many mid-sized towns throughout the region

are growing rapidly. One in four of the most rapidly growing cities in the nation is located in the Great Plains[446] (see *Society* sector). Most of these growing centers can be found in the southern parts of the region, where water resources are already seriously constrained. Urban populations, particularly the young, elderly, and economically disadvantaged, may also be disproportionately affected by heat.[447]

New opportunities

There is growing recognition that the enormous wind power potential of the Great Plains could provide new avenues for future employment and land use. Texas already produces the most wind power of any state. Wind energy production is also prominent in Oklahoma. North and South Dakota have rich wind potential.[191]

As climate change creates new environmental conditions, effective adaptation strategies become increasingly essential to ecological and socioeconomic survival. A great deal of the Great Plains' adaptation potential might be realized through agriculture. For example, plant species that mature earlier and are more resistant to disease and pests are more likely to thrive under warmer conditions.

Other emerging adaptation strategies include dynamic cropping systems and increased crop diversity. In particular, mixed cropping-livestock systems maximize available resources while minimizing the need for external inputs such as irrigation that draws down precious water supplies.[436] In many parts of the region, diverse cropping systems and improved water use efficiency will be key to sustaining crop and rangeland systems.[448] Reduced water supplies might cause some farmers to alter the intensive cropping systems currently in use.[193,219]

Adaptation: Agricultural Practices to Reduce Water Loss and Soil Erosion

Conservation of water is critical to efficient crop production in areas where water can be scarce. Following the Dust Bowl in the 1930s, Great Plains farmers implemented a number of improved farming practices to increase the effectiveness of rainfall capture and retention in the soil and protect the soil against water and wind erosion. Examples include rotating crops, retaining crop residues, increasing vegetative cover, and altering plowing techniques.

With observed and projected increases in summer temperatures and in the frequency and intensity of heavy downpours, it will become even more important to protect against increasing loss of water and soil. Across the upper Great Plains, where strong storms are projected to occur more frequently, producers are being encouraged to increase the amount of crop residue left on the soil or to plant cover crops in the fall to protect the soil in the spring before crops are planted.

Across the southern Great Plains, some farmers are returning to dryland farming rather than relying on irrigation for their crops. Preserving crop residue helps the soil absorb more moisture from rain and eases the burden on already-stressed groundwater. These efforts have been promoted by the U.S. Department of Agriculture through research and extension efforts such as Kansas State University's Center for Sustainable Agriculture and Alternative Crops.

Southwest

The Southwest region stretches from the southern Rocky Mountains to the Pacific Coast. Elevations range from the lowest in the country to among the highest, with climates ranging from the driest to some of the wettest. Past climate records based on changes in Colorado River flows indicate that drought is a frequent feature of the Southwest, with some of the longest documented "megadroughts" on Earth. Since the 1940s, the region has experienced its most rapid population and urban growth. During this time, there were both unusually wet periods (including much of 1980s and 1990s) and dry periods (including much of 1950s and 1960s).[449] The prospect of future droughts becoming more severe as a result of global warming is a significant concern, especially because the Southwest continues to lead the nation in population growth.

Human-induced climate change appears to be well underway in the Southwest. Recent warming is among the most rapid in the nation, significantly more than the global average in some areas. This is driving declines in spring snowpack and Colorado

River flow.[34,160,161] Projections suggest continued strong warming, with much larger increases under higher emissions scenarios[91] compared to lower emissions scenarios. Projected summertime temperature increases are greater than the annual average increases in some parts of the region, and are likely to be exacerbated locally by expanding urban heat island effects.[450] Further water cycle changes are projected, which, combined with increasing temperatures, signal a serious water supply challenge in the decades and centuries ahead.[34,159]

Water supplies are projected to become increasingly scarce, calling for trade-offs among competing uses, and potentially leading to conflict.

Water is, quite literally, the lifeblood of the Southwest. The largest use of water in the region is associated with agriculture, including some of the nation's most important crop-producing areas in California. Water is also an important source of hydroelectric power, and water is required for the large population growth in the region, particularly that of major cities such as Phoenix and Las Vegas. Water also plays a critical role in supporting healthy ecosystems across the region, both on land and in rivers and lakes.

Water supplies in some areas of the Southwest are already becoming limited, and this trend toward scarcity is likely to be a harbinger of future water shortages.[34,451] Groundwater pumping is lowering water tables, while rising temperatures reduce river flows in vital rivers including the Colorado.[34] Limitations imposed on water supply by projected temperature increases are likely to be made worse by substantial reductions in rain and snowfall in the spring months, when precipitation is most needed to fill reservoirs to meet summer demand.[151]

A warmer and drier future means extra care will be needed in planning the allocation of water for

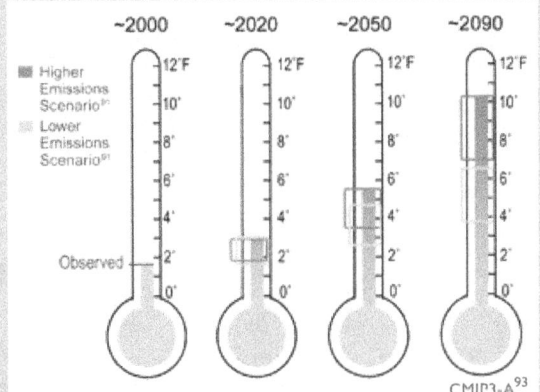

Observed and Projected Temperature Rise

~2000 ~2020 ~2050 ~2090

■ Higher Emissions Scenario[91]
▨ Lower Emissions Scenario[91]

Observed

CMIP3-A[93]

The average temperature in the Southwest has already increased roughly 1.5°F compared to a 1960-1979 baseline period. By the end of the century, average annual temperature is projected to rise approximately 4°F to 10°F above the historical baseline, averaged over the Southwest region. The brackets on the thermometers represent the likely range of model projections, though lower or higher outcomes are possible.

the coming decades. The Colorado Compact, negotiated in the 1920s, allocated the Colorado River's water among the seven basin states. It was based, however, on unrealistic assumptions about how much water was available because the observations of runoff during the early 1900s turned out to be part of the greatest and longest high-flow period of the last five centuries.[452] Today, even in normal decades, the Colorado River does not have enough water to meet the agreed-upon allocations. During droughts and under projected future conditions, the situation looks even bleaker.

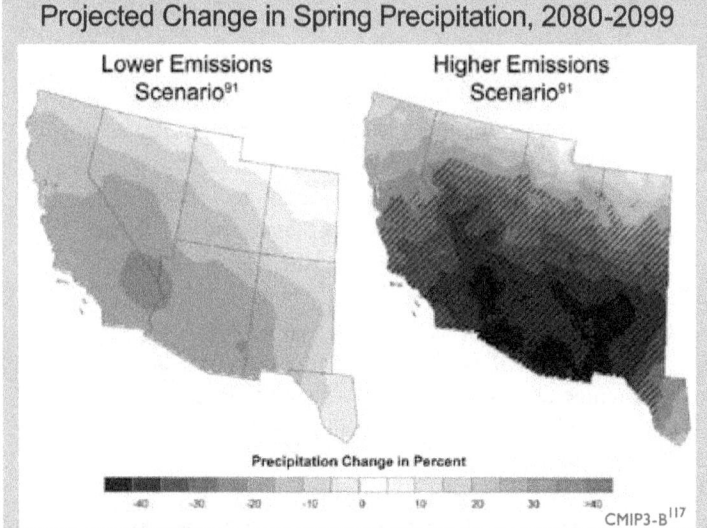

Projected Change in Spring Precipitation, 2080-2099

Percentage change in March-April-May precipitation for 2080-2099 compared to 1961-1979 for a lower emissions scenario[91] (left) and a higher emissions scenario[91] (right). Confidence in the projected changes is highest in the hatched areas.

During droughts, water designated for agriculture could provide a temporary back-up supply for urban water needs. Similarly, non-renewable groundwater could be tapped during especially dry periods. Both of these options, however, come at the cost of either current or future agricultural production.

Water is already a subject of contention in the Southwest, and climate change – coupled with rapid population growth – promises to increase the likelihood of water-related

Future of Drought in the Southwest

Droughts are a long-standing feature of the Southwest's climate. The droughts of the last 110 years pale in comparison to some of the decades-long "megadroughts" that the region has experienced over the last 2000 years.[419] During the closing decades of the 1500s, for example, major droughts gripped parts of the Southwest.[189] These droughts sharply reduced the flow of the Colorado River[452,453] and the all-important Sierra Nevada headwaters for California,[454] and dried out the region as a whole. As of 2009, much of the Southwest remains in a drought that began around 1999. This event is the most severe western drought of the last 110 years, and is being exacerbated by record warming.[455]

Over this century, projections point to an increasing probability of drought for the region.[90,115] Many aspects of these projections, including a northward shift in winter and spring storm tracks, are consistent with observed trends over recent decades.[96,456,457] Thus, the most likely future for the Southwest is a substantially drier one (although there is presently no consensus on how the region's summer monsoon [rainy season] might change in the future). Combined with the historical record of

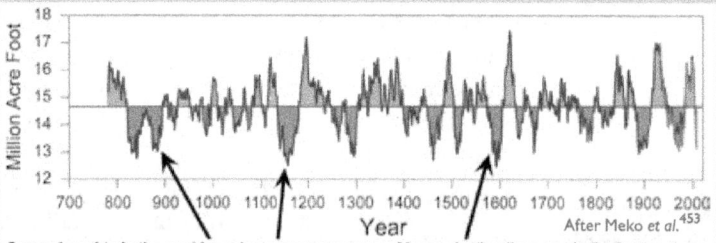

Some droughts in the past have been more severe and longer lasting than any in the last century.

severe droughts and the current uncertainty regarding the exact causes and drivers of these past events, the Southwest must be prepared for droughts that could potentially result from multiple causes. The combined effects of natural climate variability and human-induced climate change could turn out to be a devastating "one-two punch" for the region.

Colorado River flow has been reconstructed back over 1200 years based primarily on tree-ring data. These data reveal that some droughts in the past have been more severe and longer lasting than any experienced in the last 100 years. The red line indicates actual measurements of river flow during the last 100 years. Models indicate that, in the future, droughts will continue to occur, but will become hotter, and thus more severe, over time.[90]

conflict. Projected temperature increases, combined with river-flow reductions, will increase the risk of water conflicts between sectors, states, and even nations. In recent years, negotiations regarding existing water supplies have taken place among the seven states sharing the Colorado River and the two states (New Mexico and Texas) sharing the Rio Grande. Mexico and the United States already disagree on meeting their treaty allocations of Rio Grande and Colorado River water.

In addition, many water settlements between the U.S. Government and Native American tribes have yet to be fully worked out. The Southwest is home to dozens of Native communities whose status as sovereign nations means they hold rights to the water for use on their land. However, the amount of water actually available to each nation is determined through negotiations and litigation. Increasing water demand in the Southwest is driving current negotiations and litigation of tribal water rights. While several nations have legally settled their water rights, many other tribal negotiations are either currently underway or pending. Competing demands from treaty rights, rapid development, and changes in agriculture in the region, exacerbated by years of drought and climate change, have the potential to spark significant conflict over an already over-allocated and dwindling resource.

Increasing temperature, drought, wildfire, and invasive species will accelerate transformation of the landscape.

Climate change already appears to be influencing both natural and managed ecosystems of the Southwest.[455,458] Future landscape impacts are likely to be substantial, threatening biodiversity, protected areas, and ranching and agricultural lands. These changes are often driven by multiple factors, including changes in temperature and drought patterns, wildfire, invasive species, and pests.

Conditions observed in recent years can serve as indicators for future change. For example, temperature increases have made the current drought in the region more severe than the natural droughts of the last several centuries. As a result, about 4,600

square miles of piñon-juniper woodland in the Four Corners region of the Southwest have experienced substantial die-off of piñon pine trees.[455] Record wildfires are also being driven by rising temperatures and related reductions in spring snowpack and soil moisture.[458]

How climate change will affect fire in the Southwest varies according to location. In general, total area burned is projected to increase.[459] How this plays out at individual locations, however, depends on regional changes in temperature and precipitation, as well as on whether fire in the area is currently limited by fuel availability or by rainfall.[460] For example, fires in wetter, forested areas are expected to increase in frequency, while areas where fire is limited by the availability of fine fuels experience decreases.[460] Climate changes could also create subtle shifts in fire behavior, allowing more "runaway fires" – fires that are thought to have been brought under control, but then rekindle.[461] The magnitude of fire damages, in terms of economic impacts as well as direct endangerment, also increases as urban development increasingly impinges on forested areas.[460,462]

Climate-fire dynamics will also be affected by changes in the distribution of ecosystems across the Southwest. Increasing temperatures and shifting precipitation patterns will drive declines in high-elevation ecosystems such as alpine forests and tundra.[459,463] Under higher emissions scenarios,[91] high-elevation forests in California, for example, are projected to decline by 60 to 90 percent before the end of the century.[284,459] At the same time, grasslands are projected to expand, another factor likely to increase fire risk.

As temperatures rise, some iconic landscapes of the Southwest will be greatly altered as species shift their ranges northward and upward to cooler climates, and fires attack unaccustomed ecosystems which lack natural defenses. The Sonoran Desert, for example, famous for the saguaro cactus, would look very different if more woody species spread northward from Mexico into areas currently dominated by succulents (such as cacti) or native grasses.[464] The desert is already being invaded by red brome and buffle grasses that do well in high temperatures and are native to Africa and the

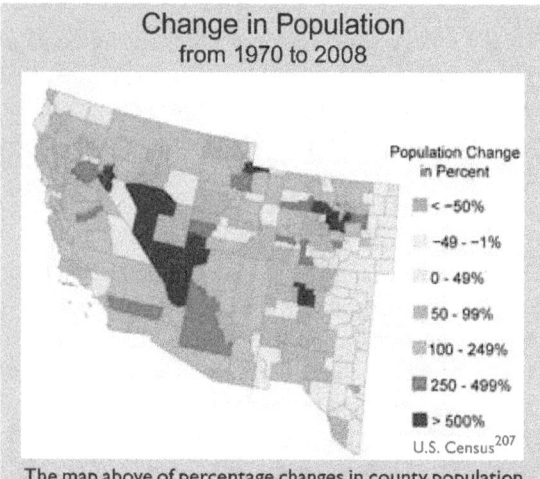

Change in Population
from 1970 to 2008

Population Change
in Percent

■ < -50%

■ -49 - -1%

□ 0 - 49%

■ 50 - 99%

■ 100 - 249%

■ 250 - 499%

■ > 500%

U.S. Census[207]

The map above of percentage changes in county population between 1970 and 2008 shows that the Southwest has experienced very rapid growth in recent decades (indicated in orange, red, and maroon).

ecosystems; those species moving southward to higher elevations might cut off future migration options as temperatures continue to increase.

The potential for successful plant and animal adaptation to coming change is further hampered by existing regional threats such as human-caused fragmentation of the landscape, invasive species, river-flow reductions, and pollution. Given the mountainous nature of the Southwest, and the associated impediments to species shifting their ranges, climate change likely places other species at risk. Some areas have already been identified as possible refuges where species at risk could continue to live if these areas were preserved for this purpose.[466] Other rapidly changing landscapes will require major adjustments, not only from plant and animal species, but also by the region's ranchers, foresters, and other inhabitants.

Mediterranean. Not only do these noxious weeds out-compete some native species in the Sonoran Desert, they also fuel hot, cactus-killing fires. With these invasive plant species and climate change, the Saguaro and Joshua Tree national parks could end up with far fewer of their namesake plants.[465] In California, two-thirds of the more than 5,500 native plant species are projected to experience range reductions up to 80 percent before the end of this century under projected warming.[466] In their search for optimal conditions, some species will move uphill, others northward, breaking up present-day

Increased frequency and altered timing of flooding will increase risks to people, ecosystems, and infrastructure.

Paradoxically, a warmer atmosphere and an intensified water cycle are likely to mean not only a greater likelihood of drought for the Southwest, but also an increased risk of flooding. Winter precipitation in Arizona, for example, is already

A Biodiversity Hotspot

The Southwest is home to two of the world's 34 designated "biodiversity hotspots." These at-risk regions have two special qualities: they hold unusually large numbers of plant and animal species that are endemic (found nowhere else), and they have already lost over 70 percent of their native vegetation.[467,468] About half the world's species of plants and land animals occur only in these 34 locations, though they cover just 2.3 percent of the Earth's land surface.

One of these biodiversity hotspots is the Madrean Pine-Oak Woodlands. Once covering 178 square miles, only isolated patches remain in the United States, mainly on mountaintops in southern Arizona, New Mexico, and West Texas. The greatest diversity of pine species in the world grows in this area: 44 of the 110 varieties,[469] as well as more than 150 species of oak.[470] Some 5,300 to 6,700 flowering plant species inhabit the ecosystem, and over 500 bird species, 23 of which are endemic. More hummingbirds are found here than anywhere else in the United States. There are 384 species of reptiles, 37 of which are endemic, and 328 species of mammals, six of which are endemic. There are 84 fish species, 18 of which are endemic. Some 200 species of butterfly thrive here, of which 45 are endemic, including the Monarch that migrates 2,500 miles north to Canada each year.[471] Ecotourism has become the economic driver in many parts of this region, but logging, land clearing for agriculture, urban development, and now climate change threaten the region's viability.

becoming more variable, with a trend toward both more frequent extremely dry and extremely wet winters.[472] Some water systems rely on smaller reservoirs being filled up each year. More frequent dry winters suggest an increased risk of these systems running short of water. However, a greater potential for flooding also means reservoirs cannot be filled to capacity as safely in years where that is possible. Flooding also causes reservoirs to fill with sediment at a faster rate, thus reducing their water-storage capacities.

On the global and national scales, precipitation patterns are already observed to be shifting, with more rain falling in heavy downpours that can lead to flooding.[90,473] Rapid landscape transformation due to vegetation die-off and wildfire as well as loss of wetlands along rivers is also likely to reduce flood-buffering capacity. Moreover, increased flood risk in the Southwest is likely to result from a combination of decreased snow cover on the lower slopes of high mountains, and an increased fraction of winter precipitation falling as rain and therefore running off more rapidly.[154] The increase in rain on snow events will also result in rapid runoff and flooding.[474]

The most obvious impact of more frequent flooding is a greater risk to human beings and their infrastructure. This applies to locations along major rivers, but also to much broader and highly vulnerable areas such as the Sacramento–San Joaquin River Delta system. Stretching from the San Francisco Bay nearly to the state capital of Sacramento, the Sacramento–San Joaquin River Delta and Suisun Marsh make up the largest estuary on the West Coast of North America. With its rich soils and rapid subsidence rates – in some locations as high as 2 or more feet per decade – the entire Delta region is now below sea level, protected by more than a thousand miles of levees and dams.[475] Projected changes in the timing and amount of river flow, particularly in winter and spring, is estimated to more than double the risk of Delta flooding events by mid-century, and result in an eight-fold increase before the end of the century.[476] Taking into account the additional risk of a major seismic event and increases in sea level due to climate change over this century, the California Bay–Delta Authority has concluded that the Delta and Suisun Marsh are

not sustainable under current practices; efforts are underway to identify and implement adaptation strategies aimed at reducing these risks.[476]

Unique tourism and recreation opportunities are likely to suffer.

Tourism and recreation are important aspects of the region's economy. Increasing temperatures will affect important winter activities such as downhill and cross-country skiing, snowshoeing, and snowmobiling, which require snow on the ground. Projections indicate later snow and less snow coverage in ski resort areas, particularly those at lower elevations and in the southern part of the region.[284] Decreases from 40 to almost 90 percent are likely in end-of-season snowpack under a higher emissions scenario[91] in counties with major ski resorts from New Mexico to California.[477] In addition to shorter seasons, earlier wet snow avalanches – more than six weeks earlier by the end of this century under a higher emissions scenario[91] – could force ski areas to shut down affected runs before the season would otherwise end.[478] Resorts require a certain number of days just to break even; cutting the season short by even a few weeks, particularly if those occur during the lucrative holiday season, could easily render a resort unprofitable.

Even in non-winter months, ecosystem degradation will affect the quality of the experience for hikers, bikers, birders, and others who enjoy the Southwest's natural beauty. Water sports that depend on the flows of rivers and sufficient water in lakes and reservoirs are already being affected, and much larger changes are expected.

Cities and agriculture face increasing risks from a changing climate.

Resource use in the Southwest is involved in a constant three-way tug-of-war among preserving natural ecosystems, supplying the needs of rapidly expanding urban areas, and protecting the lucrative agricultural sector, which, particularly in California, is largely based on highly temperature- and water-sensitive specialty crops. Urban areas are also sensitive to temperature-related impacts on air quality, electricity demand, and the health of their inhabitants.

The magnitude of projected temperature increases for the Southwest, particularly when combined with urban heat island effects for major cities such as Phoenix, Albuquerque, Las Vegas, and many California cities, represent significant stresses to health, electricity, and water supply in a region that already experiences very high summer temperatures.[284,325,450]

If present-day levels of ozone-producing emissions are maintained, rising temperatures also imply declining air quality in urban areas such as those in California which already experience some of the worst air quality in the nation (see *Society* sector).[479] Continued rapid population growth is expected to exacerbate these concerns.

With more intense, longer-lasting heat wave events projected to occur over this century, demands for air conditioning are expected to deplete electricity supplies, increasing risks of brownouts and black-outs.[325] Electricity supplies will also be affected by changes in the timing of river flows and where hydroelectric systems have limited storage capacity and reservoirs (see *Energy* sector).[480,481]

Much of the region's agriculture will experience detrimental impacts in a warmer future,

particularly specialty crops in California such as apricots, almonds, artichokes, figs, kiwis, olives, and walnuts.[482,483] These and other specialty crops require a minimum number of hours at a chilling temperature threshold in the winter to become dormant and set fruit for the following year.[482] Accumulated winter chilling hours have already decreased across central California and its coastal valleys. This trend is projected to continue to the point where chilling thresholds for many key crops would no longer be met. A steady reduction in winter chilling could have serious economic impacts on fruit and nut production in the region. California's losses due to future climate change are estimated between zero and 40 percent for wine and table grapes, almonds, oranges, walnuts, and avocadoes, varying significantly by location.[483]

Adaptation strategies for agriculture in California include more efficient irrigation and shifts in cropping patterns, which have the potential to help compensate for climate-driven increases in water demand for agriculture due to rising temperatures.[484] The ability to use groundwater and/or water designated for agriculture as backup supplies for urban uses in times of severe drought is expected to become more important in the future as climate change dries out the Southwest; however, these supplies are at risk of being depleted as urban populations swell (see *Water* sector).

Adaptation: Strategies for Fire

Living with present-day levels of fire risk, along with projected increases in risk, involves actions by residents along the urban-forest interface as well as fire and land management officials. Some basic strategies for reducing damage to structures due to fires are being encouraged by groups like National Firewise Communities, an interagency program that encourages wildfire preparedness measures such as creating defensible space around residential structures by thinning trees and brush, choosing fire-resistant plants, selecting ignition-resistant building materials and design features, positioning structures away from slopes, and working with firefighters to develop emergency plans.

Additional strategies for responding to the increased risk of fire as climate continues to change could include adding firefighting resources[461] and improving evacuation procedures and communications infrastructure. Also important would be regularly updated insights into what the latest climate science implies for changes in types, locations, timing, and potential severity of fire risks over seasons to decades and beyond; implications for related political, legal, economic, and social institutions; and improving predictions for regeneration of burnt-over areas and the implications for subsequent fire risks. Reconsideration of policies that encourage growth of residential developments in or near forests is another potential avenue for adaptive strategies.[462]

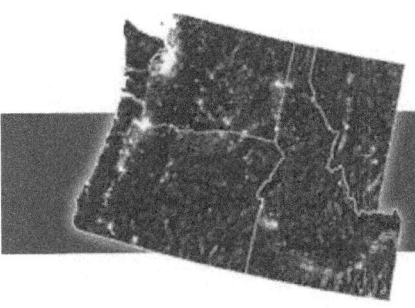

Northwest

The Northwest's rapidly growing population, as well as its forests, mountains, rivers, and coastlines, are already experiencing human-induced climate change and its impacts.[34] Regionally averaged temperature rose about 1.5°F over the past century[485] (with some areas experiencing increases up to 4°F) and is projected to increase another 3 to 10°F during this century.[486] Higher emissions scenarios would result in warming in the upper end of the projected range. Increases in winter precipitation and decreases in summer precipitation are projected by many climate models,[487] though these projections are less certain than those for temperature. Impacts related to changes in snowpack, streamflows, sea level, forests, and other important aspects of life in the Northwest are already underway, with more severe impacts expected over coming decades in response to continued and more rapid warming.

Declining springtime snowpack leads to reduced summer streamflows, straining water supplies.

The Northwest is highly dependent on temperature-sensitive springtime snowpack to meet growing, and often competing, water demands such as municipal and industrial uses, agricultural irrigation, hydropower production, navigation, recreation, and in-stream flows that protect aquatic ecosystems including threatened and endangered species. Higher cool season (October through March) temperatures cause more precipitation to fall as rain rather than snow and contribute to earlier snowmelt. April 1 snowpack, a key indicator of natural water storage available for the warm season, has already declined substantially throughout the region. The average decline in the Cascade Mountains, for example, was about 25 percent over the past 40 to 70 years, with most of this due to the 2.5°F increase in cool season temperatures over that period.[108,488] Further declines in Northwest snowpack are projected to result from additional warming over this century,

varying with latitude, elevation, and proximity to the coast. April 1 snowpack is projected to decline as much as 40 percent in the Cascades by the 2040s.[489] Throughout the region, earlier snowmelt will cause a reduction in the amount of water available during the warm season.[68]

In areas where it snows, a warmer climate means major changes in the timing of runoff: streamflow increases in winter and early spring, and then decreases in late spring, summer, and fall. This shift in streamflow timing has already been observed over the past 50 years,[252] with the peak of spring runoff shifting from a few days earlier in some places to as much as 25 to 30 days earlier in others.[157]

This trend is projected to continue, with runoff shifting 20 to 40 days earlier within this century.[157] Reductions in summer water availability will vary with the temperatures experienced in different parts of the region. In relatively warm areas on the western slopes of the Cascade Mountains, for example, reductions in warm season (April through September) runoff of 30 percent or more are projected by mid-century, whereas colder areas in the Rocky Mountains are expected to see reductions of about 10 percent. Areas dominated by rain rather than snow are not expected to see major shifts in the timing of runoff.[492]

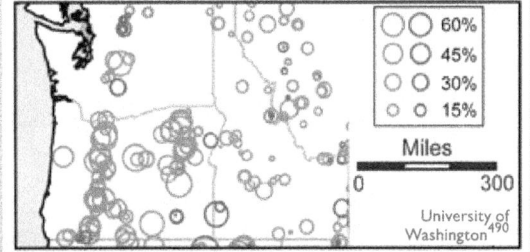

Trends in April 1 Snow Water Equivalent
1950 to 2002

60%
45%
30%
15%

Miles
0   300

University of Washington[490]

April 1 snowpack (a key indicator of natural water storage available for the warm season) has declined throughout the Northwest. In the Cascade Mountains, April 1 snowpack declined by an average of 25 percent, with some areas experiencing up to 60 percent declines. On the map, decreasing trends are in red and increasing trends are in blue.[491]

Extreme high and low streamflows also are expected to change with warming. Increasing winter rainfall (as opposed to snowfall) is expected to lead to more winter flooding in relatively warm watersheds on the west side of the Cascades. The already low flows of late summer are projected to decrease further due to both earlier snowmelt and increased evaporation and water loss from vegetation. Projected decreases in summer precipitation would exacerbate these effects. Some sensitive watersheds are projected to experience both increased flood risk in winter and increased drought risk in summer due to warming.

The region's water supply infrastructure was built based on the assumption that most of the water needed for summer uses would be stored naturally in snowpack. For example, the storage capacity in Columbia Basin reservoirs is only 30 percent of the annual runoff, and many small urban water supply systems on the west side of the Cascades store less than 10 percent of their annual flow.[493] Besides providing water supply and managing flows for hydropower, the region's reservoirs are operated for flood-protection purposes and, as such, might have to release (rather than store) large amounts of runoff during the winter and early spring to maintain enough space for flood protection. Earlier flows would thus place more of the year's runoff into the category of hazard rather than resource. An advance in the timing of snowmelt runoff would also

increase the length of the summer dry period, with important consequences for water supply, ecosystems, and wildfire management.[157]

One of the largest demands on water resources in the region is hydroelectric power production. About 70 percent of the Northwest's electricity is provided by hydropower, a far greater percentage than in any other region. Warmer summers will increase electricity demands for air conditioning and refrigeration at the same time of year that lower streamflows will lead to reduced hydropower generation. At the same time, water is needed for irrigated agriculture, protecting fish species, reservoir and river recreation, and urban uses. Conflicts between all of these water uses are expected to increase, forcing complex trade-offs between competing objectives (see *Energy* and *Water* sectors).[487,494]

Increased insect outbreaks, wildfires, and changing species composition in forests will pose challenges for ecosystems and the forest products industry.

Higher summer temperatures and earlier spring snowmelt are expected to increase the risk of forest fires in the Northwest by increasing summer moisture deficits; this pattern has already been observed in recent decades. Drought stress and higher temperatures will decrease tree growth in most low- and mid-elevation forests. They will also increase the frequency and intensity of mountain pine beetle and other insect attacks,[243] further increasing fire risk and reducing timber production, an important part of the regional economy. The mountain pine beetle outbreak in British Columbia has destroyed 33 million acres of trees so far, about 40 percent of the marketable pine trees in the province. By 2018, it is projected that the infestation will have run its course and over 78 percent of the mature pines will have been killed; this will affect more than one-third of the total area of British Columbia's forests[495] (see *Ecosystems* sector). Forest and fire management practices are also factors in these insect outbreaks.[252] Idaho's Sawtooth Mountains are also now threatened by pine beetle infestation.

In the short term, high elevation forests on the west side of the Cascade Mountains are expected to

Shift to Earlier Peak Streamflow
Quinault River (Olympic Peninsula, northern Washington)

As precipitation continues to shift from snow to rain, by the 2040s, peak flow on the Quinault River is projected to occur in December, and flows in June are projected to be reduced to about half of what they were over the past century. On the graph, the blue swath represents the range of projected streamflows based on an increase in temperature of 3.6 to 5.4°F. The other lines represent streamflows in the early and late 1900s.[487,494]

see increased growth. In the longer term, forest growth is expected to decrease as summertime soil moisture deficits limit forest productivity, with low-elevation forests experiencing these changes first. The extent and species composition of forests are also expected to change as tree species respond to climate change. There is also the potential for extinction of local populations and loss of biological diversity if environmental changes outpace species' ability to shift their ranges and form successful new ecosystems.

Agriculture, especially production of tree fruit such as apples, is also an important part of the regional economy. Decreasing irrigation supplies, increasing pests and disease, and increased competition from weeds are likely to have negative effects on agricultural production.

Salmon and other coldwater species will experience additional stresses as a result of rising water temperatures and declining summer streamflows.

Northwest salmon populations are at historically low levels due to stresses imposed by a variety of human activities including dam building, logging, pollution, and over-fishing. Climate change affects salmon throughout their life stages and poses an additional stress. As more winter precipitation falls as rain rather than snow, higher winter streamflows scour streambeds, damaging spawning nests and washing away incubating eggs. Earlier peak streamflows flush young salmon from rivers to estuaries before they are physically mature enough for the transition, increasing a variety of stresses including the risk of being eaten by predators. Lower summer streamflows and warmer water temperatures create less favorable summer stream conditions for salmon and other coldwater fish species in many parts of the Northwest. In addition, diseases and parasites that infect salmon tend to flourish in warmer water. Climate change also impacts the ocean environment, where salmon spend several years of their lives. Historically, warm periods in the coastal ocean have coincided with relatively low abundances of salmon, while cooler ocean periods have coincided with relatively high salmon numbers.[70, 563]

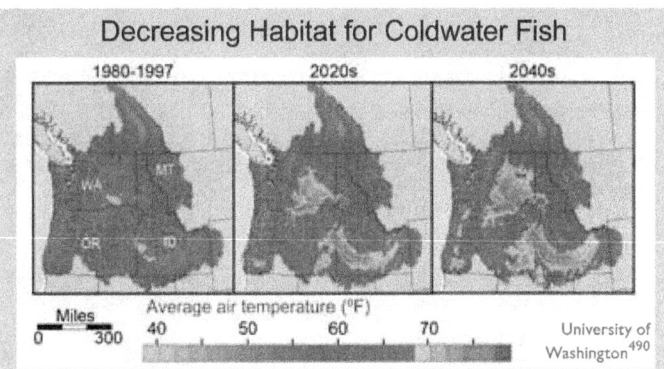

Decreasing Habitat for Coldwater Fish

1980-1997 2020s 2040s

Miles
0   300

Average air temperature (°F)
40   50   60   70

University of Washington[490]

Increasing air temperatures lead to rising water temperatures, which increase stress on coldwater fish such as trout, salmon, and steelhead. August average air temperature above 70°F is a threshold above which these fish are severely stressed. Projected temperatures for the 2020s and 2040s under a higher emissions scenario suggest that the habitat for these fish is likely to decrease dramatically.[486,497,568,569]

Most wild Pacific salmon populations are extinct or imperiled in 56 percent of their historical range in the Northwest and California,[496] and populations are down more than 90 percent in the Columbia River system. Many species are listed as either threatened or endangered under the Federal Endangered Species Act. Studies suggest that about one-third of the current habitat for the Northwest's salmon and other coldwater fish will no longer be suitable for them by the end of this century as key temperature thresholds are exceeded. Because climate change impacts on their habitat are projected to be negative, climate change is expected to hamper efforts to restore depleted salmon populations.

Sea-level rise along vulnerable coastlines will result in increased erosion and the loss of land.

Climate change is projected to exacerbate many of the stresses and hazards currently facing the coastal zone. Sea-level rise will increase erosion of the Northwest coast and cause the loss of beaches and significant coastal land areas. Among the most vulnerable parts of the coast is the heavily populated south Puget Sound region, which includes the cities of Olympia, Tacoma, and Seattle, Washington. Some climate models project changes in atmospheric pressure patterns that suggest a more southwesterly direction of future winter winds. Combined with higher sea levels, this would accelerate coastal erosion all along the Pacific Coast. Sea-level rise in the Northwest (as elsewhere) is

determined by global rates of sea-level rise, changes in coastal elevation associated with local vertical movement of the land, and atmospheric circulation patterns that influence wind-driven "pile-up" of water along the coast. A mid-range estimate of relative sea-level rise for the Puget Sound basin is about 13 inches by 2100. However, higher levels of up to 50 inches by 2100 in more rapidly subsiding (sinking) portions of the basin are also possible given the large uncertainties about accelerating rates of ice melt from Greenland and Antarctica in recent years (see *Global* and *National Climate Change* sections).[498]

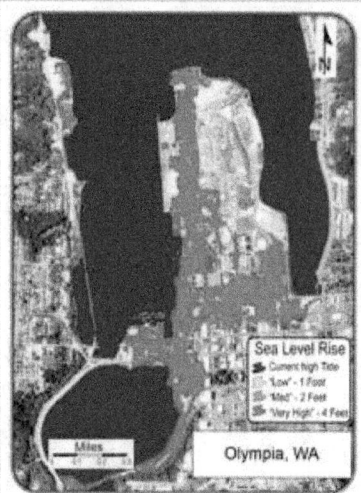

Northwest Cities at Risk to Sea-Level Rise

Petersen[499]

Highly populated coastal areas throughout Puget Sound, Washington, are vulnerable to sea-level rise. The maps show regions of Olympia and Harbor Island (both located in Puget Sound) that are likely to be lost to sea-level rise by the end of this century based on moderate and high estimates.

An additional concern is landslides on coastal bluffs. The projected heavier winter rainfall suggests an increase in saturated soils and, therefore, an increased number of landslides. Increased frequency and/or severity of landslides is expected to be especially problematic in areas where there has been intensive development on unstable slopes. Within Puget Sound, the cycle of beach erosion and bluff landslides will be exacerbated by sea-level rise, increasing beach erosion, and decreasing slope stability.

Adaptation: Improved Planning to Cope with Future Changes

States, counties, and cities in the Northwest are beginning to develop strategies to adapt to climate change. In 2007, Washington state convened stakeholders to develop adaptation strategies for water, agriculture, forests, coasts, infrastructure, and human health. Recommendations included improved drought planning, improved monitoring of diseases and pests, incorporating sea-level rise in coastal planning, and public education. An implementation strategy is under development.

In response to concerns about increasing flood risk, King County, Washington, approved plans in 2007 to fund repairs to the county's aging levee system. The county also will replace more than 57 "short-span" bridges with wider span structures that allow more debris and floodwater to pass underneath rather than backing up and causing the river to flood. The county has begun incorporating porous concrete and rain gardens into road projects to manage the effects of stormwater runoff during heavy rains, which are increasing as climate changes. King County has also published an adaptation guidebook that is becoming a model that other local governments can refer to in order to organize adaptation actions within their municipal planning processes.[500]

Concern about sea-level rise in Olympia, Washington, contributed to the city's decision to relocate its primary drinking water source from a low-lying surface water source to wells on higher ground. The city adjusted its plans for construction of a new City Hall to locate the building in an area less vulnerable to sea-level rise than the original proposed location. The building's foundation also was raised by 1 foot.

Alaska

Over the past 50 years, Alaska has warmed at more than twice the rate of the rest of the United States' average. Its annual average temperature has increased 3.4°F, while winters have warmed even more, by 6.3°F.[501] As a result, climate change impacts are much more pronounced than in other regions of the United States. The higher temperatures are already contributing to earlier spring snowmelt, reduced sea ice, widespread glacier retreat, and permafrost warming.[220,501] These observed changes are consistent with climate model projections of greater warming over Alaska, especially in winter, as compared to the rest of the country.

Climate models also project increases in precipitation over Alaska. Simultaneous increases in evaporation due to higher air temperatures, however, are expected to lead to drier conditions overall, with reduced soil moisture.[90] In the future, therefore, model projections suggest a longer summer growing season combined with an increased likelihood of summer drought and wildfires.

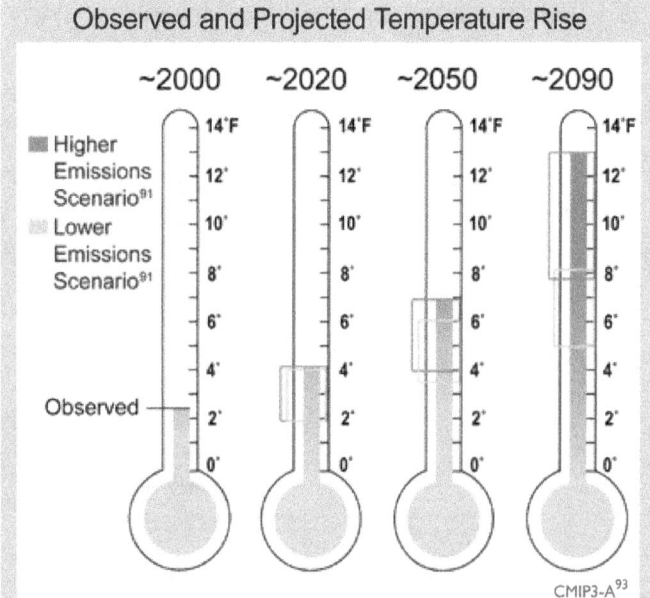

Observed and Projected Temperature Rise

Alaska's annual average temperature has increased 3.4°F over the past 50 years. The observed increase shown above compares the average temperature of 1993-2007 with a 1960s-1970s baseline, an increase of over 2°F. The brackets on the thermometers represent the likely range of model projections, though lower or higher outcomes are possible. By the end of this century, the average temperature is projected to rise by 5 to 13°F above the 1960s-1970s baseline.

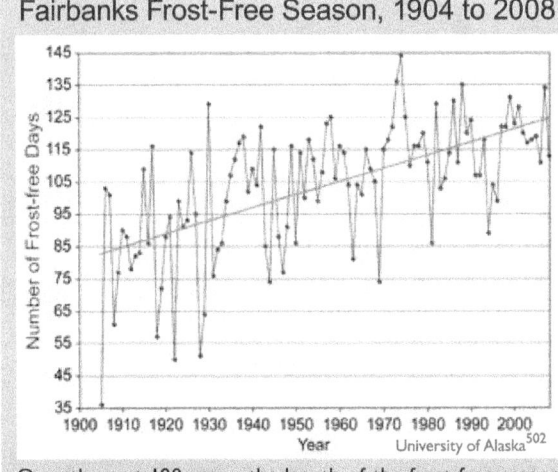

Over the past 100 years, the length of the frost-free season in Fairbanks, Alaska, has increased by 50 percent. The trend toward a longer frost-free season is projected to produce benefits in some sectors and detriments in others.

Average annual temperatures in Alaska are projected to rise about 3.5 to 7°F by the middle of this century. How much temperatures rise later in the century depends strongly on global emissions choices, with increases of 5 to 8°F projected with lower emissions, and increases of 8 to 13°F with higher emissions.[91] Higher temperatures are expected to continue to reduce Arctic sea ice coverage. Reduced sea ice provides opportunities for increased shipping and resource extraction. At the same time, it increases coastal erosion[522] and flooding associated with coastal storms. Reduced sea ice also alters the timing and location of plankton blooms, which is expected to drive major shifts of marine species such as pollock and other commercial fish stocks.[527]

Longer summers and higher temperatures are causing drier conditions, even in the absence of strong trends in precipitation.

Between 1970 and 2000, the snow-free season increased by approximately 10 days across Alaska, primarily due to earlier snowmelt in the spring.[503,504] A longer growing season has potential economic benefits, providing a longer period of outdoor and commercial activity such as tourism. However, there are also downsides. For example, white spruce forests in Alaska's interior are experiencing declining growth due to drought stress[505] and continued warming could lead to widespread death of trees.[506] The decreased soil moisture in Alaska also suggests that agriculture in Alaska might not benefit from the longer growing season.

Insect outbreaks and wildfires are increasing with warming.

Climate plays a key role in determining the extent and severity of insect outbreaks and wildfires.[506,507] During the 1990s, for example, south-central Alaska experienced the largest outbreak of spruce beetles in the world.[243,506] This outbreak occurred because rising temperatures allowed the spruce beetle to survive over the winter and to complete its life cycle in just one year instead of the normal two years. Healthy trees ordinarily defend themselves by pushing back against burrowing beetles with their pitch. From 1989 to 1997, however, the region experienced an extended drought, leaving the trees too stressed to fight off the infestation.

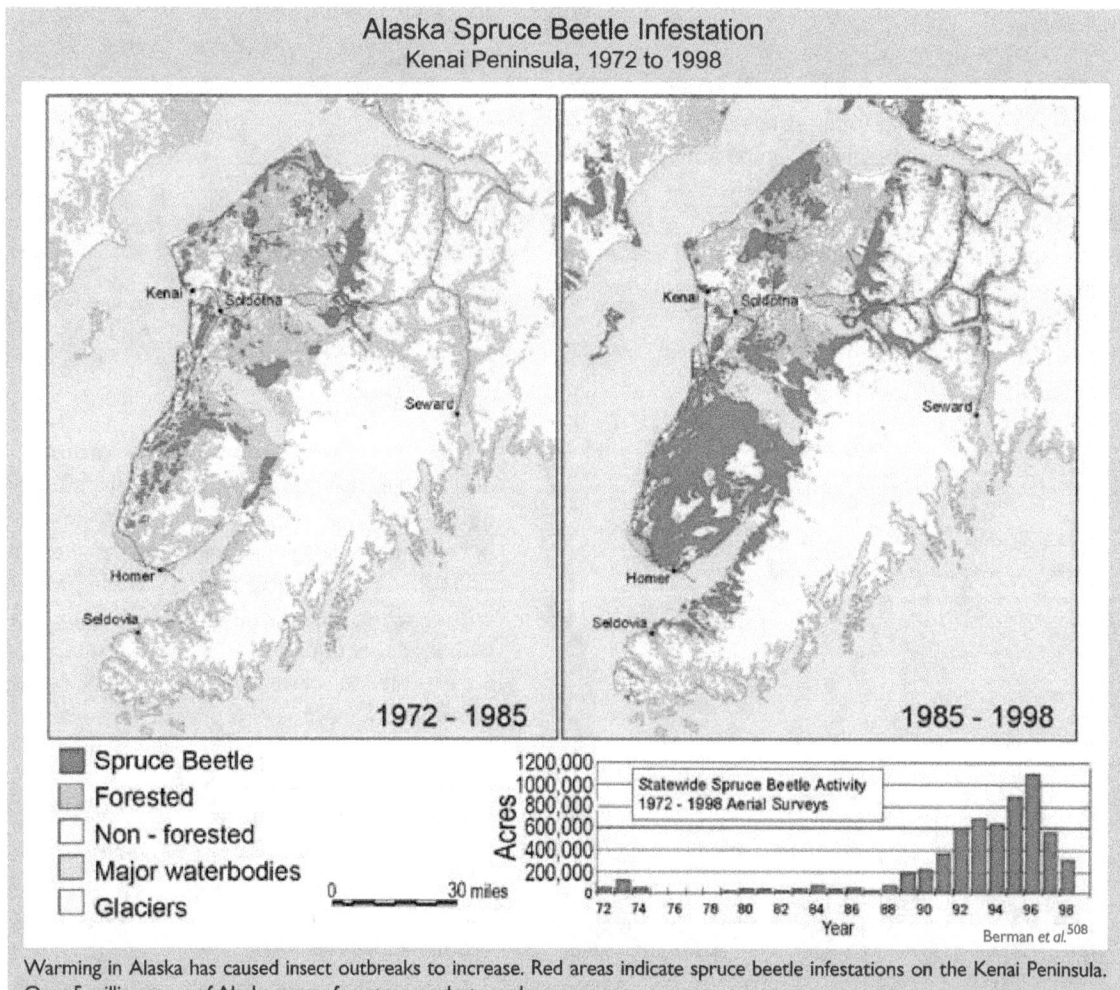

Warming in Alaska has caused insect outbreaks to increase. Red areas indicate spruce beetle infestations on the Kenai Peninsula. Over 5 million acres of Alaska spruce forests were destroyed.

Prior to 1990, the spruce budworm was not able to reproduce in interior Alaska.[506] Hotter, drier summers, however, now mean that the forests there are threatened by an outbreak of spruce budworms.[509] This trend is expected to increase in the future if summers in Alaska become hotter and drier.[506] Large areas of dead trees, such as those left behind by pest infestations, are highly flammable and thus much more vulnerable to wildfire than living trees.

The area burned in North America's northern forest that spans Alaska and Canada tripled from the 1960s to the 1990s. Two of the three most extensive wildfire seasons in Alaska's 56-year record occurred in 2004 and 2005, and half of the most severe fire years on record have occurred since 1990.[510] Under changing climate conditions, the average area burned per year in Alaska is projected to double by the middle of this century.[507] By the end of this century, area burned by fire is projected to triple under a moderate greenhouse gas emissions scenario and to quadruple under a higher emissions scenario.[91] Such increases in area burned would result in numerous impacts, including hazardous air quality conditions such as those suffered by residents of Fairbanks during the summers of 2004 and 2005, as well as increased risks to rural Native Alaskan communities because of reduced availability of the fish and game that make up their diet. This would cause them to adopt a more "Western" diet,[511] known to be associated with increased risk of cancers, diabetes, and cardiovascular disease.[512]

Lakes are declining in area.

Across the southern two-thirds of Alaska, the area of closed-basin lakes (lakes without stream inputs and outputs) has decreased over the past 50 years. This is likely due to the greater evaporation and thawing of permafrost that result from warming.[513,514] A continued decline in the area of surface water would present challenges for the management of natural resources and ecosystems on National Wildlife Refuges in Alaska. These refuges, which cover over 77 million acres (21 percent of Alaska) and comprise 81 percent of the U.S. National Wildlife Refuge System, provide breeding habitat for millions of waterfowl and shorebirds that winter in the lower 48 states. Wetlands are

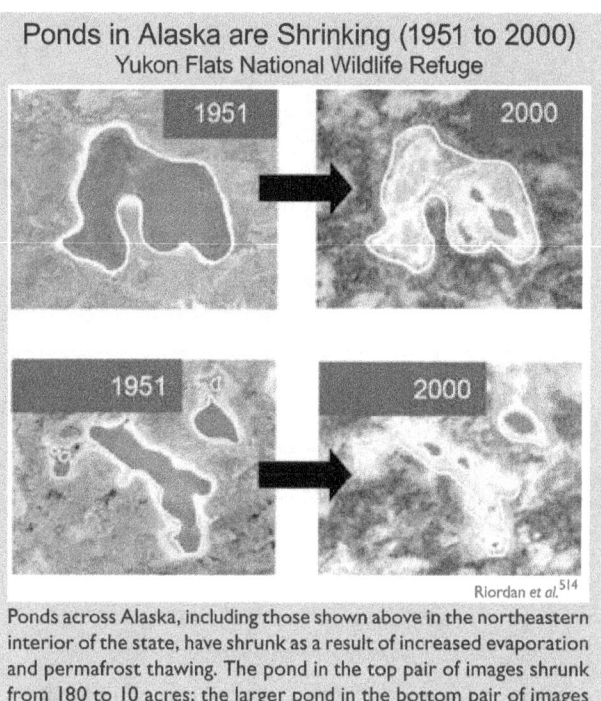

Ponds in Alaska are Shrinking (1951 to 2000)
Yukon Flats National Wildlife Refuge

Riordan et al.[514]

Ponds across Alaska, including those shown above in the northeastern interior of the state, have shrunk as a result of increased evaporation and permafrost thawing. The pond in the top pair of images shrunk from 180 to 10 acres; the larger pond in the bottom pair of images shrunk from 90 to 4 acres.

also important to Native peoples who hunt and fish for their food in interior Alaska. Many villages are located adjacent to wetlands that support an abundance of wildlife resources. The sustainability of these traditional lifestyles is thus threatened by a loss of wetlands.

Thawing permafrost damages roads, runways, water and sewer systems, and other infrastructure.

Permafrost temperatures have increased throughout Alaska since the late 1970s.[149] The largest increases have been measured in the northern part of the state.[515] While permafrost in interior Alaska so far has experienced less warming than permafrost in northern Alaska, it is more vulnerable to thawing during this century because it is generally just below the freezing point, while permafrost in northern Alaska is colder.

Land subsidence (sinking) associated with the thawing of permafrost presents substantial challenges to engineers attempting to preserve infrastructure in Alaska.[516] Public infrastructure at risk for damage includes roads, runways, and water and sewer systems. It is estimated that thawing

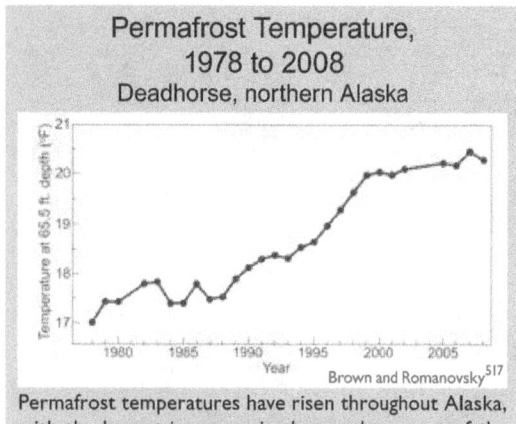

Permafrost Temperature, 1978 to 2008
Deadhorse, northern Alaska

Brown and Romanovsky[517]

Permafrost temperatures have risen throughout Alaska, with the largest increases in the northern part of the state.

Changing Permafrost Distribution
Moderate Warming Scenario

Current   Seward Peninsula   Projected: Late this Century

☐ Continuous (90 to 100% of land area frozen)
▨ Discontinuous (10 to 90% of land area frozen)
▇ Thawing/Permafrost Free

Busey et al.[518]

The maps show projected thawing on the Seward Peninsula by the end of this century under a moderate warming scenario approximately half-way between the lower and higher emissions scenarios[91] described on page 23.

permafrost would add between $3.6 billion and $6.1 billion (10 to 20 percent) to future costs for publicly owned infrastructure by 2030 and between $5.6 billion and $7.6 billion (10 to 12 percent) by 2080.[230] Analyses of the additional costs of permafrost thawing to private property have not yet been conducted.

Thawing ground also has implications for oil and gas drilling. As one example, the number of days per year in which travel on the tundra is allowed under Alaska Department of Natural Resources standards has dropped from more than 200 to about 100 days in the past 30 years. This results in a 50 percent reduction in days that oil and gas exploration and extraction equipment can be used.[220,245]

Thawing permafrost can push natural ecosystems across thresholds. Some forests in Alaska are literally toppling over as the permafrost beneath them thaws, undermining the root systems of trees (see photo next page).

Coastal storms increase risks to villages and fishing fleets.

Alaska has more coastline than the other 49 states combined. Frequent storms in the Gulf of Alaska and the Bering, Chukchi, and Beaufort Seas already affect the coasts during much of the year. Alaska's coastlines, many of which are low in elevation, are increasingly threatened by a combination of the loss of their protective sea ice buffer, increasing storm activity, and thawing coastal permafrost.

Increasing storm activity in autumn in recent years[520] has delayed or prevented barge operations

Adaptation: Keeping Soil Around the Pipeline Cool

When permafrost thaws, it can cause the soil to sink or settle, damaging structures built upon or within that soil. A warming climate and burial of supports for the Trans-Alaska Pipeline System both contribute to thawing of the permafrost around the pipeline. In locations on the pipeline route where soils were ice-rich, a unique above-ground system was developed to keep the ground cool. Thermal siphons were designed to disperse heat to the air that would otherwise be transferred to the soil, and these siphons were placed on the pilings that support the pipeline. While this unique technology added significant expense to the pipeline construction, it helps to greatly increase the useful lifetime of this structure.[519]

Leaning trees in this Alaska forest tilt because the ground beneath them, which used to be permanently frozen, has thawed. Forests like this are named "drunken forests."

Projected Coastal Erosion, 2007 to 2027
Newtok, western Alaska

Year
2027
2022
2017
2012
2007

N

Image dated June 2004

0   250   500   750   1000 Feet

U.S. Army Corps of Engineers[521]

Many of Alaska's coastlines are eroding rapidly; the disappearance of coastal land is forcing communities to relocate. The 2007 line on the image indicates where Newtok, Alaska's shoreline had eroded to by 2007. The other lines are projected assuming a conservative erosion rate of 36 to 83 feet per year; however, Newtok residents reported a July 2003 erosion rate of 110 feet per year.

that supply coastal communities with fuel. Commercial fishing fleets and other marine traffic are also strongly affected by Bering Sea storms. High-wind events have become more frequent along the western and northern coasts. The same regions are experiencing increasingly long sea-ice-free seasons and hence longer periods during which coastal areas are especially vulnerable to wind and wave damage. Downtown streets in Nome, Alaska, have flooded in recent years. Coastal erosion is causing the shorelines of some areas to retreat at average rates of tens of feet per year. The ground beneath several native communities is literally crumbling into the sea, forcing residents to confront difficult and expensive choices between relocation and engineering strategies that require continuing investments despite their uncertain effectiveness (see *Society*

sector). The rate of erosion along Alaska's northeastern coastline has doubled over the past 50 years.[522]

Over this century, an increase of sea surface temperatures and a reduction of ice cover are likely to lead to northward shifts in the Pacific storm track and increased impacts on coastal Alaska.[523,524] Climate models project the Bering Sea to experience the largest decreases in atmospheric pressure in the Northern Hemisphere, suggesting an increase in storm activity in the region.[90] In addition, the longer ice-free season is likely to make more heat and moisture available for storms in the Arctic Ocean, increasing their frequency and/or intensity.

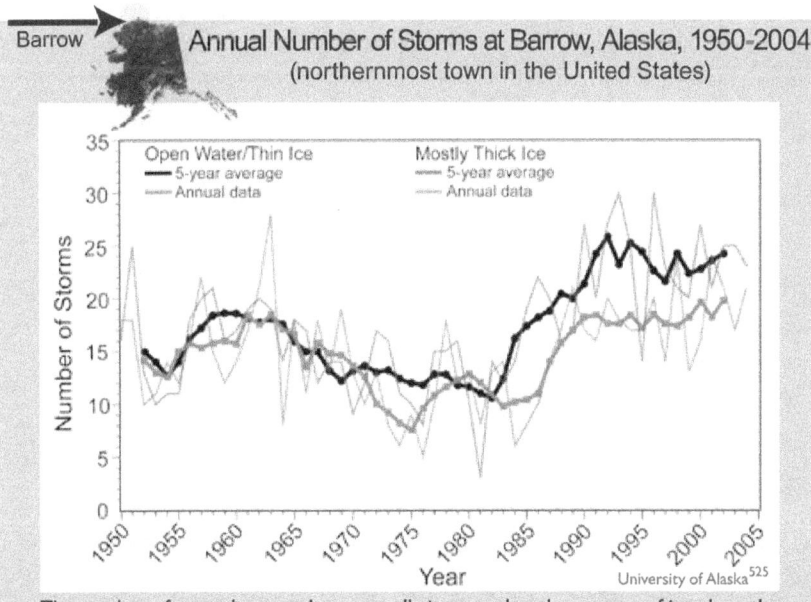

Barrow

Annual Number of Storms at Barrow, Alaska, 1950-2004
(northernmost town in the United States)

University of Alaska[525]

The number of coastal storms has generally increased as the amount of ice along the coast has decreased. This increase threatens commercial activity and communities in Alaska. The blue line indicates the annual number of open-water storms, those occurring in primarily ice-free water (July to December). The purple line indicates the number of storms occurring when thick sea ice is present (January to June). The black and green lines are smoothed using 5-year averages.

Displacement of marine species will affect key fisheries.

Alaska leads the United States in the value of its commercial fishing catch. Most of the nation's salmon, crab, halibut, and herring come from Alaska. In addition, many Native communities depend on local harvests of fish, walruses, seals, whales, seabirds, and other marine species for their food supply. Climate change causes significant alterations in marine ecosystems with important implications for fisheries. Ocean acidification associated with a rising carbon dioxide concentration represents an additional threat to coldwater marine ecosystems[23,526] (see *Ecosystems* sector and *Coasts* region).

One of the most productive areas for Alaska fisheries is the northern Bering Sea off Alaska's west coast. The world's largest single fishery is the Bering Sea pollock fishery, which has undergone major declines in recent years. Over much of the past decade, as air and water temperatures rose, sea ice in this region declined sharply. Populations of fish, seabirds, seals, walruses, and other species depend on plankton blooms that are regulated by

the extent and location of the ice edge in spring. As the sea ice retreats, the location, timing, and species composition of the plankton blooms changes, reducing the amount of food reaching the living things on the ocean floor. This radically changes the species composition and populations of fish and other marine life forms, with significant repercussions for fisheries[527] (see *Ecosystems* sector).

Over the course of this century, changes already observed on the shallow shelf of the northern Bering Sea are likely to affect a much broader portion of the Pacific-influenced sector of the Arctic Ocean. As such changes occur, the most productive commercial fisheries are likely to become more distant from existing fishing ports and processing infrastructure, requiring either relocation or greater investment in transportation time and fuel costs. These changes will also affect the ability of Native Peoples to successfully hunt and fish for the food they need to survive. Coastal communities are already noticing a displacement of walrus and seal populations. Bottom-feeding walrus populations are threatened when their sea ice platform retreats from the shallow coastal feeding grounds on which they depend.[528]

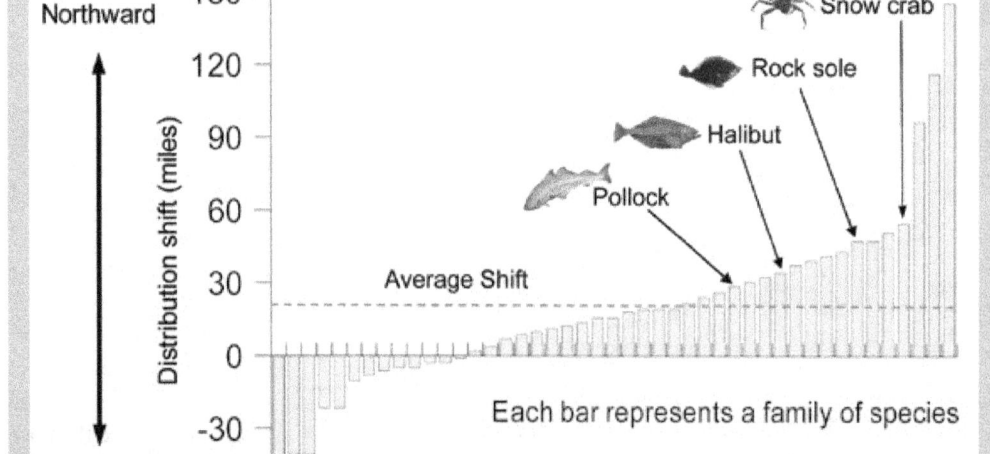

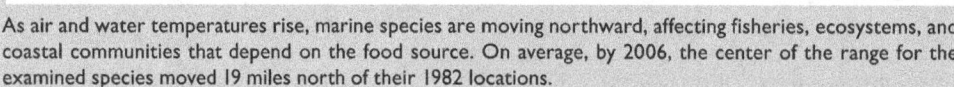

As air and water temperatures rise, marine species are moving northward, affecting fisheries, ecosystems, and coastal communities that depend on the food source. On average, by 2006, the center of the range for the examined species moved 19 miles north of their 1982 locations.

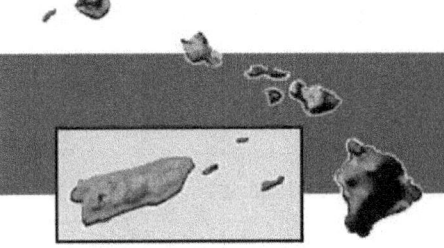

Islands

Climate change presents the Pacific and Caribbean islands with unique challenges. The U.S. affiliated Pacific Islands are home to approximately 1.7 million people in the Hawaiian Islands; Palau; the Samoan Islands of Tutuila, Manua, Rose, and Swains; and islands in the Micronesian archipelago, the Carolines, Marshalls, and Marianas.[530] These include volcanic, continental, and limestone islands, atolls, and islands of mixed geologies.[530] The degree to which climate change and variability will affect each of the roughly 30,000 islands in the Pacific depends upon a variety of factors, including the island's geology, area, height above sea level, extent of reef formation, and the size of its freshwater aquifer.[531]

In addition to Puerto Rico and the U.S. Virgin Islands, there are 40 island nations in the Caribbean that are home to approximately 38 million people.[532] Population growth, often concentrated in coastal areas, escalates the vulnerability of both Pacific and Caribbean island communities to the effects of climate change, as do weakened traditional support systems. Tourism and fisheries, both of which are climate-sensitive, play a large economic role in these communities.[530]

Small islands are considered among the most vulnerable to climate change because extreme events have major impacts on them. Changes in weather patterns and the frequency and intensity of extreme events, sea-level rise, coastal erosion, coral reef bleaching, ocean acidification, and contamination of freshwater resources by salt water are among the impacts small islands face.[533]

Islands have experienced rising temperatures and sea levels in recent decades. Projections for the rest of this century suggest:

- Increases in air and ocean surface temperatures in both the Pacific and Caribbean;[90]
- An overall decrease in rainfall in the Caribbean; and
- An increased frequency of heavy downpours and increased rainfall during summer months (rather than the normal rainy season in winter months) for the Pacific (although the range of projections regarding rainfall in the Pacific is still quite large).

The number of heavy rain events is very likely to increase.[90] Hurricane (typhoon) wind speeds and rainfall rates are likely to increase with continued

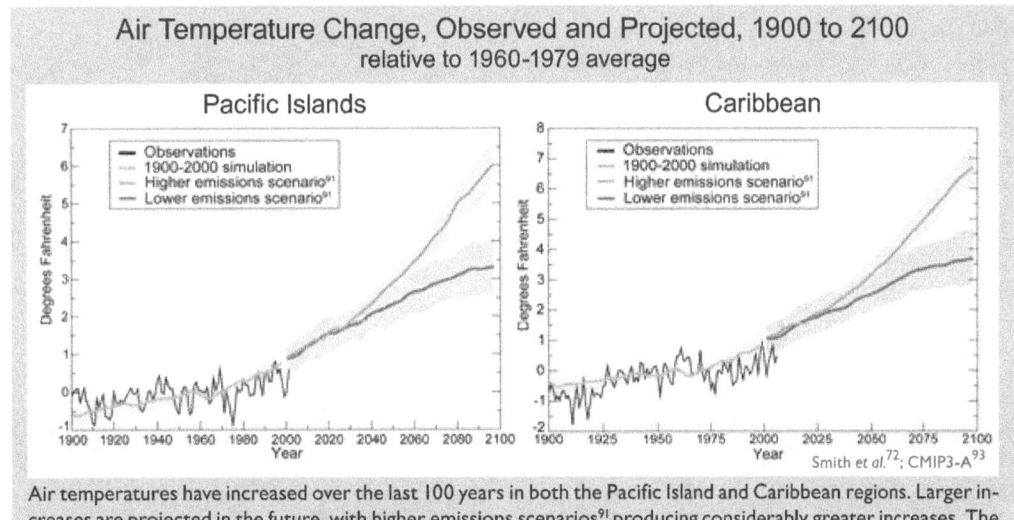

Air Temperature Change, Observed and Projected, 1900 to 2100
relative to 1960-1979 average

Pacific Islands

- Observations
- 1900-2000 simulation
- Higher emissions scenario[91]
- Lower emissions scenario[91]

Caribbean

- Observations
- 1900-2000 simulation
- Higher emissions scenario[91]
- Lower emissions scenario[91]

Smith et al.[72]; CMIP3-A[93]

Air temperatures have increased over the last 100 years in both the Pacific Island and Caribbean regions. Larger increases are projected in the future, with higher emissions scenarios[91] producing considerably greater increases. The shaded areas show the likely ranges while the lines show the central projections from a set of climate models.

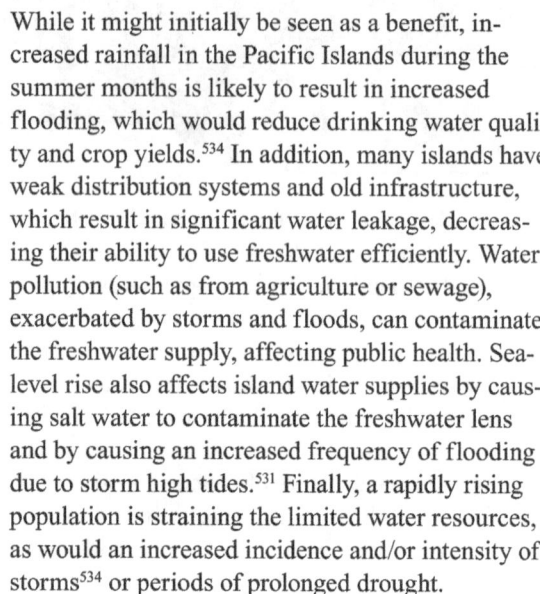

Freshwater Lens

Adapted from Burns[534]

Many island communities depend on freshwater lenses, which are recharged by precipitation. The amount of water a freshwater lens contains is determined by the size of the island, the amount of rainfall, rates of water withdrawal, the permeability of the rock beneath the island, and salt mixing due to storm- or tide-induced pressure. Freshwater lenses can be as shallow as 4 to 8 inches or as deep as 65 feet.[534]

warming.[68] Islands and other low-lying coastal areas will be at increased risk from coastal inundation due to sea-level rise and storm surge, with major implications for coastal communities, infrastructure, natural habitats, and resources.

The availability of freshwater is likely to be reduced, with significant implications for island communities, economies, and resources.

Most island communities in the Pacific and the Caribbean have limited sources of the freshwater needed to support unique ecosystems and biodiversity, public health, agriculture, and tourism. Conventional freshwater resources include rainwater collection, groundwater, and surface water.[534] For drinking and bathing, smaller Pacific islands primarily rely on individual rainwater catchment systems, while groundwater from the freshwater lens is used for irrigation. The size of freshwater lenses in atolls is influenced by factors such as rates of recharge (through precipitation), rates of use, and extent of tidal inundation.[531] Since rainfall triggers the formation of the freshwater lens, changes in precipitation, such as the significant decreases projected for the Caribbean, can significantly affect the availability of water. Because tropical storms replenish water supplies, potential changes in these storms are a great concern.

While it might initially be seen as a benefit, increased rainfall in the Pacific Islands during the summer months is likely to result in increased flooding, which would reduce drinking water quality and crop yields.[534] In addition, many islands have weak distribution systems and old infrastructure, which result in significant water leakage, decreasing their ability to use freshwater efficiently. Water pollution (such as from agriculture or sewage), exacerbated by storms and floods, can contaminate the freshwater supply, affecting public health. Sea-level rise also affects island water supplies by causing salt water to contaminate the freshwater lens and by causing an increased frequency of flooding due to storm high tides.[531] Finally, a rapidly rising population is straining the limited water resources, as would an increased incidence and/or intensity of storms[534] or periods of prolonged drought.

Island communities, infrastructure, and ecosystems are vulnerable to coastal inundation due to sea-level rise and coastal storms.

Sea-level rise will have enormous effects on many island nations. Flooding will become more frequent due to higher storm tides, and coastal land will be permanently lost as the sea inundates low-lying areas and the shorelines erode. Loss of land

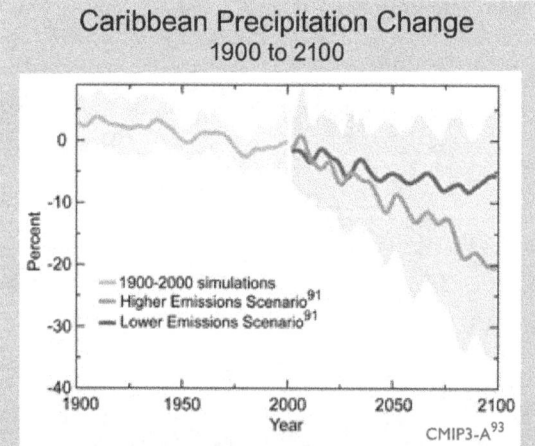

Caribbean Precipitation Change
1900 to 2100

— 1900-2000 simulations
— Higher Emissions Scenario[91]
— Lower Emissions Scenario[91]

CMIP3-A[93]

Total annual precipitation has declined in the Caribbean and climate models project stronger declines in the future, particularly under higher emission scenarios.[91] Such decreases threaten island communities that rely on rainfall for replenishing their freshwater supplies. The shaded areas show the likely ranges while the lines show the central projections from a set of climate models.

will reduce freshwater supplies[531] and affect living things in coastal ecosystems. For example, the Northwestern Hawaiian Islands, which are low-lying and therefore at great risk from increasing sea level, have a high concentration of endangered and threatened species, some of which exist nowhere else.[535] The loss of nesting and nursing habitat is expected to threaten the survival of already vulnerable species.[535]

In addition to gradual sea-level rise, extreme high water level events can result from a combination of coastal processes.[271] For example, the harbor in Honolulu, Hawaii, experienced the highest daily average sea level ever recorded in September 2003. This resulted from the combination of long-term sea-level rise, normal seasonal heating (which causes the volume of water to expand and thus the level of the sea to rise), seasonal high tide, and an ocean circulation event which temporarily raised local sea level.[536] The interval between such extreme events has decreased from more than 20 years to approximately 5 years as average sea level has risen.[536]

Hurricanes, typhoons, and other storm events, with their intense precipitation and storm surge, cause major impacts to Pacific and Caribbean island com-

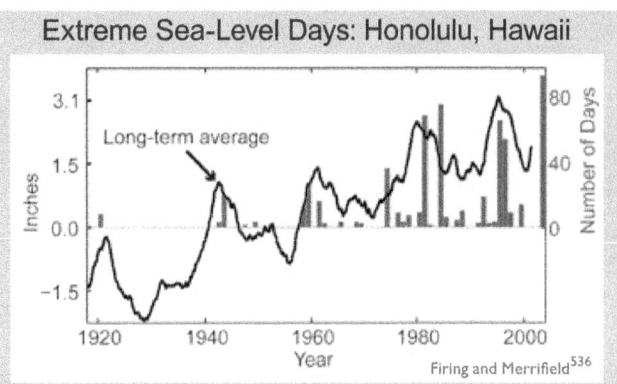

Extreme Sea-Level Days: Honolulu, Hawaii

Firing and Merrifield[536]

Sea-level rise will result in permanent land loss and reductions in freshwater supplies, as well as threaten coastal ecosystems. "Extreme" sea-level days (with a daily average of more than 6 inches above the long-term average[90]) can result from the combined effects of gradual sea-level rise due to warming and other phenomena, including seasonal heating and high tides.

munities, including loss of life, damage to infrastructure and property, and contamination of freshwater supplies.[537] As the climate continues to warm, the peak wind intensities and near-storm precipitation from future tropical cyclones are likely to increase,[90] which, combined with sea-level rise, is expected to cause higher storm surge levels. If such events occur frequently, communities would face challenges in recovering between events, resulting in long-term deterioration of infrastructure, freshwater and agricultural resources, and other impacts.[246]

Adaptation: Securing Water Resources

In the islands, "water is gold." Effective adaptation to climate-related changes in the availability of freshwater is thus a high priority. While island communities cannot completely counter the threats to water supplies posed by global warming, effective adaptation approaches can help reduce the damage.

When existing resources fall short, managers look to unconventional resources, such as desalinating seawater, importing water by ship, and using treated wastewater for non-drinking uses. Desalination costs are declining, though concerns remain about the impact on marine life, the

A billboard on Pohnpei, in the Federated States of Micronesia, encourages water conservation in preparation for the 1997 to 1998 El Niño.

disposal of concentrated brines that may contain chemical waste, and the large energy use (and associated carbon footprint) of the process.[146] With limited natural resources, the key to successful water resource management in the islands will continue to be "conserve, recover, and reuse."[530]

Pacific Island communities are also making use of the latest science. This effort started during the 1997 to 1998 El Niño, when managers began using seasonal forecasts to prepare for droughts by increasing public awareness and encouraging water conservation. In addition, resource managers can improve infrastructure, such as by fixing water distribution systems to minimize leakage and by increasing freshwater storage capacity.[530]

Coastal houses and an airport in the U.S.-affiliated Federated States of Micronesia rely on mangroves' protection from erosion and damage due to rising sea level, waves, storm surges, and wind.

Critical infrastructure, including homes, airports, and roads, tends to be located along the coast. Flooding related to sea-level rise and hurricanes and typhoons negatively affects port facilities and harbors, and causes closures of roads, airports, and bridges.[538] Long-term infrastructure damage would affect social services such as disaster risk management, health care, education, management of freshwater resources, and economic activity in sectors such as tourism and agriculture.

Climate changes affecting coastal and marine ecosystems will have major implications for tourism and fisheries.

Marine and coastal ecosystems of the islands are particularly vulnerable to the impacts of climate change. Sea-level rise, increasing water temperatures, rising storm intensity, coastal inundation and flooding from extreme events, beach erosion, ocean acidification, increased incidences of coral disease, and increased invasions by non-native species are among the threats that endanger the ecosystems that provide safety, sustenance, economic viability, and cultural and traditional values to island communities.[539]

Tourism is a vital part of the economy for many islands. In 1999, the Caribbean had tourism-based gross earnings of $17 billion, providing 900,000 jobs and making the Caribbean one of the most tourism dependent regions in the world.[532] In the South Pacific, tourism can contribute as much as 47 percent of gross domestic product.[540] In Hawaii, tourism generated $12.4 billion for the state in 2006, with over 7 million visitors.[541]

Sea-level rise can erode beaches, and along with increasing water temperatures, can destroy or degrade natural resources such as mangroves and coral reef ecosystems that attract tourists.[246] Extreme weather events can affect transportation systems

and interrupt communications. The availability of freshwater is critical to sustaining tourism, but is subject to the climate-related impacts described on the previous page. Public health concerns about diseases would also negatively affect tourism.

Coral reefs sustain fisheries and tourism, have biodiversity value, scientific and educational value, and form natural protection against wave erosion.[542] For Hawaii alone, net benefits of reefs to the economy are estimated at $360 million annually, and the overall asset value is conservatively estimated to be nearly $10 billion.[542] In the Caribbean, coral reefs provide annual net benefits from fisheries, tourism, and shoreline protection services of between $3.1 billion and $4.6 billion. The loss of income by 2015 from degraded reefs is conservatively estimated at several hundred million dollars annually.[532,543]

Coral reef ecosystems are particularly susceptible to the impacts of climate change, as even small increases in water temperature can cause coral bleaching,[544] damaging and killing corals. Ocean acidification due to a rising carbon dioxide concentration poses an additional threat (see *Ecosystems* sector and *Coasts* region). Coral reef ecosystems are also especially vulnerable to invasive species.[545] These impacts, combined with changes in the occurrence and intensity of El Niño events, rising sea level, and increasing storm damage,[246] will have major negative effects on coral reef ecosystems.

Fisheries feed local people and island economies. Almost all communities within the Pacific Islands derive over 25 percent of their animal protein from fish, with some deriving up to 69 percent.[546] For island fisheries sustained by healthy coral reef and marine ecosystems, climate change impacts exacerbate stresses such as overfishing,[246] affecting both fisheries and tourism that depend on abundant and diverse reef fish. The loss of live corals results in local extinctions and a reduced number of reef fish species.[547]

Nearly 70 percent of the world's annual tuna harvest, approximately 3.2 million tons, comes from the Pacific Ocean.[548] Climate change is projected to cause a decline in tuna stocks and an eastward shift in their location, affecting the catch of certain countries.[246]

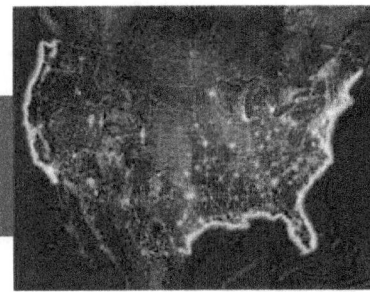

Coasts

Approximately one-third of all Americans live in counties immediately bordering the nation's ocean coasts.[549,550] In addition to accommodating major cities, the coasts and the exclusive economic zone extending 200 miles offshore provide enjoyment, recreation, seafood, transportation of goods, and energy. Coastal and ocean activities contribute more than $1 trillion to the nation's gross domestic product and the ecosystems hold rich biodiversity and provide invaluable services.[551] However, intense human uses have taken a toll on coastal environments and their resources. Many fish stocks have been severely diminished by over-fishing, large "dead zones" depleted of oxygen have developed as a result of pollution by excess nitrogen runoff, toxic blooms of algae are increasingly frequent, and coral reefs are badly damaged or becoming overgrown with algae. About half of the nation's coastal wetlands have been lost – and most of this loss has occurred during the past 50 years.

Global climate change imposes additional stresses on coastal environments. Rising sea level is already eroding shorelines, drowning wetlands, and threatening the built environment.[43,224] The destructive potential of Atlantic tropical storms and hurricanes has increased since 1970 in association with increasing Atlantic sea surface temperatures, and it is likely that hurricane rainfall and wind speeds will increase in response to global warming.[112] Coastal water temperatures have risen by about 2°F in several regions, and the geographic distributions of marine species have shifted.[37,68,347] Precipitation increases on land have increased river runoff, polluting coastal waters with more nitrogen and phosphorous, sediments, and other contaminants. Furthermore, increasing acidification resulting from the uptake of carbon dioxide by ocean waters threatens corals, shellfish, and other living things that form their shells and skeletons from calcium carbonate[23] (see *Ecosystems* sector). All of these forces converge and interact at the coasts, making these areas particularly sensitive to the impacts of climate change.

Significant sea-level rise and storm surge will adversely affect coastal cities and ecosystems around the nation; low-lying and subsiding areas are most vulnerable.

The rise in sea level relative to the land surface in any given location is a function of both the amount of global average sea-level rise and the degree to which the land is rising or falling. During the past century in the United States, relative sea level changes ranged from falling several inches to rising as much as 2 feet.[225] High rates of relative sea-level rise, coupled with cutting off the supply of sediments from the Mississippi River and other human alterations, have resulted in the loss of 1,900 square miles of Louisiana's coastal wetlands during the past century, weakening their capacity

Multiple Stresses Confront Coastal Regions

Various forces of climate change at the coasts pose a complex array of management challenges and adaptation requirements. For example, relative sea level is expected to rise at least 2 feet in Chesapeake Bay (located between Maryland and Virginia) where the land is subsiding, threatening portions of cities, inhabited islands, most tidal wetlands, and other low-lying regions. Climate change also will affect the volume of the bay, its salinity distribution and circulation, as will changes in precipitation and freshwater runoff. These changes, in turn, will affect summertime oxygen depletion and efforts to reduce the agricultural nitrogen runoff that causes it. Meanwhile the warming of the bay's waters will make survival there difficult for northern species such as eelgrass and soft clams, while allowing southern species and invaders riding in ships' ballast water to move in and change the mix of species that are caught and must be managed. Additionally, more acidic waters resulting from rising carbon dioxide levels will make it difficult for oysters to build their shells and will complicate the recovery of this key species.[553]

Projected Sea-Level Rise by 2100

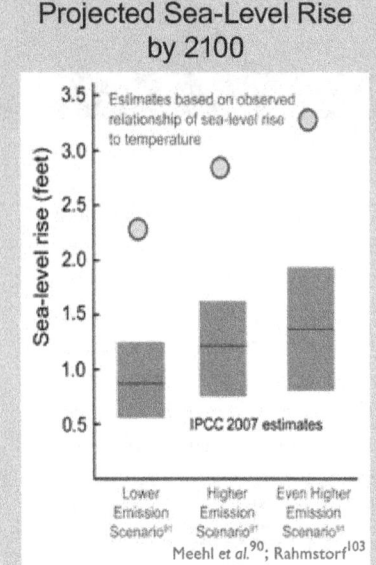

Estimates of sea-level rise by the end of the century for three emissions scenarios.[91] Intergovernmental Panel on Climate Change 2007 projections (range shown as bars) exclude changes in ice sheet flow.[90] Light blue circles represent more recent, central estimates derived using the observed relationship of sea-level rise to temperature.[103] Areas where coastal land is sinking, for example by as much as 1.5 feet in this century along portions of the Gulf Coast, would experience that much additional sea-level rise relative to the land.[128]

to absorb the storm surge of hurricanes such as Katrina.[552] Shoreline retreat is occurring along most of the nation's exposed shores.

The amount of sea-level rise likely to be experienced during this century depends mainly on the expansion of the ocean volume due to warming and the response of glaciers and polar ice sheets. Complex processes control the discharges from polar ice sheets and some are already producing substantial additions of water to the ocean.[554] Because these processes are not well understood, it is difficult to predict their future contributions to sea-level rise.[90,555]

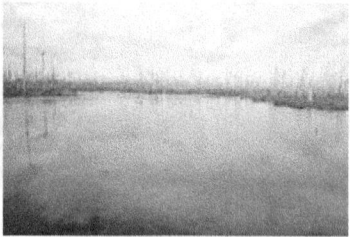

A "ghost swamp" in south Louisiana shows the effects of saltwater intrusion.

As discussed in the *Global Climate Change* section, recent estimates of global sea-level rise substantially exceed the IPCC estimates, suggesting sea-level rise between 3 and 4 feet in this century. Even a 2-foot rise in relative sea level over a century would result in the loss of a large portion of the nation's remaining coastal wetlands, as they are not able to build new soil at a fast enough rate.[164] Accelerated sea-level rise would affect sea-grasses, coral reefs, and other important habitats. It would also fragment barrier islands, and place into jeopardy existing homes, businesses, and infrastructure, including roads, ports, and water and sewage systems. Portions of major cities, including Boston and New York, would be subject to inundation by ocean water during storm surges or even during regular high tides.[234]

More spring runoff and warmer coastal waters will increase the seasonal reduction in oxygen resulting from excess nitrogen from agriculture.

Coastal dead zones in places such as the northern Gulf of Mexico[556] and the Chesapeake Bay[557] are likely to increase in size and intensity as warming increases unless efforts to control runoff of agricultural fertilizers are redoubled. Greater spring runoff into East Coast estuaries and the Gulf of Mexico would flush more nitrogen into coastal waters stimulating harmful blooms of algae and the excess production of microscopic plants that settle near the seafloor and deplete oxygen supplies as they decompose. In addition, all else being equal, greater runoff reduces salinity, which when coupled with warmer surface water increases the difference in density between surface and bottom waters, thus preventing the replacement of oxygen in the deeper waters. As dissolved oxygen levels decline below a certain level, living things cannot survive. They leave the area if they can, and die if they cannot.

Dead Zones in the Chesapeake Bay

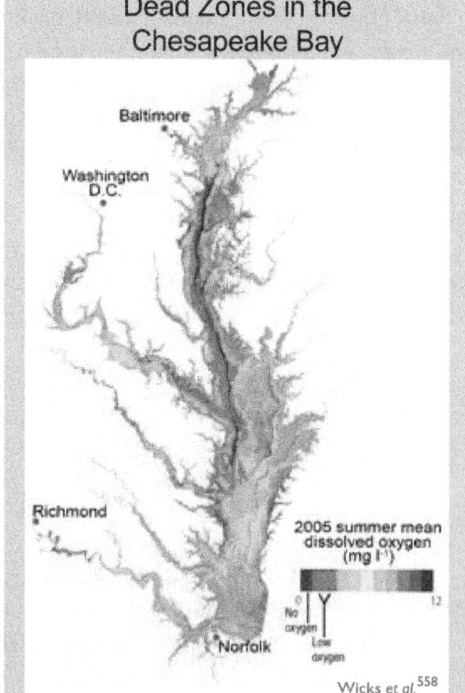

Climate change is likely to expand and intensify "dead zones," areas where bottom water is depleted of dissolved oxygen because of nitrogen pollution, threatening living things.

Coastal waters are very likely to continue to warm by as much 4 to 8°F in this century, both in summer and winter.[234] This will result in a northward shift in the geographic distribution of marine life along the coasts; this is already being observed.[70,347] The shift occurs because some species cannot tolerate the higher temperatures and others are out-competed by species from farther south moving in.[270] Warming also opens the door to invasion by species that humans are intentionally or unintentionally transporting around the world, for example in the ballast water carried by ships. Species that were previously unable to establish populations because of cold winters are likely to find the warmer conditions more welcoming and gain a foothold,[567] particularly as native species are under stress from climate change and other human activities. Non-native clams and small crustaceans have already had major effects on the San Francisco Bay ecosystem and the health of its fishery resources.[559]

Higher water temperatures and ocean acidification due to increasing atmospheric carbon dioxide will present major additional stresses to coral reefs, resulting in significant die-offs and limited recovery.

In addition to carbon dioxide's heat-trapping effect, the increase in its concentration in the atmosphere is gradually acidifying the ocean. About one-third of the carbon dioxide emitted by human activities has been absorbed by the ocean, resulting in a decrease in the ocean's pH. Since the beginning of the industrial era, ocean pH has declined demonstrably and is projected to decline much more by 2100 if current emissions trends continue. Further declines in pH are very

likely to continue to affect the ability of living things to create and maintain shells or skeletons of calcium carbonate. This is because at a lower pH less of the dissolved carbon is available as carbonate ions (see *Global Climate Change*).[70,259]

Ocean acidification will affect living things including important plankton species in the open ocean, mollusks and other shellfish, and corals.[22,23,70,259] The effects on reef-building corals are likely to be particularly severe during this century. Coral calcification rates are likely to decline by more than 30 percent under a doubling of atmospheric carbon dioxide concentrations, with erosion outpacing reef formation at even lower concentrations.[22] In addition, the reduction in pH also affects photosynthesis, growth, and reproduction. The upwelling of deeper ocean water, deficient in carbonate, and thus potentially detrimental to the food chains supporting juvenile salmon has recently been observed along the U.S. West Coast.[259]

Acidification imposes yet another stress on reef-building corals, which are also subject to bleaching – the expulsion of the microscopic algae that live inside the corals

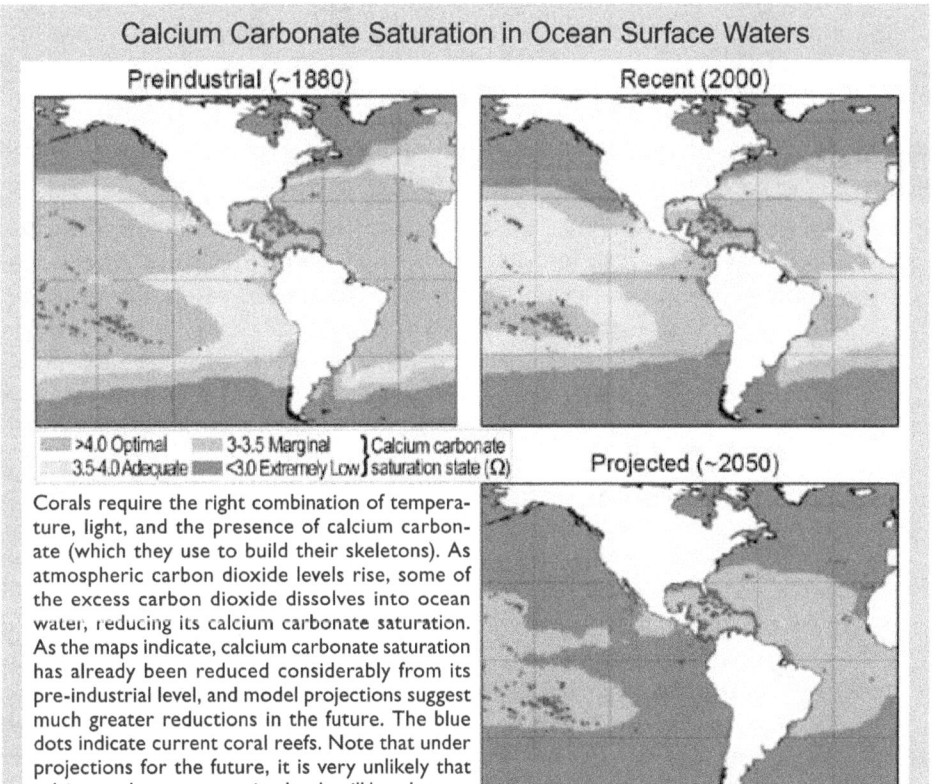

Calcium Carbonate Saturation in Ocean Surface Waters

Preindustrial (~1880)   Recent (2000)

>4.0 Optimal   3-3.5 Marginal ⎫ Calcium carbonate
3.5-4.0 Adequate   <3.0 Extremely Low ⎭ saturation state (Ω)

Projected (~2050)

Corals require the right combination of temperature, light, and the presence of calcium carbonate (which they use to build their skeletons). As atmospheric carbon dioxide levels rise, some of the excess carbon dioxide dissolves into ocean water, reducing its calcium carbonate saturation. As the maps indicate, calcium carbonate saturation has already been reduced considerably from its pre-industrial level, and model projections suggest much greater reductions in the future. The blue dots indicate current coral reefs. Note that under projections for the future, it is very unlikely that calcium carbonate saturation levels will be adequate to support coral reefs in any U.S. waters.[219]

NAST[219]

Pacific Coast "Dead Zone"
2006 to 2007

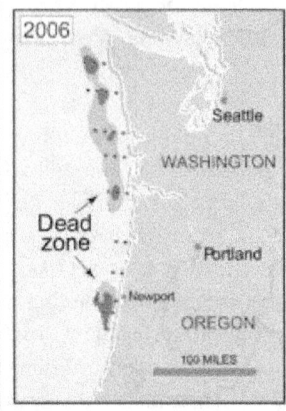

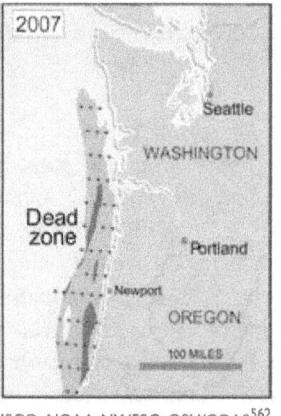

PISCO; NOAA-NWFSC; OSU/COAS[562]

Climate change affects coastal currents that moderate ocean temperatures and the productivity of ecosystems. As such, it is believed to be a factor in the low-oxygen "dead zone" that has appeared along the coast of Washington and Oregon in recent years.[561] In the maps above, blue indicates low-oxygen areas and purple shows areas that are the most severely oxygen depleted.

and are essential to their survival – as a result of heat stress[70] (see *Ecosystems* sector and *Islands* region). As a result of these and other stresses, the corals that form the reefs in the Florida Keys, Puerto Rico, Hawaii, and the Pacific Islands are projected to be lost if carbon dioxide concentrations continue to rise at their current rate.[560]

Changing ocean currents will affect coastal ecosystems.

Because it affects the distribution of heat in the atmosphere and the oceans, climate change will affect winds and currents that move along the nation's coasts, such as the California Current that bathes the West Coast from British Columbia to Baja California.[70] In this area, wind-driven upwelling of deeper ocean water along the coast is vital to moderation of temperatures and the high productivity of Pacific Coast ecosystems. Coastal currents are subject to periodic variations caused by the El Niño-Southern Oscillation and the Pacific Decadal Oscillation, which have substantial effects on the success of salmon and other fishery resources. Climate change is expected to affect such coastal currents, and possibly the larger scale natural oscillations as well, though these effects are not yet well understood. The recent emergence of oxygen-depletion events on the continental shelf off Oregon and Washington (a dead zone not directly caused by agricultural runoff and waste discharges such as those in the Gulf of Mexico or Chesapeake Bay) is one example.[561]

Adaptation: Coping with Sea-Level Rise

Adaptation to sea-level rise is already taking place in three main categories: (1) protecting the coastline by building hard structures such as levees and seawalls (although hard structures can, in some cases, actually increase risks and worsen beach erosion and wetland retreat), (2) accommodating rising water by elevating or redesigning structures, enhancing wetlands, or adding sand from elsewhere to beaches (the latter is not a permanent solution, and can encourage development in vulnerable locations), and (3) planned retreat from the coastline as sea level rises.[269]

Several states have laws or regulations that require setbacks for construction based on the planned life of the development and observed erosion rates.[371] North Carolina, Rhode Island, and South Carolina are using such a moving baseline to guide planning. Maine's Coastal Sand Dune Rules prohibit buildings of a certain size that are unlikely to remain stable with a sea-level rise of 2 feet. The Massachusetts Coastal Hazards Commission is preparing a 20-year infrastructure and protection plan to improve hazards management and the Maryland Commission on Climate Change has recently made comprehensive recommendations to reduce the state's vulnerability to sea-level rise and coastal storms by addressing building codes, public infrastructure, zoning, and emergency preparedness. Governments and private interests are beginning to take sea-level rise into account in planning levees and bridges, and in the siting and design of facilities such as sewage treatment plants (see Adaptation box in *Northeast* region).

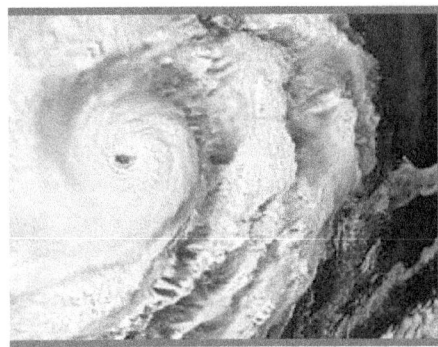

An Agenda for Climate Impacts Science

Both mitigation and adaptation decisions are becoming increasingly necessary. Advancing our knowledge in the many aspects of science that affect the climate system has already contributed greatly to decision making on climate change issues. Further advances in climate science including better understanding and projections regarding rainfall, storm tracks, storm intensity, heat waves, and sea-level rise will improve decision making capabilities.

The focus below, however, is on advancing our knowledge specifically on climate change impacts and those aspects of climate change responsible for these impacts in order to continue to guide decision making.

Recommendation 1:
Expand our understanding of climate change impacts.

There is a clear need to increase understanding of how ecosystems, social and economic systems, human health, and infrastructure will be affected by climate change in the context of other stresses. New understanding will come from a mix of activities including sustained and systematic observations, field and laboratory experiments, model development, and integrated impact assessments. These will incorporate shared learning among researchers, practitioners (such as engineers and water managers), and local stakeholders.

Ecosystems
Ecosystem changes, in response to changes in climate and other environmental conditions, have already been documented. These include changes in the chemistry of the atmosphere and precipitation, vegetation patterns, growing season length, plant productivity, animal species distributions, and the frequency and severity of pest outbreaks and fires. In the marine environment,

changes include the health of corals and other living things due to temperature stress and ocean acidification. These observations not only document climate-change impacts, but also provide critical input to understanding how and why these changes occur, and how changes in ecosystems in turn affect climate. In this way, records of observed changes can improve projections of future impacts related to various climate change scenarios.

In addition to observations, large-scale, whole-ecosystem experiments are essential for improving projections of impacts. Ecosystem-level experiments that vary multiple factors, such as temperature, moisture, ground-level ozone, and atmospheric carbon dioxide, would provide process-level understanding of the ways ecosystems could respond to climate change in the context of other environmental stresses. Such experiments are particularly important for ecosystems with the greatest potential to experience massive change due to the crossing of thresholds or tipping points.

Insights regarding ecosystem responses to climate change gained from both observations and experiments are the essential building blocks of ecosystem simulation models. These models, when rigorously developed and tested, provide powerful tools for exploring the ecosystem consequences of alternative future climates. The incorporation of ecosystem models into an integrated assessment framework that includes socioeconomic, atmospheric and ocean chemistry, and atmosphere-ocean general circulation models should be a major goal of impacts research. This knowledge can provide a base for research studies into ways to manage critical ecosystems in an environment that is continually changing.

Economic systems, human health, and the built environment
As natural systems experience variations due to a changing climate, social and economic systems will

be affected. Food production, water resources, forests, parks, and other managed systems provide life support for society. Their sustainability will depend on how well they can adapt to a future climate that is different from historical experience.

At the same time, climate change is exposing human health and the built environment to increasing risks. Among the likely impacts are an expansion of the ranges of insects and other animals that carry diseases and a greater incidence of health-threatening air pollution events compounded by unusually hot weather associated with climate change. In coastal areas, sea-level rise and storm surge threaten infrastructure including homes, roads, ports, and oil and gas drilling and distribution facilities. In other parts of the country, floods, droughts, and other weather and climate extremes pose increasing threats.

Careful observations along with climate and Earth system models run with a range of emissions scenarios can help society evaluate these risks and plan actions to minimize them. Work in this area would include assessments of the performance of delivery systems, such as those for regional water and electricity supply, so that climate change impacts and costs can be evaluated in terms of changes in risk to system performance. It will be particularly important to understand when the effects on these systems are extremely large and/or rapid, similar to tipping points and thresholds in ecosystems.

In addition, the climate change experienced outside the United States will have implications for our nation. A better understanding of these international linkages, including those related to trade, security, and large-scale movements of people in response to climate change, is desirable.

Recommendation 2:
Refine ability to project climate change, including extreme events, at local scales.

One of the main messages to emerge from the past decade of synthesis and assessments is that while  climate change is a global issue, it has a great deal of regional variability. There is an indisputable need to improve understanding of climate system effects at these smaller scales, because these are often the scales of decision making in society. Understanding impacts at

local scales will also help to target finite resources for adaptation measures. Although much progress has been made in understanding important aspects of this variability, uncertainties remain. Further work is needed on how to quantify cumulative uncertainties across spatial scales and the uncertainties associated with complex, intertwined natural and social systems.

Because region-specific climate changes will occur in the context of other environmental and social changes that are also region-specific, it is important to continue to refine our understanding of regional details, especially those related to precipitation and soil moisture. This would be aided by further testing of models against observations using established metrics designed to evaluate and improve the realism of regional model simulations.

Continued development of improved, higher resolution global climate models, increased computational capacity, extensive climate model experiments, and improved downscaling methods will increase the value of geographically specific climate projections for decision makers in government, business, and the general population.

Extreme weather and climate events are a key component of regional climate. Additional attention needs to be focused on improved observations (made on the relevant time and space scales to capture high-impact extreme events) and associated research and analysis of the potential for future changes in extremes. Impacts analyses indicate that extreme weather and climate events often play a major role in determining climate-change consequences.

Recommendation 3:
Expand capacity to provide decision makers and the public with relevant information on climate change and its impacts.

The United States has tremendous potential to create more comprehensive measurement, archive, and data-access systems and to convey needed information that could provide great benefit to society. There are several aspects to fulfilling this goal: defining what is most relevant, gathering the needed information, expanding the capacity to deliver information, and improving the tools for decision makers to use this information to the best advantage. All of these aspects should involve an interactive and iterative process of continual learning between those who provide information and those who use it. Through such a process, monitoring systems, distribution networks, and tools for using information can all be refined to meet user needs.

For example, tools used by researchers that could also be useful to decision makers include those that analyze and display the probability of occurrence of a range of outcomes to help in assessing risks.

Improved climate monitoring can be efficiently achieved by following the Climate Monitoring Principles recommended by the National Academy of Sciences and the Climate Change Science Strategic Plan in addition to integrating current efforts of governments at all levels. Such a strategy complements a long-term commitment to the measurement of the set of essential climate variables identified by both the Climate Change Science Program and the Global Climate Observing System. Attention must be placed on the variety of time and space scales critical for decision making.

Improved impacts monitoring would include information on the physical and economic effects of extreme events (such as floods and droughts), available, for example, from emergency preparedness and resource management authorities. It would also include regular archiving of information about impacts.

Improved access to data and information archives could substantially enhance society's ability to respond to climate change. While many data related to climate impacts are already freely and readily available to a broad range of users, other data, such as damage costs, are not, and efforts should be made to make them available. Easily accessible information should include a set of agreed-upon baseline indicators and measures of environmental conditions that can be used to track the effects of changes in climate. Services that provide reliable, well-documented, and easily used climate information, and make this information available to support users, are important.

Recommendation 4:
Improve understanding of thresholds likely to lead to abrupt changes in climate or ecosystems.

Paleoclimatic data show that climate can and has changed quite abruptly when certain thresholds are crossed. Similarly, there is evidence that ecological and human systems can undergo abrupt change when tipping points are reached.

Within the climate system there are a number of key risks to society for which understanding is still quite limited. Additional research is needed in some key areas, for example, identifying thresholds that lead to rapid changes in ice sheet dynamics. Sea-level rise is a major concern and improved understanding of the sensitivity of the major ice sheets to sustained warming requires improved observing capability, analysis, and modeling of the ice sheets and their interactions with nearby oceans. Estimates of sea-level rise in previous assessments, such as the recent Intergovernmental Panel on Climate Change 2007 report, did not fully quantify the magnitude and rate of future sea-level rise due to inadequate scientific understanding of potential instabilities of the Greenland and Antarctic ice sheets.

Tipping points in biological systems include the temperature thresholds above which insects survive winter, and can complete two life cycles instead of one in a single growing season, contributing to infestations that kill large numbers of trees. The devastation caused by bark beetles in Canada, and increasingly in the U.S. West, provides an example of how crossing such a threshold can set off massive destruction in an ecosystem with far-reaching consequences.

Similarly, there is increasing concern about the acidification of the world's oceans due to rising atmospheric carbon dioxide levels. There are ocean acidity thresholds beyond which corals and other living things, including some that form the base of important marine food chains, will no longer be able to form the shells and other body structures they need to survive. Improving understanding of such thresholds is an important goal for future research.

Recommendation 5:
Improve understanding of the most effective ways to reduce the rate and magnitude of climate change, as well as unintended consequences of such activities.

This report underscores the importance of reducing the concentrations of heat-trapping gases in the atmosphere. Impacts of climate change during this century and beyond are projected to be far larger and more rapid in scenarios in which greenhouse gas concentrations continue to grow rapidly compared to scenarios in which concentrations grow more slowly. Additional research will help identify the desired mix of mitigation options necessary to control the rate and magnitude of climate change.

In addition to their intended reduction of atmospheric concentrations of greenhouse gases, mitigation options also have the potential for unintended consequences, which should also be examined in future research. For example, the production, transportation, and use of biofuels could lead to increases in water and fertilizer use as well as in some air pollutants. It could also create competition among land uses for food production, biofuels production, and natural ecosystems that provide many benefits to society. Improved understanding of such unintended consequences, and identification of those options that carry the largest negative impacts, can help decision makers make more informed choices regarding the possible trade-offs inherent in various mitigation strategies.

Recommendation 6:
Enhance understanding of how society can adapt to climate change.

There is currently limited knowledge about the ability of communities, regions, and sectors to adapt to future climate change. It is important to improve understanding of how to enhance society's capacity to adapt to a changing climate in the context of other environmental stresses. Interdisciplinary research on adaptation that takes into account the interconnectedness of the Earth system and the complex nature of the social, political, and economic environment in which adaptation decisions must be made would be central to this effort.

The potential exists to provide insights into the possible effectiveness and limits of adaptation options that might be considered in the future. To realize this potential, new research would be helpful to document past responses to climate variability and other environmental changes, analyze the underlying reasons for them, and explain how individual and institutional decisions were made. However, human-induced climate change is projected to be larger and more rapid than any experienced by modern society so there are limits to what can be learned from the past.

A major difficulty in the analysis of adaptation strategies in this report has been the lack of information about the potential costs of adaptation measures, their effectiveness under various scenarios of climate change, the time horizons required for their implementation, and unintended consequences. These types of information should be systematically gathered and shared with decision makers as they consider a range of adaptation options. It is also clear that there is a substantial gap between the available information about climate change and the development of new guidelines for infrastructure such as housing, transportation, water systems, commercial buildings, and energy systems. There are also social and institutional obstacles to appropriate action, even in the face of adequate knowledge. These obstacles need to be better understood so that they can be reduced or eliminated.

Finally, it is important to carry out regular assessments of adaptation measures that address combined scenarios of future climate change, population growth, and economic development paths. This is an important opportunity for shared learning in which researchers, practitioners, and stakeholders collaborate using observations, models, and dialogue to explore adaptation as part of long-term, sustainable development planning.

Concluding Thoughts

Responding to changing conditions

Human-induced climate change is happening now, and impacts are already apparent. Greater impacts are projected, particularly if heat-trapping gas emissions continue unabated. Previous assessments have established these facts, and this report confirms, solidifies, and extends these conclusions for the United States. It reports the latest understanding of how climate change is already affecting important sectors and regions. In particular, it reports that some climate change impacts appear to be increasing faster than previous assessments had suggested. This report represents a significant update to previous work, as it draws from the U.S. Climate Change Science Program's Synthesis and Assessment Products and other recent studies that examine how climate change and its effects are projected to continue to increase over this century and beyond.

Climate choices

Choices about emissions now and in the coming years will have far-reaching consequences for climate change impacts. A consistent finding of this assessment is that the rate and magnitude of future climate change and resulting impacts depend critically on the level of global atmospheric heat-trapping gas concentrations as well as the types and concentrations of atmospheric particles (aerosols). Lower emissions of heat-trapping gases will delay the appearance of climate change impacts and lessen their magnitude. Unless the rate of emissions is substantially reduced, impacts are expected to become increasingly severe for more people and places.

Similarly, there are choices to be made about adaptation strategies that can help to reduce or avoid some of the undesirable impacts of climate change. There is much to learn about the effectiveness of the various types of adaptation responses and how they will interact with each other and with mitigation actions.

Responses to the climate change challenge will almost certainly evolve over time as society learns by doing. Determining and refining societal responses will be an iterative process involving scientists, policymakers, and public and private decision makers at all levels. Implementing these response strategies will require careful planning and continual feedback on the impacts of mitigation and adaptation policies for government, industry, and society.

The value of assessments

Science has revolutionized our ability to observe and model the Earth's climate and living systems, to understand how they are changing, and to project future changes in ways that were not possible in prior generations. These advances have enabled the assessment of climate change, impacts, vulnerabilities, and response strategies. Assessments serve a very important function in providing the scientific underpinnings of informed policy. They can identify advances in the underlying science, provide critical analysis of issues, and highlight key findings and key unknowns that can guide decision making. Regular assessments also serve as progress reports to evaluate and improve policy making and other types of decision making related to climate change.

Impacts and adaptation research includes complex human dimensions, such as economics, management, governance, behavior, and equity. Comprehensive assessments provide an opportunity to evaluate the social implications of climate change within the context of larger questions of how communities and the nation as a whole create sustainable and environmentally sound development paths.

A vision for future U.S. assessments

Over the past decade, U.S. federal agencies have undertaken two coordinated, national-scale efforts to evaluate the impacts of global climate change on this country. Each effort produced a report to the nation – *Climate Change Impacts on the United States*, published in 2000, and this report, *Global Climate Change Impacts in the United States*, published in 2009. A unique feature of the first report was that in addition to reporting the current state of the science, it created a national discourse on climate change that involved hundreds of scientists and thousands of stakeholders including farmers, ranchers, resource managers, city planners, business people, and local and regional government officials. A notable feature of the second report is the incorporation of information from the 21 topic-specific Synthesis and Assessment Products, many motivated by stakeholder interactions.

A vision for future climate change assessments includes both sustained, extensive stakeholder involvement, and targeted, scientifically rigorous reports that address concerns in a timely fashion. The value of stakeholder involvement includes helping scientists understand what information society wants and needs. In addition, the problem-solving abilities of stakeholders will be essential to designing, initiating, and evaluating mitigation and adaptation strategies and their interactions. The best decisions about these strategies will come when there is widespread understanding of the complex issue of climate change – the science and its many implications for our nation.

Federal Advisory Committee Authors

David M. Anderson is the Director for the World Data Center for Paleoclimatology, Chief of the Paleoclimatology Branch of NOAA's National Climatic Data Center, and an Associate Professor at the University of Colorado.

Donald F. Boesch is President of the University of Maryland Center for Environmental Science. His area of expertise is biological oceanography.

Virginia R. Burkett is the Chief Scientist for Global Change Research at the U.S. Geological Survey. Her areas of expertise are coastal ecology, wetland management, and forestry.

Lynne M. Carter is the Director of the Adaptation Network, a non-profit organization, and a project of the Earth Island Institute. Through assessment and action, she works to build resilience in communities and ecosystems in the face of a changing climate.

Stewart J. Cohen is senior researcher with the Adaptation and Impacts Research Division of Environment Canada, and an Adjunct Professor with the Department of Forest Resources Management of the University of British Columbia.

Nancy B. Grimm is a Professor of Life Sciences at Arizona State University. She studies how human-environment interactions and climate variability influence biogeochemical processes in both riverine and urban ecosystems.

Jerry L. Hatfield is the Laboratory Director of the USDA-ARS National Soil Tilth Laboratory in Ames, Iowa. His expertise is in the quantifications of spatial and temporal interactions across the soil-plant-atmosphere continuum.

Katharine Hayhoe is a Research Associate Professor in the Department of Geosciences at Texas Tech University and Principal Scientist and CEO of ATMOS Research & Consulting. Her research examines the potential impacts of human activities on the global environment.

Anthony C. Janetos is the Director of the Joint Global Change Research Institute, a joint venture between the Pacific Northwest National Laboratory and the University of Maryland. His area of expertise is biology.

Thomas R. Karl, (Co-Chair), is the Director of NOAA's National Climatic Data Center. His areas of expertise include monitoring for climate change and changes in extreme climate and weather events. He is also president of the American Meteorological Society.

Jack A. Kaye currently serves as Associate Director for Research of the Earth Science Division within NASA's Science Mission Directorate. He is responsible for NASA's research and data analysis programs in Earth System Science.

Jay H. Lawrimore is Chief of the Climate Analysis Branch at NOAA's National Climatic Data Center. He has led a team of scientists that monitors the Earth's climate on an operational basis.

James J. McCarthy is Alexander Agassiz Professor of Biological Oceanography at Harvard University. His areas of expertise are biology and oceanography. He is also President of the American Association for the Advancement of Science.

A. David McGuire is a Professor of Ecology in the U.S. Geological Survey's Alaska Cooperative Fish and Wildlife Research Unit located at the University of Alaska Fairbanks. His areas of expertise are ecosystem ecology and terrestrial feedbacks to the climate system.

Jerry M. Melillo, (Co-Chair), is the Director of The Ecosystems Center at the Marine Biological Laboratory in Woods Hole. He specializes in understanding the impacts of human activities on the biogeochemistry of ecological systems.

Edward L. Miles is the Virginia and Prentice Bloedel Professor of Marine Studies and Public Affairs at the University of Washington. His fields of specialization are international science and technology policy, marine policy and ocean management, and the impacts of climate variability and change.

Evan Mills is currently a Staff Scientist at the U.S. Department of Energy's Lawrence Berkeley National Laboratory. His areas of expertise are energy systems and risk management in the context of climate change.

Jonathan T. Overpeck is a climate system scientist at the University of Arizona, where he is also the Director of the Institute of the Environment, as well as a Professor of Geosciences and a Professor of Atmospheric Sciences.

Federal Advisory Committee Authors

Jonathan A. Patz is a Professor & Director of Global Environmental Health at the University of Wisconsin in Madison. He has earned medical board certification in both Occupational/Environmental Medicine and Family Medicine.

Thomas C. Peterson, (Co-Chair), is a physical scientist at NOAA's National Climatic Data Center in Asheville, North Carolina. His areas of expertise include data fidelity, international data exchange and global climate analysis using both *in situ* and satellite data.

Roger S. Pulwarty is a physical scientist and the Director of the National Integrated Drought Information System Program at NOAA in Boulder, Colorado. His interests are in climate risk assessment and adaptation.

Benjamin D. Santer is an atmospheric scientist at Lawrence Livermore National Laboratory. His research focuses on climate model evaluation, the use of statistical methods in climate science, and identification of "fingerprints" in observed climate records.

Michael J. Savonis has 25 years of experience in transportation policy, with extensive expertise in air quality and emerging environmental issues. He currently serves as a Senior Policy Advisor at the Federal Highway Administration.

H. Gerry Schwartz, Jr. is an internationally known expert in environmental and civil engineering. He is past-president of both the Water Environment Federation and the American Society of Civil Engineers, a member of the National Academy of Engineering, and a private consultant.

Eileen L. Shea serves as Director of the NOAA Integrated Data and Environmental Applications Center and Chief of the Climate Monitoring and Services Division, National Climatic Data Center, NOAA/NESDIS. Her educational experience focused on marine science, environmental law, and resource management.

John M.R. Stone is an Adjunct Research Professor in the Department of Geography and Environmental Studies at Carleton University. He has spent the last 20 years managing climate research in Canada and helping to influence the dialogue between science and policy.

Bradley H. Udall is the Director of the University of Colorado Western Water Assessment. He was formerly a consulting engineer at Hydrosphere Resource Consultants. His expertise includes water and policy issues of the American West and especially the Colorado River. He is an affiliate of NOAA's Earth System Research Laboratory.

John E. Walsh is a President's Professor of Global Change at the University of Alaska, Fairbanks and Professor Emeritus of Atmospheric Sciences at the University of Illinois. His research interests include the climate of the Arctic, extreme weather events as they relate to climate, and climate-cryosphere interactions.

Michael F. Wehner is a member of the Scientific Computing Group at the Lawrence Berkeley National Laboratory in Berkeley, California. He has been active in both the design of global climate models and in the analysis of their output.

Thomas J. Wilbanks is a Corporate Research Fellow at the Oak Ridge National Laboratory and leads the Laboratory's Global Change and Developing Country Programs. He conducts research on such issues as sustainable development and responses to concerns about climate change.

Donald J. Wuebbles is the Harry E. Preble Professor of Atmospheric Sciences at the University of Illinois. His research emphasizes the study of chemical and physical processes of the atmosphere towards improved understanding of the Earth's climate and atmospheric composition.

PRIMARY SOURCES OF INFORMATION

CCSP Goal 1: Improve knowledge of the Earth's past and present climate and environment, including its natural variability, and improve understanding of the causes of observed variability and change.	

CCSP 1.1 Temperature Trends	*Temperature Trends in the Lower Atmosphere: Steps for Understanding and Reconciling Differences*
	Thomas R. Karl, NOAA; Susan J. Hassol, STG Inc.; Christopher D. Miller, NOAA; William L. Murray, STG Inc.
CCSP 1.2 Past Climate	*Past Climate Variability and Change in the Arctic and at High Latitudes*
	Richard B. Alley, Pennsylvania State Univ.; Julie Brigham-Grette, Univ. of Massachusetts; Gifford H. Miller, Univ. of Colorado; Leonid Polyak, Ohio State Univ.; James W.C. White, Univ. of Colorado; Joan J. Fitzpatrick, USGS
CCSP 1.3 Reanalysis	*Re-Analysis of Historical Climate Data for Key Atmospheric Features: Implications for Attribution of Causes of Observed Change*
	Randall M. Dole, Martin P. Hoerling, Siegfried Schubert, NOAA

CCSP Goal 2: Improve quantification of the forces bringing about changes in the Earth's climate and related systems.	

CCSP 2.1 GHG Emissions	Part A: *Scenarios of Greenhouse Gas Emissions and Atmospheric Concentrations* Part B: *Global-Change Scenarios: Their Development and Use*
	Leon E. Clarke, James A. Edmonds, Hugh M. Pitcher, Pacific Northwest National Lab.; Henry D. Jacoby, MIT; John M. Reilly, MIT; Richard G. Richels, Electric Power Research Institute; Edward A. Parson, Univ. of Michigan; Virginia R. Burkett, USGS; Karen Fisher-Vanden, Dartmouth College; David W. Keith, Univ. of Calgary; Linda O. Mearns, NCAR; Cynthia E. Rosenzweig, NASA; Mort D. Webster, MIT; John C. Houghton DOE/Office of Biological and Environmental Research
CCSP 2.2 Carbon Cycle	*The First State of the Carbon Cycle Report (SOCCR) North American Carbon Budget and Implications for the Global Carbon Cycle*
	Anthony W. King, ORNL; Lisa Dilling, Univ. of Colorado/NCAR; Gregory P. Zimmerman, ORNL; David Fairman, Consensus Building Institute Inc.; Richard A. Houghton, Woods Hole Research Center; Gregg Marland, ORNL; Adam Z. Rose, Pennsylvania State Univ. and Univ. Southern California; Thomas J. Wilbanks, ORNL
CCSP 2.3 Aerosol Impacts	*Atmospheric Aerosol Properties and Climate Impacts*
	Mian Chin, NASA; Ralph A. Kahn, NASA; Stephen E. Schwartz, DOE/BNL; Lorraine A. Remer, NASA/GSFC; Hogbin Yu, NASA/GSFC/UMBC; David Rind, NASA/GISS; Graham Feingold, NOAA/ESRL; Patricia K. Quinn, NOAA/PMEL; David G. Streets, DOE/ANL; Philip DeCola, NASA HQ; Rangasayi Halthore, NASA HQ/NRL
CCSP 2.4 Ozone Trends	*Trends in Emissions of Ozone-Depleting Substances, Ozone Layer Recovery, & Implications for Ultraviolet Radiation Exposure*
	A.R. Ravishankara, NOAA; Michael J. Kurylo, NASA; Christine Ennis, NOAA/ESRL

CCSP Goal 3: *Reduce uncertainty in projections of how the Earth's climate and related systems may change in the future.*	

CCSP 3.1 — Climate Models	*Climate Models: An Assessment of Strengths and Limitations*
	David C. Bader and Curt Covey, Lawrence Livermore National Lab.; William J. Gutowski Jr., Iowa State Univ.; Isaac M. Held, NOAA/GFDL; Kenneth E. Kunkel, Illinois State Water Survey; Ronald L. Miller, NASA/GISS; Robin T. Tokmakian, Naval Postgraduate School; Minghua H. Zhang, State Univ. of New York Stony Brook; Anjuli S. Bamzai, U.S. DOE
CCSP 3.2 — Climate Projections	*Climate Projections Based on Emissions Scenarios for Long-Lived and Short-Lived Radiatively Active Gases and Aerosols*
	Hiram Levy II, NOAA/GFDL; Drew Shindell, NASA/GISS; Alice Gilliland, NOAA /ARL; M. Daniel Schwarzkopf, NOAA/GFDL; Larry W. Horowitz, NOAA/GFDL; Anne M. Waple, STG Inc.
CCSP 3.3 — Extremes	*Weather and Climate Extremes in a Changing Climate:* *Regions of Focus: North America, Hawaii, Caribbean, and U.S. Pacific Islands*
	Thomas R. Karl, NOAA; Gerald A. Meehl, NCAR; Christopher D. Miller, NOAA; Susan J. Hassol, STG Inc.; Anne M. Waple, STG Inc.; William L. Murray, STG Inc.
CCSP 3.4 — Abrupt Climate Change	*Abrupt Climate Change*
	John P. McGeehin, USGS; John A. Barron, USGS; David M. Anderson, NOAA; David J. Verardo, NSF; Peter U. Clark, Oregon State Univ.; Andrew J. Weaver, Univ. of Victoria; Konrad Steffen, Univ. of Colorado; Edward R. Cook, Columbia Univ.; Thomas L. Delworth, NOAA; Edward Brook, Oregon State Univ.

CCSP Goal 4: *Understand the sensitivity and adaptability of different natural and managed ecosystems and human systems to climate and related global changes.*	

CCSP 4.1 — Sea-Level Rise	*Coastal Sensitivity to Sea-Level Rise:* *A Focus on the Mid-Atlantic Region*
	James G. Titus, U.S. EPA; K. Eric Anderson, USGS; Donald R. Cahoon, USGS; Dean B. Gesch, USGS; Stephen K. Gill, NOAA; Benjamin T. Gutierrez, USGS; E. Robert Thieler, USGS; S. Jeffress Williams, USGS
CCSP 4.2 — Ecosystem Thresholds	*Thresholds of Climate Change in Ecosystems*
	Daniel B. Fagre, USGS; Colleen W. Charles, USGS
CCSP 4.3 — Impacts	*The Effects of Climate Change on Agriculture, Land Resources, Water Resources and Biodiversity in the United States*
	Peter Backlund, NCAR; Anthony Janetos, PNNL/Univ. of Maryland; David Schimel, National Ecological Observatory Network; Margaret Walsh, USDA
CCSP 4.4 — Ecosystem Adaptation	*Preliminary Review of Adaptation Options for Climate-Sensitive Ecosystems and Resources*
	Susan Herrod Julius, U.S. EPA; Jordan M. West, U.S. EPA; Jill S. Baron, USGS and Colorado State Univ.; Linda A. Joyce, USDA Forest Service; Brad Griffith, USGS; Peter Kareiva, The Nature Conservancy; Brian D. Keller, NOAA; Margaret Palmer, Univ. of Maryland; Charles Peterson, Univ. of North Carolina; J. Michael Scott, USGS and Univ. of Idaho
CCSP 4.5 — Energy	*Effects of Climate Change on Energy Production and Use in the United States*
	Thomas J. Wilbanks, ORNL; Vatsal Bhatt, Brookhaven National Lab.; Daniel E. Bilello, National Renewable Energy Lab.; Stanley R. Bull, National Renewable Energy Lab.; James Ekmann, National Energy Technology Lab.; William C. Horak, Brookhaven National Lab.; Y. Joe Huang, Mark D. Levine, Lawrence Berkeley National Lab.; Michael J. Sale, ORNL; David K. Schmalzer, Argonne National Lab.; Michael J. Scott, Pacific Northwest National Lab.

CCSP 4.6 Health	*Analyses of the Effects of Global Change on Human Health and Welfare and Human Systems*
	Janet L. Gamble, U.S. EPA; Kristie L. Ebi, ESS LLC.; Anne E. Grambsch, U.S. EPA; Frances G. Sussman, Environmental Economics Consulting; Thomas J. Wilbanks, ORNL
CCSP 4.7 Transportation	*Impacts of Climate Variability and Change on Transportation Systems and Infrastructure -- Gulf Coast Study*
	Michael J. Savonis, Federal Highway Administration; Virginia R. Burkett, USGS; Joanne R. Potter, Cambridge Systematics

CCSP Goal 5: *Explore the uses and identify the limits of evolving knowledge to manage risks and opportunities related to climate variability and change.*

CCSP 5.1 Data Uses & Limitations	*Uses and Limitations of Observations, Data, Forecasts, and Other Projections in Decision Support for Selected Sectors and Regions*
	John Haynes, NASA; Fred Vukovich, SAIC; Molly K. Macauley, RFF; Daewon W. Byun, Univ. of Houston; David Renne, NREL; Gregory Glass, Johns Hopkins School of Public Health; Holly Hartmann, Univ. of Arizona
CCSP 5.2 Uncertainty	*Best Practice Approaches for Characterizing, Communicating and Incorporating Scientific Uncertainty in Climate Decision Making*
	M. Granger Morgan, Dept. of Engineering and Public Policy, Carnegie Mellon Univ.; Hadi Dowlatabadi, Inst. for Resources, Environment and Sustainability, Univ. of British Columbia; Max Henrion, Lumina Decision Systems; David Keith, Dept. of Chemical and Petroleum Engineering and Dept. of Economics, Univ. of Calgary; Robert Lempert, The RAND Corp.; Sandra McBride, Duke Univ.; Mitchell Small, Dept. of Engineering and Public Policy, Carnegie Mellon Univ.; Thomas Wilbanks, Environmental Science Division, ORNL
CCSP 5.3 Decision Support	*Decision Support Experiments and Evaluations using Seasonal-to-Interannual Forecasts and Observational Data: A Focus on Water Resources*
	Nancy Beller-Simms, NOAA; Helen Ingram, Univ. of Arizona; David Feldman, Univ. of California; Nathan Mantua, Climate Impacts Group, Univ. of Washington; Katharine L. Jacobs, Arizona Water Institute; Anne M. Waple, STG Inc.

Other Assessments Referenced

IPCC WG-1	Working Group I - *Climate Change 2007: The Physical Science Basis*
	Susan Solomon, Dahe Qin, Martin Manning, Zhenlin Chen, Melinda Marquis, Kristen B. Averyt, Melina M.B. Tignor, Henry LeRoy Miller, Jr.
IPCC WG-2	Working Group II - *Climate Change 2007: Impacts, Adaptation and Vulnerability*
	Martin L. Parry, Osvalda F. Canziani, Jean P. Palutikof, Paul J. van der Linden, Clair E. Hanson
IPCC WG-3	Working Group III - *Climate Change 2007: Mitigation of Climate Change*
	Bert Metz, Ogunlade R. Davidson, Peter R. Bosch, Rutu Dave, Leo A. Meyer
IPCC Emissions Scenarios	*Special Report on Emissions Scenarios*
	Nebojsa Nakicenovic, Robert Swart

IPCC Water	*Climate Change and Water*
	Bryson Bates, Zbigniew W. Kundzewicz, Shaohong Wu, Jean P. Palutikof

NRC Transportation Impacts	*Potential Impacts of Climate Change on U.S. Transportation*
	Henry G. Schwartz, Jr., Alan C. Clark, G. Edward Dickey, George C. Eads, Robert E. Gallamore, Genevieve Giuliano, William J. Gutowski, Jr., Randell H. Iwasaki, Klaus H. Jacob, Thomas R. Karl, Robert J. Lempert, Luisa M. Paiewonsky, S. George H. Philander, Christopher R. Zeppie

NAST U.S. Impacts	*Climate Change Impacts on the United States: The Potential Consequences of Climate Variability and Change*
	Jerry M. Melillo, Anthony C. Janetos, Thomas R. Karl, Eric J. Barron, Virginia Rose Burkett, Thomas F. Cecich, Robert W. Corell, Katharine L. Jacobs, Linda A. Joyce, Barbara Miller, M. Granger Morgan, Edward A. Parson, Richard G. Richels, David S. Schimel

ACIA Arctic Impacts	*Impacts of a Warming Arctic, Arctic Climate Impact Assessment*
	Robert W. Corell, Susan J. Hassol, Pål Prestrud, Patricia A. Anderson, Snorri Baldursson, Elizabeth Bush, Terry V. Callaghan, Paul Grabhorn, Gordon McBean, Michael MacCracken, Lars-Otto Reiersen, Jan Idar Solbakken, Gunter Weller

ACRONYMS AND ABBREVIATIONS

ARS:	Agricultural Research Service	**NOAA:**	National Oceanic and Atmospheric Administration
CCSP:	Climate Change Science Program		
CIESIN:	Center for International Earth Science Information Network	**NRCS:**	Natural Resources Conservation Service
		NSIDC:	National Snow and Ice Data Center
CIRES:	Cooperative Institute for Research in Environmental Sciences	**NWS:**	National Weather Service
		NWFSC:	Northwest Fisheries Science Center
CMIP:	Coupled Model Intercomparison Project	**PISCO:**	Partnership for Interdisciplinary Studies of Coastal Oceans
DOE:	Department of Energy		
EIA:	Energy Information Administration	**PLJV:**	Playa Lakes Joint Venture
IARC:	International Arctic Research Center	**SAP:**	Synthesis and Assessment Product
IPCC:	Intergovernmental Panel on Climate Change	**SRH:**	Southern Regional Headquarters
NASA:	National Aeronautics and Space Administration	**USACE:**	United States Army Corps of Engineers
NASS:	National Agricultural Statistics Service	**USBR:**	States Bureau of Reclamation
NAST:	National Assessment Synthesis Team	**USDA:**	United States Department of Agriculture
NCDC:	National Climatic Data Center	**U.S. EPA:**	United States Environmental Protection Agency
NESDIS:	National Environmental Satellite, Data, and Information Service	**USFS:**	United States Forest Service
		USGS:	United States Geological Survey

REFERENCES

1 CCSP, 2009: *Best Practice Approaches for Characterizing, Communicating, and Incorporating Scientific Uncertainty in Decisionmaking*. [Morgan, G., H. Dowlatabadi, M. Henrion, D. Keith, R. Lempert, S. McBrid, M. Small, and T. Wilbanks (eds.)]. Synthesis and Assessment Product 5.2. National Oceanic and Atmospheric Administration, Washington DC.

2 Historical data: Lüthi, D., M. Le Floch, B. Bereiter, T. Blunier, J.-M. Barnola, U. Siegenthaler, D. Raynaud, J. Jouzel, H. Fischer, K. Kawamura, and T.F. Stocker, 2008: High-resolution carbon dioxide concentration record 650,000-800,000 years before present. *Nature*, **453(7193)**, 379-382.

   1959-2008 data: Tans, P., 2008: *Trends in Atmospheric Carbon Dioxide: Mauna Loa*. NOAA Earth System Research Laboratory (ESRL). [Web site] <http://www.esrl.noaa.gov/gmd/ccgg/trends/> Data available at <ftp://ftp.cmdl.noaa.gov/ccg/co2/trends/co2_annmean_mlo.txt>

   2100 projected data: International Institute for Applied System Analysis (IIASA) GGI Scenario Database, 2008. <http://www.iiasa.ac.at/Research/GGI/DB/>

3 Forster, P., V. Ramaswamy, P. Artaxo, T. Berntsen, R. Betts, D.W. Fahey, J. Haywood, J. Lean, D.C. Lowe, G. Myhre, J. Nganga, R. Prinn, G. Raga, M. Schulz, and R. Van Dorland, 2007: Changes in atmospheric constituents and in radiative forcing. In: *Climate Change 2007: The Physical Basis*. Contribution of Working Group I to the Fourth Assessment Report of the Intergovernmental Panel on Climate Change [Solomon, S., D. Qin, M. Manning, Z. Chen, M. Marquis, K.B. Averyt, M. Tignor, and H.L. Miller (eds.)]. Cambridge University Press, Cambridge, UK, and New York, pp. 129-234.

4 Denman, K.L., G. Brasseur, A. Chidthaisong, P. Ciais, P.M. Cox, R.E. Dickinson, D. Hauglustaine, C. Heinze, E. Holland, D. Jacob, U. Lohmann, S Ramachandran, P.L. da Silva Dias, S.C. Wofsy, and X. Zhang, 2007: Couplings between changes in the climate system and biogeochemistry. In: *Climate Change 2007: The Physical Science Basis*. Contribution of Working Group I to the Fourth Assessment Report of the Intergovernmental Panel on Climate Change [Solomon, S., D. Qin, M. Manning, Z. Chen, M. Marquis, K.B. Averyt, M. Tignor and H.L. Miller (eds.)]. Cambridge University Press, Cambridge, UK, and New York, pp. 499-587.

5 Ko, M., J.S. Daniel, J.R. Herman, P.A. Newman, and V. Ramaswamy, 2008: The future and recovery. In: *Trends in Emissions of Ozone-Depleting Substances, Ozone Layer Recovery, and Implications for Ultraviolet Radiation Exposure*. [Ravishankara, A.R., M.J. Kurylo, and C.A. Ennis (eds.)]. Synthesis and Assessment Product 2.4. NOAA's National Climatic Data Center, Asheville, NC, pp. 133-154.

6 Ravishankara, A.R., M.J. Kurylo, and A.-M. Schmoltner, 2008: Introduction. In: *Trends in Emissions of Ozone-Depleting Substances, Ozone Layer Recovery, and Implications for Ultraviolet Radiation Exposure*. [Ravishankara, A.R., M.J. Kurylo, and C.A. Ennis (eds.)]. Synthesis and Assessment Product 2.4. NOAA's National Climatic Data Center, Asheville, NC, pp. 23-28.

7 Blasing, T.J., 2008: *Recent Greenhouse Gas Concentrations*. Carbon Dioxide Information Analysis Center, Oak Ridge National Laboratory. <http://cdiac.ornl.gov/pns/current_ghg.html>

8 Fahey, D.W. (lead author), 2007: *Twenty Questions and Answers about the Ozone Layer: 2006 Update*. World Meteorological Organization, Geneva, Switzerland, 50 pp. <http://www.esrl.noaa.gov/csd/assessments/2006/twentyquestions.html>

9 Thompson, D.W.J. and S. Solomon, 2002: Interpretation of recent Southern Hemisphere climate change. *Science*, **296(5569)**, 895-899.

10 Kahn, R.A., H. Yu, S.E. Schwartz, M. Chin, G. Feingold, L.A. Remer, D. Rind, R. Halthore, and P. DeCola, 2009: Introduction. In: *Atmospheric Aerosol Properties and Climate Impacts*. [Chin, M., R.A. Kahn, and S.E. Schwartz (eds.)]. Synthesis and Assessment Product 2.3. National Aeronautics and Space Administration, Washington, DC, pp. 9-20.

11 Solomon, S., G.-K. Plattner, R. Knutti, and P. Friedlingstein, 2009: Irreversible climate change because of carbon dioxide emissions. *Proceedings of the National Academy of Sciences*, **106(6)**, 1704-1709.

12 Archer, D., 2005: Fate of fossil fuel CO_2 in geologic time. *Journal of Geophysical Research*, **110**, C09S05, doi:10.1029/2004JC002625.

13 Shindell, D.T., H. Levy II, A. Gilliland, M.D. Schwarzkopf, and L.W. Horowitz, 2008: Climate change from short-lived emissions due to human activities. In: *Climate Projections Based on Emissions Scenarios for Long-Lived and Short-Lived Radiatively Active Gases and Aerosols*. [Levy II, H., D.T. Shindell, A. Gilliland, M.D. Schwarzkopf, and L.W. Horowitz, (eds.)]. Synthesis and Assessment Product 3.2. U.S. Climate Change Science Program, Washington, DC, pp. 27-60.

14 Santer, B.D., J.E. Penner, and P.W. Thorne, 2006: How well can the observed vertical temperature changes be reconciled with our understanding of the causes of these changes? In: *Temperature Trends in the Lower Atmosphere: Steps for Understanding and Reconciling Differences* [Karl, T.R., S.J. Hassol, C.D. Miller, and W.L. Murray (eds.)]. Synthesis and Assessment Product 1.1 U.S. Climate Change Science Program, Washington, DC, pp. 89-118.

15 Hansen, J., M. Sato, R. Ruedy, L. Nazarenko, A. Lacis, G.A. Schmidt, G. Russell, I. Aleinov, M. Bauer, S. Bauer, N. Bell, B. Cairns, V. Canuto, M. Chandler, Y. Cheng, A. Del Genio, G. Faluveg E. Fleming, A. Friend, T. Hall, C. Jackman, M. Kelley, N. Kiang, D. Koch, J. Lean, J. Lerner, K. Lo, S. Menon, R. Miller, P. Minnis, T. Novakov, V. Oinas, Ja. Perlwitz, Ju. Perlwitz, D. Rind, A. Romanou, D. Shindell, P. Stone, S. Sun, N. Tausnev, D. Thresher, B. Wielicki, T. Wong, M. Yao, and S. Zhang 2005: Efficacy of climate forcings. *Journal of Geophysical Research*, **110**, D18104, doi:10.1029/2005JD005776.

16 National Research Council, 2005: *Radiative Forcing of Climate Change: Expanding the Concept and Addressing Uncertainties*. National Academies Press, Washington DC, 207 pp.

17 Hansen, J., M. Sato, R. Ruedy, A. Lacis, and V. Oinas, 2000: Global warming in the twenty-first century: an alternative scenario. *Proceedings of the National Academy of Sciences*, **97(18)**, 9875-9880.

18 Field, C.B., J. Sarmiento, and B. Hales, 2007: The carbon cycle of North America in a global context. In: *The First State of the Carbon Cycle Report (SOCCR): The North American Carbon Budget and Implications for the Global Carbon Cycle* [King, A.W., L. Dilling, G.P. Zimmerman, D.M. Fairman, R.A. Houghton, G. Marland, A.Z. Rose, and T.J.

Wilbanks (eds.)]. Synthesis and Assessment Product 2.2. NOAA's National Climatic Data Center, Asheville, NC, pp. 21-28.

19 Tarnocai, C., C.-L. Ping, and J. Kimble, 2007: Carbon cycles in the permafrost region of North America. In: *The First State of the Carbon Cycle Report (SOCCR): The North American Carbon Budget and Implications for the Global Carbon Cycle* [King, A.W., L. Dilling, G.P. Zimmerman, D.M. Fairman, R.A. Houghton, G. Marland, A.Z. Rose, and T.J. Wilbanks (eds.)]. Synthesis and Assessment Product 2.2. NOAA's National Climatic Data Center, Asheville, NC, pp. 127-138.

20 Jansen, E., J. Overpeck, K.R. Briffa, J.-C. Duplessy, F. Joos, V. Masson-Delmotte, D. Olago, B. Otto-Bliesner, W.R. Peltier, S. Rahmstorf, R. Ramesh, D. Raynaud, D. Rind, O. Solomina, R. Villalba and D. Zhang, 2007: Palaeoclimate. In: *Climate Change 2007: The Physical Science Basis*. Contribution of Working Group I to the Fourth Assessment Report of the Intergovernmental Panel on Climate Change [Solomon, S., D. Qin, M. Manning, Z. Chen, M. Marquis, K.B. Averyt, M. Tignor, and H.L. Miller (eds.)]. Cambridge University Press, Cambridge, UK, and New York, pp. 433-497.

21 Canadell, J.G., C. Le Quéré, M.R. Raupach, C.B. Field, E.T. Buitenhuis, P. Ciais, T.J. Conway, N.P. Gillett, R.A. Houghton, and G. Marland, 2007: Contributions to accelerating atmospheric CO_2 growth from economic activity, carbon intensity, and efficiency of natural sinks. *Proceedings of the National Academy of Sciences*, **104(47)**, 18866-18870.

22 Royal Society, 2005: *Ocean Acidification Due to Increasing Atmospheric Carbon Dioxide*. Policy Document 12/05. Royal Society, London, 60 pp.

23 Orr, J.C., V.J. Fabry, O. Aumont, L. Bopp, S.C. Doney, R.A. Feely, A. Gnanadesikan, N. Gruber, A. Ishida, F. Joos, R.M. Key, K. Lindsay, E. Maier-Reimer, R. Matear, P. Monfray, A. Mouchet, R.G. Najjar, G.-K. Plattner, K.B. Rodgers, C.L. Sabine, J.L. Sarmiento, R. Schlitzer, R.D. Slater, I.J. Totterdell, M.-F. Weirig, Y. Yamanaka, and A. Yool, 2005: Anthropogenic ocean acidification over the twenty-first century and its impact on calcifying organisms. *Nature*, **437(7059)**, 681-686.

24 Allen, M.R., 2003: Liability for climate change. *Nature*, **421(6926)**, 891-892.

25 Clarke, L., J. Edmonds, H. Jacoby, H. Pitcher, J. Reilly, and R. Richels, 2007: *Scenarios of Greenhouse Gas Emissions and Atmospheric Concentrations*. Sub-report 2.1A of Synthesis and Assessment Product 2.1. U.S. Department of Energy, Office of Biological & Environmental Research, Washington, DC, 154 pp.

26 The spatial average of annual-average surface air temperatures around the globe is commonly referred to as the global average surface air temperature.

27 Meier, M.F., M.B. Dyurgerov, U.K. Rick, S. O'Neel, W.T. Pfeffer, R.S. Anderson, S.P. Anderson, and A.F. Glazovsky, 2007: Glaciers dominate eustatic sea-level rise in the 21st century. *Science*, **317(5841)**, 1064-1067.

28 Trenberth, K.E., P.D. Jones, P. Ambenje, R. Bojariu, D. Easterling, A. Klein Tank, D. Parker, F. Rahimzadeh, J.A. Renwick, M. Rusticucci, B. Soden, and P. Zhai, 2007: Observations: surface and atmospheric climate change. In: *Climate Change 2007: The Physical Basis*. Contribution of Working Group I to the Fourth Assessment Report of the Intergovernmental Panel on Climate Change [Solomon, S., D. Qin, M. Manning, Z. Chen, M. Marquis, K.B. Averyt, M. Tignor, and H.L. Miller (eds.)]. Cambridge University Press, Cambridge, UK, and New York, pp. 235-335.

29 Steffen, K., P.U. Clark, J.G. Cogley, D. Holland, S. Marshall, E. Rignot, and R. Thomas, 2008: Rapid changes in glaciers and ice sheets and their impacts on sea level. In: *Abrupt Climate Change*.

Synthesis and Assessment Product 3.4. U.S. Geological Survey, Reston, VA, pp. 60-142.

30 Lanzante, J.R., T.C. Peterson, F.J. Wentz, and K.Y. Vinnikov, 2006: What do observations indicate about the change of temperatures in the atmosphere and at the surface since the advent of measuring temperatures vertically? In: *Temperature Trends in the Lower Atmosphere: Steps for Understanding and Reconciling Differences* [Karl, T.R., S.J. Hassol, C.D. Miller, and W.L. Murray (eds.)]. Synthesis and Assessment Product 1.1 U.S. Climate Change Science Program, Washington, DC, pp. 47-70.

31 Santer, B.D., P.W. Thorne, L. Haimberger, K.E. Taylor, T.M.L. Wigley, J.R. Lanzante, S. Solomon, M. Free, P.J. Gleckler, P.D. Jones, T.R. Karl, S.A. Klein, C. Mears, D. Nychka, G.A. Schmidt, S.C. Sherwood, and F.J. Wentz, 2008: Consistency of modelled and observed temperature trends in the tropical troposphere. *International Journal of Climatology*, **28(13)**, 1703-1722.

32 Uncertainties in the data are an order of magnitude smaller than the trend according to Karl, T.R., J.R. Christy, R.A. Clarke, G.V. Gruza, J. Jouzel, M.E. Mann, J. Oerlemans, M.J. Salinger, and S.-W. Wang, 2001: Observed climate variability and change. In: *Climate Change 2001: The Scientific Basis*. Contribution of Working Group I to the Third Assessment Report of the Intergovernmental Panel on Climate Change [Houghton, J.T., Y. Ding, D.J. Griggs, M. Noguer, P.J. van der Linden, X. Dai, K. Maskell, and C.A. Johnson (eds.)]. Cambridge University Press, Cambridge, UK, and New York, pp. 99-181.

Temperature data:

Smith, T.M. and R.W. Reynolds, 2004: Improved extended reconstruction of SST (1854–1997). *Journal of Climate*, **17(12)**, 2466-2477.

Jones, P.D., M. New, D.E. Parker, S. Martin, and I.G. Rigor, 1999: Surface air temperature and its changes over the past 150 years. *Reviews of Geophysics*, **37(2)**, 173-199.

Carbon dioxide data:

Data from 1974 to present: Tans, P., 2008: *Trends in Atmospheric Carbon Dioxide: Mauna Loa*. NOAA Earth System Research Laboratory (ESRL). [Web site] <http://www.esrl.noaa.gov/gmd/ccgg/trends/> Data available at <ftp://ftp.cmdl.noaa.gov/ccg/co2/trends/co2_annmean_mlo.txt>

1958-1974 data are from the Scripps Institution of Oceanography (Keeling) Mauna Loa Observatory record. <http://scrippsco2.ucsd.edu/>

Pre-1958 values are annual points taken from a smooth fit to the Law Dome data: Etheridge, D.M., L.P. Steele, R.L. Langenfelds, R.J. Francey, J.-M. Barnola, and V.I. Morgan, 1996: Natural and anthropogenic changes in atmospheric CO_2 over the last 1000 years from air in Antarctic ice and firn. *Journal of Geophysical Research*, **101(D2)**, 4115-4128.

33 Easterling, D. and M. Wehner, 2009: Is the climate warming or cooling? *Geophysical Research Letters*, **36**, L08706, doi:10.1029/2009GL037810.

34 Barnett, T.P., D.W. Pierce, H.G. Hidalgo, C. Bonfils, B.D. Santer, T. Das, G. Bala, A.W. Wood, T. Nozawa, A.A. Mirin, D.R. Cayan, and M.D. Dettinger, 2008: Human-induced changes in the hydrology of the western United States. *Science*, **319(5866)**, 1080-1083.

35 Willett, K.M., N.P. Gillett, P.D. Jones, and P.W. Thorne, 2007: Attribution of observed surface humidity changes to human influence. *Nature*, **449(7163)**, 710-712.

36 Santer, B.D., C. Mears, F.J. Wentz, K.E. Taylor, P.J. Gleckler, T.M.L. Wigley, T.P. Barnett, J.S. Boyle, W. Brüggemann, N.P. Gillett, S.A. Klein, G.A. Meehl, T. Nozawa, D.W. Pierce, P.A. Stott, W.M. Washington, and M.F. Wehner, 2007: Identification of human-induced changes in atmospheric moisture content. *Proceedings of the National Academy of Sciences*, **104(39)**, 15248-15253.

[37] Bindoff, N.L., J. Willebrand, V. Artale, A. Cazenave, J. Gregory, S. Gulev, K. Hanawa, C. Le Quéré, S. Levitus, Y. Nojiri, C.K. Shum, L.D. Talley, and A. Unnikrishnan, 2007: Observations: oceanic climate change and sea level. In: *Climate Change 2007: The Physical Science Basis*. Contribution of Working Group I to the Fourth Assessment Report of the Intergovernmental Panel on Climate Change [Solomon, S., D. Qin, M. Manning, Z. Chen, M. Marquis, K.B. Averyt, M. Tignor, and H.L. Miller (eds.)]. Cambridge University Press, Cambridge, UK, and New York, pp. 385-432.

[38] Barnett, T.P., D.W. Pierce, K.M. AchutaRao, P.J. Gleckler, B.D. Santer, J.M. Gregory, and W.M. Washington, 2005: Penetration of human-induced warming into the world's oceans. *Science*, **309(5732)**, 284-287.

[39] Pierce, D.W., T.P. Barnett, K.M. AchutaRao, P.J. Gleckler, J.M. Gregory, and W.M. Washington, 2006: Anthropogenic warming of the oceans: observations and model results. *Journal of Climate*, **19(10)**, 1873-1900.

[40] Lemke, P., J. Ren, R.B. Alley, I. Allison, J. Carrasco, G. Flato, Y. Fujii, G. Kaser, P. Mote, R.H. Thomas, and T. Zhang, 2007: Observations: changes in snow, ice and frozen ground. In: *Climate Change 2007: The Physical Basis*. Contribution of Working Group I to the Fourth Assessment Report of the Intergovernmental Panel on Climate Change [Solomon, S., D. Qin, M. Manning, Z. Chen, M. Marquis, K.B. Averyt, M. Tignor, and H.L. Miller (eds.)]. Cambridge University Press, Cambridge, UK, and New York, pp. 337-383.

[41] Luthcke, S.B., H.J. Zwally, W. Abdalati, D.D. Rowlands, R.D. Ray, R.S. Nerem, F.G. Lemoine, J.J. McCarthy, and D.S. Chinn, 2006: Recent Greenland ice mass loss by drainage system from satellite gravity observations. *Science*, **314(5803)**, 1286-1289.

[42] Pfeffer, W.T., J.T. Harper, and S. O'Neel, 2008: Kinematic constraints on glacier contributions to 21st-century sea-level rise. *Science*, **321(5894)**, 1340-1343.

[43] Williams, S.J., B.T. Gutierrez, J.G. Titus, S.K. Gill, D.R. Cahoon, E.R. Thieler, K.E. Anderson, D. FitzGerald, V. Burkett, and J. Samenow, 2009: Sea-level rise and its effects on the coast. In: *Coastal Elevations and Sensitivity to Sea-level Rise: A Focus on the Mid-Atlantic Region* [J.G. Titus (coordinating lead author), K.E. Anderson, D.R. Cahoon, D.B. Gesch, S.K. Gill, B.T. Gutierrez, E.R. Thieler, and S.J. Williams (lead authors)]. Synthesis and Assessment Product 4.1. U.S. Environmental Protection Agency, Washington, DC, pp. 11-24.

[44] IPCC (Intergovernmental Panel on Climate Change), 1996: Summary for policy makers. In: *Climate Change 1995: The Science of Climate Change*. Contribution of Working Group I to the Second Assessment Report of the Intergovernmental Panel on Climate Change [Houghton, J.T., L.G. Meiro Filho, B.A. Callander, N. Harris, A. Kattenberg, and K. Maskell (eds.)]. Cambridge University Press, Cambridge, UK, and New York, pp. 1-7.

[45] IPCC (Intergovernmental Panel on Climate Change), 2007: Summary for policymakers. In: *Climate Change 2007: The Physical Science Basis*. Contribution of Working Group I to the Fourth Assessment Report of the Intergovernmental Panel on Climate Change [Solomon, S., D. Qin, M. Manning, Z. Chen, M. Marquis, K.B. Averyt, M. Tignor, and H.L. Miller (eds.)]. Cambridge University Press, Cambridge, UK, and New York, pp. 1-18.

[46] Wigley, T.M.L., V. Ramaswamy, J.R. Christy, J.R. Lanzante, C.A. Mears, B.D. Santer, and C.K. Folland, 2006: Executive summary. In: *Temperature Trends in the Lower Atmosphere: Steps for Understanding and Reconciling Differences* [Karl, T.R., S.J. Hassol, C.D. Miller, and W.L. Murray (eds.)]. Synthesis and Assessment Product 1.1 U.S. Climate Change Science Program, Washington, DC, pp. 1-15.

[47] National Research Council, 2006: *Surface Temperature Reconstructions for the Last 2,000 Years*. National Academies Press, Washington DC, 196 pp.

[48] Mann, M.E., Z. Zhang, M.K. Hughes, R.S. Bradley, S.K. Miller, S. Rutherford, and F. Ni, 2008: Proxy-based reconstructions of hemispheric and global surface temperature variations over the past two millennia. *Proceedings of the National Academy of Sciences*, **105(36)**, 13252-13257.

[49] Hegerl, G.C., F.W. Zwiers, P. Braconnot, N.P. Gillett, Y. Luo, J.A. Marengo Orsini, N. Nicholls, J.E. Penner, and P.A. Stott, 2007: Understanding and attributing climate change. In: *Climate Change 2007: The Physical Science Basis*. Contribution of Working Group I to the Fourth Assessment Report of the Intergovernmental Panel on Climate Change [Solomon, S., D. Qin, M. Manning, Z. Chen, M. Marquis, K.B. Averyt, M. Tignor, and H.L. Miller (eds.)]. Cambridge University Press, Cambridge, UK, and New York, pp. 663-745.

[50] LeTreut, H., R. Somerville, U. Cubasch, Y. Ding, C. Mauritzen, A. Mokssit, T. Peterson, and M. Prather, 2007: Historical overview of climate changes science. In: *Climate Change 2007: The Physical Basis*. Contribution of Working Group I to the Fourth Assessment Report of the Intergovernmental Panel on Climate Change [Solomon, S., D. Qin, M. Manning, Z. Chen, M. Marquis, K.B. Averyt, M. Tignor, and H.L. Miller (eds.)]. Cambridge University Press, Cambridge, UK, and New York, pp. 93-127.

[51] Santer, B.D., T.M.L. Wigley, T.P. Barnett, and E. Anyamba, 1996: Detection of climate change, and attribution of causes. In: *Climate Change 1995: The Science of Climate Change*. Contribution of Working Group I to the Second Assessment Report of the Intergovernmental Panel on Climate Change [Houghton, J.T., L.G. Meiro Filho, B.A. Callander, N. Harris, A. Kattenberg, and K. Maskell (eds.)]. Cambridge University Press, Cambridge, UK, and New York, pp. 407-443.

[52] Mitchell, J.F.B., D.J. Karoly, G.C. Hegerl, F.W. Zwiers, M.R. Allen, and J. Marengo, 2001: Detection of climate change and attribution of causes. In: *Climate Change 2001: The Scientific Basis*. Contribution of Working Group I to the Third Assessment Report of the Intergovernmental Panel on Climate Change [Houghton, J.T., Y. Ding, D.J. Griggs, M. Noguer, P.J. van der Linden, X. Dai, K. Maskell, and C.A. Johnson (eds.)]. Cambridge University Press, Cambridge, UK, and New York, pp. 695-738.

[53] Santer, B.D., M.F. Wehner, T.M.L. Wigley, R. Sausen, G.A. Meehl, K.E. Taylor, C. Ammann, J. Arblaster, W.M. Washington, J.S. Boyle, and W. Brüggemann, 2003: Contributions of anthropogenic and natural forcing to recent tropopause height changes. *Science*, **301(5632)**, 479-483.

[54] Zhang, X., F.W. Zwiers, G.C. Hegerl, F.H. Lambert, N.P. Gillett, S. Solomon, P.A. Stott and T. Nozawa, 2007: Detection of human influence on twentieth-century precipitation trends. *Nature*, **448(7152)**, 461-465.

[55] Burke, E.J., S.J. Brown, and N. Christidis, 2006: Modeling the recent evolution of global drought and projections for the twenty-first century with the Hadley Centre climate model. *Journal of Hydrometeorology*, **7(5)**, 1113-1125.

[56] Gillett, N.P., F.W. Zwiers, A.J. Weaver, and P.A. Stott, 2003: Detection of human influence on sea level pressure. *Nature*, **422(6929)**, 292-294.

[57] Gedney, N., P.M. Cox, R.A. Betts, O. Boucher, C. Huntingford, and P.A. Stott, 2006: Detection of a direct carbon dioxide effect in continental river runoff records. *Nature*, **439(7078)**, 835-838.

[58] Dole, R. and M. Hoerling, 2008: Introduction. In: *Reanalysis of Historical Climate Data for Key Atmospheric Features: Implications for Attribution of Causes of Observed Change*. [Dole, R., M. Hoerling, and S. Schubert (eds.)]. Synthesis and Assessment Prod-

uct 1.3. NOAA's National Climatic Data Center, Asheville, NC, pp. 5-10.

59 The temperature data for the globe is the standard NOAA/NCDC temperature product.

The solar data are a composite of 3 different data sets:

Fröhlich, C. and J. Lean, 2004: Solar radiative output and its variability: evidence and mechanisms. *Astronomy and Astrophysics Review*, **12(4)**, 273-320.

Willson, R.C. and A.V. Mordvinov, 2003: Secular total solar irradiance trend during solar cycles 21023. *Geophysical Research Letters*, **30(5)**, 1199, doi:10.1029/2002GL016038.

Dewitte, S., D. Crommelynck, S. Mekaoui, and A. Jouko, 2004: Measurement and uncertainty of the long-term total solar irradiance trend. *Solar Physics*, **224(1-2)**, 209-216.

60 Lean, J.L. and D.H. Rind, 2008: How natural and anthropogenic influences alter global and regional surface temperatures: 1889 to 2006. *Geophysical Research Letters*, **35**, L18701, doi:10.1029/2008GL034864.

61 Min, S.-K., X. Zhang, F.W. Zwiers, and T. Agnew, 2008: Human influence on Arctic sea ice detectable from early 1990s onwards. *Geophysical Research Letters*, **35**, L21701, doi:10.1029/2008GL035725.

62 Gillett, N.P., D.A. Stone, P.A. Stott, T. Nozawa, A.Y. Karpechko, G.C. Hegerl, M.F. Wehner, and P.D. Jones, 2008: Attribution of polar warming to human influence. *Nature Geoscience*, **1(11)**, 750-754.

63 Ramaswamy, V., J.W. Hurrell, G.A. Meehl, A. Phillips, B.D. Santer, M.D. Schwarzkopf, D.J. Seidel, S.C. Sherwood, and P.W. Thorne, 2006: Why do temperatures vary vertically (from the surface to the stratosphere) and what do we understand about why they might vary and change over time? In: *Temperature Trends in the Lower Atmosphere: Steps for Understanding and Reconciling Differences* [Karl, T.R., S.J. Hassol, C.D. Miller, and W.L. Murray (eds.)]. Synthesis and Assessment Product 1.1. U.S. Climate Change Science Program, Washington, DC, pp. 15-28.

64 Stott, P.A., 2003: Attribution of regional-scale temperature changes to anthropogenic and natural causes. *Geophysical Research Letters*, **30(14)**, 1724, doi:10.1029/2003GL017324.

65 Zwiers, F.W. and X. Zhang, 2003: Towards regional-scale climate change detection. *Journal of Climate*, **16(5)**, 793-797.

66 Santer, B.D., T.M.L. Wigley, P.J. Glecker, C. Bonfils, M.F. Wehner, K. AchutaRao, T.P. Barnett, J.S. Boyle, W. Brüggemann, M. Fiorino, N.P. Gillett, J.E. Hansen, P.D. Jones, S.A. Klein, G.A. Meehl, S.C.B. Raper, R.W. Reynolds, K.E. Taylor, and W.M. Washington, 2006: Forced and unforced ocean temperature changes in Atlantic and Pacific tropical cyclogenesis regions. *Proceedings of the National Academy of Sciences*, **103(38)**, 13905-13910.

67 Gillett, N.P., P.A. Stott, and B.D. Santer, 2008: Attribution of cyclogenesis region sea surface temperature change to anthropogenic influence. *Geophysical Research Letters*, **35**, L09707, doi:10.1029/2008GL033670.

68 Gutowski, W.J., G.C. Hegerl, G.J. Holland, T.R. Knutson, L.O. Mearns, R.J. Stouffer, P.J. Webster, M.F. Wehner, and F.W. Zwiers, 2008: Causes of observed changes in extremes and projections of future changes. In: *Weather and Climate Extremes in a Changing Climate: Regions of Focus: North America, Hawaii, Caribbean, and U.S. Pacific Islands* [Karl, T.R., G.A. Meehl, C.D. Miller, S.J. Hassol, A.M. Waple, and W.L. Murray (eds.)]. Synthesis and Assessment Product 3.3. U.S. Climate Change Science Program, Washington, DC, pp. 81-116.

69 Root, T.L., D.P. MacMynowski, M.D. Mastrandrea, and S.H. Schneider, 2005: Human-modified temperatures induce species changes: joint attribution. *Proceedings of the National Academy of Sciences*, **102(21)**, 7465-7469.

70 Janetos, A., L. Hansen, D. Inouye, B.P. Kelly, L. Meyerson, B. Peterson, and R. Shaw, 2008: Biodiversity. In: *The Effects of Climate Change on Agriculture, Land Resources, Water Resources, and Biodiversity in the United States* [Backlund, P., A. Janetos, D. Schimel, J. Hatfield, K. Boote, P. Fay, L. Hahn, C. Izaurralde, B.A. Kimball, T. Mader, J. Morgan, D. Ort, W. Polley, A. Thomson, D. Wolfe, M.G. Ryan, S.R. Archer, R. Birdsey, C. Dahm, L. Heath, J. Hicke, D. Hollinger, T. Huxman, G. Okin, R. Oren, J. Randerson, W. Schlesinger, D. Lettenmaier, D. Major, L. Poff, S. Running, L. Hansen, D. Inouye, B.P. Kelly, L. Meyerson, B. Peterson, and R. Shaw (eds.)]. Synthesis and Assessment Product 4.3. U.S. Department of Agriculture, Washington, DC, pp. 151-181.

71 CCSP, 2006: *Temperature Trends in the Lower Atmosphere: Steps for Understanding and Reconciling Differences* [Karl, T.R., S.J. Hassol, C.D. Miller, and W.L. Murray (eds.)]. Synthesis and Assessment Product 1.1. U.S. Climate Change Science Program, Washington, DC, 164 pp.

72 Smith, T.M., R.W. Reynolds, T.C. Peterson, and J. Lawrimore, 2008: Improvements to NOAA's historical merged land–ocean surface temperature analysis (1880–2006). *Journal of Climate*, **21(10)**, 2283-2296.

73 Haimberger, L., C. Tavolato, and S. Sperka, 2008: Toward elimination of the warm bias in historic radiosonde temperature records – some new results from a comprehensive intercomparison of upper air data. *Journal of Climate*, **21(18)**, 4587-4606.

74 Sherwood, S.C., C.L. Meyer, R.J. Allen, and H.A. Titchner, 2008: Robust tropospheric warming revealed by iteratively homogenized radiosonde data. *Journal of Climate*, **21(20)**, 5336-5352.

75 Titchner, H.A., P.W. Thorne, M.P. McCarthy, S.F.B. Tett, L. Haimberger, and D.E. Parker, 2008: Critically reassessing tropospheric temperature trends from radiosondes using realistic validation experiments. *Journal of Climate*, **22(3)**, 465-485.

76 Delworth, T.L., P.U. Clark, M. Holland, W.E. Johns, T. Kuhlbrodt, J. Lynch-Stieglitz, C. Morrill, R. Seager, A.J. Weaver, and R. Zhang, 2008: The potential for abrupt change in the Atlantic Meridional Overturning Circulation. In: *Abrupt Climate Change*. Synthesis and Assessment Product 3.4. U.S. Geological Survey, Reston, VA, pp. 258-359.

77 Wentz, F.J., L. Ricciardulli, K. Hilburn, and C. Mears, 2007: How much more rain will global warming bring? *Science*, **317(5835)**, 233-235.

78 Stott, P.A., D.A. Stone, and M.R. Allen, 2004: Human contribution to the European heatwave of 2003. *Nature*, **432(7017)**, 610-614.

79 CCSP, 2008: Introduction. In: *Climate Models: An Assessment of Strengths and Limitations* [Bader, D.C., C. Covey, W.J. Gutowski Jr., I.M. Held, K.E. Kunkel, R.L. Miller, R.T. Tokmakian, and M.H. Zhang (authors)]. Synthesis and Assessment Product 3.1 U.S. Department of Energy, Office of Biological and Environmental Research, Washington, DC, pp. 7-12.

80 Randall, D.A., R.A. Wood, S. Bony, R. Coman, T. Fichefet, J. Fyfe, V. Kattsov, A. Pitman, J. Shukla, J. Srinivasan, R.J. Stouffer, A. Sumi, and K.E. Taylor, 2007: Climate models and their evaluation. In: *Climate Change 2007: The Physical Basis*. Contribution of Working Group I to the Fourth Assessment Report of the Intergovernmental Panel on Climate Change [Solomon, S., D. Qin, M. Manning, Z. Chen, M. Marquis, K.B. Averyt, M. Tignor, and H.L. Miller (eds.)]. Cambridge University Press, Cambridge, UK, and New York, pp. 589-662.

81 Nakićenović, N. and R. Swart (eds.), 2000: *Special Report on Emissions Scenarios*. A special report of Working Group III of the Intergovernmental Panel on Climate Change. Cambridge University Press, Cambridge, UK, and New York, 599 pp. <http://www.ipcc.ch/ipccreports/sres/emission/index.htm>

[82] Raupach, M.R., G. Marland, P. Ciais, C. Le Quéré, J.G. Canadell, G. Klepper, and C.B. Field, 2007: Global and regional drivers of accelerating CO_2 emissions. *Proceedings of the National Academy of Sciences*, **104(24)**, 10288-10293.

[83] O'Neill, B.C. and M. Oppenheimer, 2004: Climate change impacts are sensitive to the concentration stabilization path. *Proceedings of the National Academy of Sciences*, **101(47)**, 16411-16416.

[84] Schneider, S.H. and M.D. Mastrandrea, 2005: Probabilistic assessment of "dangerous" climate change and emissions pathways. *Proceedings of the National Academy of Sciences*, **102(44)**, 15728-15735.

[85] Lenton, T.M., H. Held, E. Kriegler, J.W. Hall, W. Lucht, S. Rahmstorf, and H.J. Schellnhuber, 2008: Tipping elements in the Earth's climate system. *Proceedings of the National Academy of Sciences*, **105(6)**, 1786-1793.

[86] Hansen, J., M. Sato, R. Ruedy, P. Kharecha, A. Lacis, R. Miller, L. Nazarenko, K. Lo, G.A. Schmidt, G. Russell, I. Aleinov, S. Bauer, E. Baum, B. Cairns, V. Canuto, M. Chandler, Y. Cheng, A. Cohen, A. Del Genio, G. Faluvegi, E. Fleming, A. Friend, T. Hall, C. Jackman, J. Jonas, M. Kelley, N.Y. Kiang, D. Koch, G. Labow, J. Lerner, S. Menon, T. Novakov, V. Oinas, Ja. Perlwitz, Ju. Perlwitz, D. Rind, A. Romanou, R. Schmunk, D. Shindell, P. Stone, S. Sun, D. Streets, N. Tausnev, D. Thresher, N. Unger, M. Yao, and S. Zhang, 2007: Dangerous human-made interference with climate: a GISS modelE study. *Atmospheric Chemistry and Physics*, **7(9)**, 2287-2312.

[87] Ramanathan, V. and Y. Feng, 2008: On avoiding dangerous anthropogenic interference with the climate system: formidable challenges ahead. *Proceedings of the National Academy of Sciences*, **105(38)**, 14245-14250.

[88] Meinshausen, M., 2006: What does a 2°C target mean for greenhouse gas concentrations? - A brief analysis based on multi-gas emission pathways and several climate sensitivity uncertainty estimates. In: *Avoiding Dangerous Climate Change* [Schellnhuber, J.S., W. Cramer, N. Nakićenović, T.M.L. Wigley, and G. Yohe (eds.)]. Cambridge University Press, Cambridge, UK, and New York, pp. 265-280.

[89] Meinshausen, M., B. Hare, T.M.L. Wigley, D. van Vuuren, M.G.J. den Elzen, and R. Swart, 2006: Multi-gas emission pathways to meet climate targets. *Climatic Change*, **75(1)**, 151-194.

[90] Meehl, G.A., T.F. Stocker, W.D. Collins, P. Friedlingstein, A.T. Gaye, J.M. Gregory, A. Kitoh, R. Knutti, J.M. Murphy, A. Noda, S.C.B. Raper, I.G. Watterson, A.J. Weaver, and Z.-C. Zhao, 2007: Global climate projections. In: *Climate Change 2007: The Physical Basis*. Contribution of Working Group I to the Fourth Assessment Report of the Intergovernmental Panel on Climate Change [Solomon, S., D. Qin, M. Manning, Z. Chen, M. Marquis, K.B. Averyt, M. Tignor, and H.L. Miller (eds.)]. Cambridge University Press, Cambridge, UK, and New York, pp. 747-845.

[91] Refer to the description of the emissions scenarios in the *Global Climate Change* section on pages 22-25. "Lower emissions scenario" refers to IPCC SRES B1, "higher emissions scenario" refers to A2 and "even higher emissions scenario" refers to A1FI.

[92] IPCC Emissions Scenarios (Even Higher, Higher Emission Scenario, Lower Emission Scenario): Nakićenović, N. and R. Swart (eds.), 2000: Appendix VII: Data tables. In: *Special Report on Emissions Scenarios*. A special report of Working Group III of the Intergovernmental Panel on Climate Change. Cambridge University Press, Cambridge, UK, and New York. <http://www.grida.no/publications/other/ipcc_sr/?src=/climate/ipcc/emission/> Emission trajectories are spline fits as per Raupach, M.R., G. Marland, P. Ciais, C. Le Quéré, J.G. Canadell, G. Klepper, and C.B. Field, 2007: Global and regional drivers of accelerating CO_2 emissions. *Proceedings of the National Academy of Sciences*, **104(24)**, 10288-10293.

Stabilization scenario (450 ppm): CCSP 2.1a Scenario Information 070707 data file. From: Clarke, L., J. Edmonds, H. Jacoby, H. Pitcher, J. Reilly, and R. Richels, 2007: *Scenarios of Greenhouse Gas Emissions and Atmospheric Concentrations*. Sub-report 2.1A of Synthesis and Assessment Product 2.1. U.S. Department of Energy, Office of Biological & Environmental Research, Washington, DC. The emissions and concentrations shown were from MINICAM 1 and 2. See CCSP 2.1A Executive summary for more information. Spread sheet available at <http://www.climatescience.gov/Library/sap/sap2-1/finalreport/default.htm>

Observations of CO_2 emissions (Fossil Fuel CO_2 Emissions graphic) are updates to: Marland, G., B. Andres, T. Boden, 2008: *Global CO_2 Emissions from Fossil-Fuel Burning, Cement Manufacture, and Gas Flaring: 1751-2005*. Carbon Dioxide Information Analysis Center, Oak Ridge National Laboratory, Oak Ridge, TN. <http://cdiac.ornl.gov/ftp/ndp030/global.1751_2005.ems>

Observations of CO_2 concentrations (Atmospheric CO_2 Concentrations graphic): Tans, P., 2008: *Trends in Atmospheric Carbon Dioxide: Mauna Loa*. NOAA Earth System Research Laboratory (ESRL). [Web site] <http://www.esrl.noaa.gov/gmd/ccgg/trends/> Data available at <ftp://ftp.cmdl.noaa.gov/ccg/co2/trends/co2_annmean_mlo.txt>

[93] CMIP3-A: This analysis uses 15 models simulations from the WCRP CMIP3 that were available at resolutions finer than 4 degrees (CCSM3.0, CSIRO, UKMO-HadCM3, IPSL, ECHAM5/MPI, CGCM3.1(T47), GFDL2.0, UKMO-HadGEM1, MIROC3.2(medres), MRI-CGCM2.3.2a, CNRM, GFDL2.1, INM-CM3, ECHO-G, PCM). See Wehner, M., 2005: Changes in daily precipitation and surface air temperature extremes in the IPCC AR4 models. *US CLIVAR Variations*, **3(3)**, 5-9.

Hatching indicates at least two out of three models agree on the sign of the projected change in precipitation.

We acknowledge the modeling groups, the Program for Climate Model Diagnosis and Intercomparison (PCMDI) and the WCRP's Working Group on Coupled Modelling (WGCM) for their roles in making available the WCRP CMIP3 multi-model dataset, <http://www-pcmdi.llnl.gov/projects/cmip/index.php>. Support of this dataset is provided by the Office of Science, U.S. Department of Energy. For an overview and documentation of the CMIP3 modelling activity, see Meehl, G.A., C. Covey, T. Delworth, M. Latif, B. McAvaney, J.F.B. Mitchell, R.J. Stouffer, and K.E. Taylor, 2007: The WCRP CMIP3 multi-model dataset: a new era in climate change research. *Bulletin of the American Meteorological Society*, **88(9)**, 1383-1394.

[94] Hare, B. and M. Meinshausen, 2006: How much warming are we committed to and how much can be avoided? *Climatic Change*, **75(1)**, 111-149.

[95] den Elzen, M.G.J. and M. Meinshausen, 2006: Multi-gas emission pathways for meeting the EU 2°C climate target. In: *Avoiding Dangerous Climate Change* [Schellnhuber, J.S., W. Cramer, N. Nakićenović, T.M.L. Wigley and G. Yohe (eds.)]. Cambridge University Press, Cambridge, UK, and New York, pp. 299-310.

[96] Seidel, D.J., Q. Fu, W.J. Randel, and T.J. Reichler, 2008: Widening of the tropical belt in a changing climate. *Nature Geoscience*, **1(1)**, 21-24.

[97] Cook, E.R., P.J. Bartlein, N. Diffenbaugh, R. Seager, B.N. Shuman, R.S. Webb, J.W. Williams, and C. Woodhouse, 2008: Hydrological variability and change. In: *Abrupt Climate Change*. Synthesis and Assessment Product 3.4. U.S. Geological Survey, Reston, VA, pp. 143-257.

[98] Emanuel, K., 2005: Increasing destructiveness of tropical cyclones over the past 30 years. *Nature*, **436(7051)**, 686-688.

99 Vecchi, G.A., K.L. Swanson, and B.J. Soden, 2008: Whither hurricane activity? *Science*, **322(5902)**, 687-689.

100 Emanuel, K., R. Sundararajan, and J. Williams, 2008: Hurricanes and global warming: Results from downscaling IPCC AR4 simulations. *Bulletin of the American Meteorological Society*, **89(3)**, 347-367.

101 Vecchi, G.A. and B.J. Soden, 2007: Effect of remote sea surface temperature change on tropical cyclone potential intensity. *Nature*, **450(7172)**, 1077-1070.

102 Alley, R.B., P.U. Clark, P. Huybrechts, and I. Joughin, 2005: Ice-sheet and sea-level changes ice-sheet and sea-level changes. *Science*, **310(5747)**, 456-460.

103 Rahmstorf, S., 2007: A semi-empirical approach to projecting future sea-level rise. *Science*, **315(5810)**, 368-370.

104 Mitrovica, J.X., N. Gomez, and P.U. Clark, 2009: The sea-level fingerprint of West Antarctic collapse. *Science*, **323(5915)**, 753.

105 Clark, P.U., A.J. Weaver, E. Brook, E.R. Cook, T.L. Delworth, and K. Steffen, 2008: Introduction: Abrupt changes in the Earth's climate system. In: *Abrupt Climate Change*. Synthesis and Assessment Product 3.4. U.S. Geological Survey, Reston, VA, pp. 19-59.

106 Brook, E., D. Archer, E. Dlugokencky, S. Frolking, and D. Lawrence, 2008: Potential for abrupt changes in atmospheric methane. In: *Abrupt Climate Change*. Synthesis and Assessment Product 3.4. U.S. Geological Survey, Reston, VA, pp. 360-452.

107 Temperatures for the contiguous U.S. are based on data from the U.S. Historical Climatology Network Version 2 (Menne *et al.* 2008). Temperatures for Alaska, Hawaii, and Puerto Rico are based on data from the Cooperative Observers Network adjusted to remove non-climatic influences such as changes in instruments and observer practices and changes in the station environment (Menne and Williams, 2008).

U.S. time series on page 27 is calculated with data for the contiguous US, Alaska, and Hawaii. US map on page 28 lower left includes observed temperature change in Puerto Rico. Winter temperature trend map in the agriculture section, page 76, is for the contiguous US only.

References for this endnote:
Menne, M.J., C.N. Williams, and R.S. Vose, 2009: The United States Historical Climatology Network Monthly Temperature Data - Version 2. *Bulletin of the American Meteorological Society*, Early online release, 25 February 2009, doi:10.1175/2008BAMS2613.1

Menne, M.J. and C.N. Williams Jr., 2008: Homogenization of temperature series via pairwise comparisons. *Journal of Climate*, **22(7)**, 1700-1717.

108 Christensen, J.H., B. Hewitson, A. Busuioc, A. Chen, X. Gao, I. Held, R. Jones, R.K. Kolli, W.-T. Kwon, R. Laprise, V. Magaña Rueda, L. Mearns, C.G. Menéndez, J. Räisänen, A. Rinke, A. Sarr, and P. Whetton, 2007: Regional climate projections. In: *Climate Change 2007: The Physical Basis*. Contribution of Working Group I to the Fourth Assessment Report of the Intergovernmental Panel on Climate Change [Solomon, S., D. Qin, M. Manning, Z. Chen, M. Marquis, K.B. Averyt, M. Tignor, and H.L. Miller (eds.)]. Cambridge University Press, Cambridge, UK, and New York, pp. 847-940.

109 CMIP3-C: Analysis for the contiguous U.S. was based on methods described in: Hayhoe, K., D. Cayan, C.B. Field, P.C. Frumhoff, E.P. Maurer, N.L. Miller, S.C. Moser, S.H. Schneider, K.N. Cahill, E.E. Cleland, L. Dale, R. Drapek, R.M. Hanemann, L.S. Kalkstein, J. Lenihan, C.K. Lunch, R.P. Neilson, S.C. Sheridan, and J.H. Verville, 2004: Emission pathways, climate change, and impacts on California. *Proceedings of the National Academy of Sciences*, **101(34)**, 12422-12427; and Hayhoe, K., C. Wake, B. Anderson, X.-Z. Liang, E. Maurer, J. Zhu, J. Bradbury, A. DeGaetano, A.M. Stoner, and D. Wuebbles, 2008: Regional climate change projections for the Northeast USA. *Mitigation and Adaptation Strategies for Global Change*, **13(5-6)**, 425-436. This analysis uses 16 models simulations from the WCRP CMIP3. Where models had multiple runs, only the first run available from each model was used. See <http://gdo-dcp.ucllnl.org/downscaled_cmip3_projections/dcpInterface.html> for more information.

The Alaskan projections are based on 14 models that best captured the present climate of Alaska; see Walsh, J.E., W.L. Chaman, V. Romanovsky, J.H. Christensen, and M. Stendel, 2008: Global climate model performance over Alaska and Greenland. *Journal of Climate*, **21(23)**, 6156-6174.

Caribbean and Pacific islands analyses use 15 models simulations from the WCRP CMIP3 that were available at resolutions finer than 4 degrees (CCSM3.0, CSIRO, UKMO-HadCM3, IPSL, ECHAM5/MPI, CGCM3.1(T47), GFDL2.0, UKMO-HadGEM1, MIROC3.2(medres), MRI-CGCM2.3.2a, CNRM, GFDL2.1, INM-CM3, ECHO-G, PCM). See Wehner, M., 2005: Changes in daily precipitation and surface air temperature extremes in the IPCC AR4 models. *US CLIVAR Variations*, **3(3)**, 5-9.

We acknowledge the modeling groups, the Program for Climate Model Diagnosis and Intercomparison (PCMDI) and the WCRP's Working Group on Coupled Modelling (WGCM) for their roles in making available the WCRP CMIP3 multi-model dataset, <http://www-pcmdi.llnl.gov/projects/cmip/index.php>. Support of this dataset is provided by the Office of Science, U.S. Department of Energy. For an overview and documentation of the CMIP3 modelling activity, see Meehl, G.A., C. Covey, T. Delworth, M. Latif, B. McAvaney, J.F.B. Mitchell, R.J. Stouffer, and K.E. Taylor, 2007: The WCRP CMIP3 multi-model dataset: a new era in climate change research. *Bulletin of the American Meteorological Society*, **88(9)**, 1383-1394.

110 Detailed local-scale projections about temperature and precipitation changes displayed in this report were generated using well-documented "statistical downscaling" techniques [Wood *et al.*, 2002] for the contiguous U.S. and Alaska. These techniques use statistical relationships between surface observations and climate simulations of the past to develop modifications for the global model results. These modifications are then applied to the climate projections for the future scenarios. The approach is also used to drive daily simulations by a well-established hydrological modeling framework for the contiguous U.S. [Liang *et al.*, 1994]. This method, which modifies global climate model simulations to better account for landscape variations and other features affecting climate at the regional to local scale, has been previously applied to generate high-resolution regional climate projections for the Northeast, Midwest, Northwest, and Southwest [Wood *et al.*, 2004; Hayhoe *et al.*, 2004; Hayhoe *et al.*, 2008; Cayan *et al.*, 2008; Cherkauer *et al.*, 2009]. Comparison of these methods with dynamically downscaled projections generated using regional climate model simulations provide strong justification for the use of such techniques [Wood *et al.*, 2004; Hayhoe *et al.*, 2008].

References for this endnote:
Cayan, D., E. Maurer, M. Dettinger, M. Tyree, and K. Hayhoe, 2008: Climate change scenarios for the California region. *Climatic Change*, **87(Supplement 1)**, S21-S42.

Cherkauer, K. and T. Sinha, 2009: Hydrologic impacts of projected future climate change in the Lake Michigan region. *Journal of Great Lakes Research*, in press.

Hayhoe, K., D. Cayan, C.B. Field, P.C. Frumhoff, E.P. Maurer, N.L. Miller, S.C. Moser, S.H. Schneider, K.N. Cahill, E.E. Cleland, L. Dale, R. Drapek, R.M. Hanemann, L.S. Kalkstein, J. Lenihan, C.K. Lunch, R.P. Neilson, S.C. Sheridan, and J.H. Verville, 2004: Emission pathways, climate change, and impacts on California. *Proceedings of the National Academy of Sciences*, **101(34)**, 12422-12427.

Hayhoe, K., C. Wake, B. Anderson, X.-Z. Liang, E. Maurer, J. Zhu, J. Bradbury, A. DeGaetano, A.M. Stoner, and D. Wuebbles, 2008: Regional climate change projections for the Northeast USA. *Mitigation and Adaptation Strategies for Global Change*, **13(5-6)**, 425-436.

Liang, X., D. Lettenmaier, E. Wood, and S. Burges, 1994: A simple hydrologically-based model of land surface water and energy fluxes for general circulation models. *Journal of Geophysical Research*, **99(D7)**, 14415-14428.

Maurer, E.P., A.W. Wood, J.C. Adam, D.P. Lettenmaier, and B. Nijssen, 2002: A long-term hydrologically-based data set of land surface fluxes and states for the conterminous United States. *Journal of Climate*, **15(22)**, 3237-3251.

Wood, A.W., L.R. Leung, V. Sridhar, and D.P. Lettenmaier, 2004: Hydrologic implications of dynamical and statistical approaches to downscaling climate model outputs. *Climatic Change*, **62(1-3)**, 189-216.

Wood, A.W., E.P. Maurer, A. Kumar, and D.P. Lettenmaier, 2002: Long range experimental hydrologic forecasting for the eastern U.S. *Journal of Geophysical Research*, **107(D20)**, 4429, doi:10.1029/2001JD000659.

[111] NOAA's National Climatic Data Center, 2008: *The USHCN Version 2 Serial Monthly Dataset*. [Web site] <http://www.ncdc.noaa.gov/oa/climate/research/ushcn/>

[112] Kunkel, K.E., P.D. Bromirski, H.E. Brooks, T. Cavazos, A.V. Douglas, D.R. Easterling, K.A. Emanuel, P.Ya. Groisman, G.J. Holland, T.R. Knutson, J.P. Kossin, P.D. Komar, D.H. Levinson, and R.L. Smith, 2008: Observed changes in weather and climate extremes. In: *Weather and Climate Extremes in a Changing Climate: Regions of Focus: North America, Hawaii, Caribbean, and U.S. Pacific Islands* [Karl, T.R., G.A. Meehl, C.D. Miller, S.J. Hassol, A.M. Waple, and W.L. Murray (eds.)]. Synthesis and Assessment Product 3.3. U.S. Climate Change Science Program, Washington, DC, pp. 35-80.

[113] Groisman, P.Ya., R.W. Knight, T.R. Karl, D.R. Easterling, B. Sun, and J.H. Lawrimore, 2004: Contemporary changes of the hydrological cycle over the contiguous United States, trends derived from *in situ* observations. *Journal of Hydrometeorology*, **5(1)**, 64-85. The climate regions are different than those used in this article but the methodology is identical.

[114] Karl, T.R., G.A. Meehl, T.C. Peterson, K.E. Kunkel, W.J. Gutowski Jr., and D.R. Easterling, 2008: Executive summary. In: *Weather and Climate Extremes in a Changing Climate. Regions of Focus: North America, Hawaii, Caribbean, and U.S. Pacific Islands* [Karl, T.R., G.A. Meehl, C.D. Miller, S.J. Hassol, A.M. Waple, and W.L. Murray (eds.)]. Synthesis and Assessment Product 3.3. U.S. Climate Change Science Program, Washington, DC, pp. 1-9.

[115] Seager, R., M. Ting, I. Held, Y. Kushnir, J. Lu, G. Vecchi, H.-P. Huang, N. Harnik, A. Leetmaa, N.-C. Lau, C. Li, J. Velez, and N. Naik, 2007: Model projections of an imminent transition to a more arid climate in southwestern North America. *Science*, **316(5828)**, 1181-1184.

[116] USGS, 2005: *Changes in Streamflow Timing in the Western United States in Recent Decades*. USGS fact sheet 2005-3018. U.S. Geological Survey, National Streamflow Information Program, La Jolla, CA, 4 pp. <http://pubs.usgs.gov/fs/2005/3018/>

[117] CMIP3-B: Analysis for the contiguous U.S. was based on methods described in: Hayhoe, K., D. Cayan, C.B. Field, P.C. Frumhoff, E.P. Maurer, N.L. Miller, S.C. Moser, S.H. Schneider, K.N. Cahill, E.E. Cleland, L. Dale, R. Drapek, R.M. Hanemann, L.S. Kalkstein, J. Lenihan, C.K. Lunch, R.P. Neilson, S.C. Sheridan, and J.H. Verville, 2004: Emission pathways, climate change, and

impacts on California. *Proceedings of the National Academy of Sciences*, **101(34)**, 12422-12427; and Hayhoe, K., C. Wake, B. Anderson, X.-Z. Liang, E. Maurer, J. Zhu, J. Bradbury, A. DeGaetano, A.M. Stoner, and D. Wuebbles, 2008: Regional climate change projections for the Northeast USA. *Mitigation and Adaptation Strategies for Global Change*, **13(5-6)**, 425-436. This analysis uses 16 models simulations from the WCRP CMIP3. Where models had multiple runs, only the first run available from each model was used. See <http://gdo-dcp.ucllnl.org/downscaled_cmip3_projections/dcpInterface.html> for more information.

We acknowledge the modeling groups, the Program for Climate Model Diagnosis and Intercomparison (PCMDI) and the WCRP's Working Group on Coupled Modelling (WGCM) for their roles in making available the WCRP CMIP3 multi-model dataset, <http://www-pcmdi.llnl.gov/projects/cmip/index.php>. Support of this dataset is provided by the Office of Science, U.S. Department of Energy. For an overview and documentation of the CMIP3 modelling activity, see Meehl, G.A., C. Covey, T. Delworth, M. Latif, B. McAvaney, J.F.B. Mitchell, R.J. Stouffer, and K.E. Taylor, 2007: The WCRP CMIP3 multi-model dataset: a new era in climate change research. *Bulletin of the American Meteorological Society*, **88(9)**, 1383-1394.

[118] Swanson, K.L., 2008: Nonlocality of Atlantic tropical cyclone intensities. *Geochemistry, Geophysics, Geosystems*, **9**, Q04V01, doi:10.1029/2007GC001844.

[119] Knutson, T.R., J.J. Sirutis, S.T. Garner, G.A. Vecchi, and I. Held, 2008: Simulated reduction in Atlantic hurricane frequency under twenty-first-century warming conditions. *Nature Geoscience*, **1(6)**, 359-364.

[120] Emanuel, K., 2007: Environmental factors affecting tropical cyclone power dissipation. *Journal of Climate*, **20(22)**, 5497-5509.

[121] Number of strongest hurricanes, number of landfalling strongest hurricanes, and number of landfalling hurricanes are based on data obtained from NOAA's Oceanographic and Meteorological Laboratory: <http://www.aoml.noaa.gov/hrd/hurdat/ushurrlist18512005-gt.txt> with updates. The total number of named storms are adjusted to account for missing tropical storms and hurricanes in the pre-satellite era using the method of Vecchi and Knutson (2008). Basin and landfalling totals are displayed in 5-year increments (pentads) from 1881 through 2010. The final 5-year period was standardized to a comparable 5-year period assuming the level of activity from 2006 to 2008 persists through 2010.

Vecchi, G.A. and T.R. Knutson, 2008: On estimates of historical North Atlantic tropical cyclone activity. *Journal of Climate*, **21(14)**, 3580-3600.

[122] Elsner, J.B., J.P. Kossin, and T.H. Jagger, 2008: The increasing intensity of the strongest tropical cyclones. *Nature*, **455(7209)**, 92-95.

[123] Bell, G.D. and M. Chelliah, 2006: Leading tropical modes associated with interannual and multidecadal fluctuations in North Atlantic hurricane activity. *Journal of Climate*, **19(4)**, 590-612.

[124] Levinson, D.H. and J. Lawrimore (eds.), 2008: State of the climate in 2007. *Bulletin of the American Meteorological Society*, **89(7, Supplement)**, S1-S179.

[125] Kossin, J.P., K.R. Knapp, D.J. Vimont, R.J. Murnane, and B.A. Harper, 2007: A globally consistent reanalysis of hurricane variability and trends. *Geophysical Research Letters*, **34**, L04815, doi:10.1029/2006GL028836.

[126] Rahmstorf, S., A. Cazenave, J.A. Church, J.E. Hansen, R.F. Keeling, D.E. Parker, and R.C.J. Somerville, 2007: Recent climate observations compared to projections. *Science*, **316(5825)**, 709.

[127] Zervas, C., 2001: *Sea Level Variations of the United States 1985-1999*. NOAA technical report NOS CO-OPS 36. National Oceanic and Atmospheric Administration, Silver Spring, MD, 66 pp.

<http://tidesandcurrents.noaa.gov/publications/techrpt36doc.pdf> Trends were calculated for locations that had at least 10 months of data per year and at least 41 years of data during the 51-year period.

[128] Sea-level rise numbers are calculated based on an extrapolation of NOAA tide gauge stations with records exceeding 50 years, as reported in Zervas, C., 2001: *Sea Level Variations of the United States 1985-1999*. NOAA technical report NOS CO-OPS 36. National Oceanic and Atmospheric Administration, Silver Spring, MD, 66 pp. <http://tidesandcurrents.noaa.gov/publications/techrpt36doc.pdf>

[129] Kunkel, K.E., N.E. Westcott, and D.A.R. Knistovich, 2002: Assessment of potential effects of climate changes on heavy lake-effect snowstorms near Lake Erie. *Journal of Great Lakes Research*, **28(4)**, 521-536.

[130] Burnett, A.W., M.E. Kirby, H.T. Mullins, and W.P. Patterson, 2003: Increasing Great Lake-effect snowfall during the twentieth century: a regional response to global warming? *Journal of Climate*, **16(21)**, 3535-3542.

[131] Trapp, R.J., N.S. Diffenbaugh, H.E. Brooks, M.E. Baldwin, E.D. Robinson, and J.S. Pal, 2007: Changes in severe thunderstorm environment frequency during the 21st century caused by anthropogenically enhanced global radiative forcing. *Proceedings of the National Academy of Sciences*, **104(50)**, 19719-19723.

[132] ACIA, 2005: *Arctic Climate Impact Assessment*. Cambridge University Press, Cambridge, UK, and New York, 1042 pp. <http://www.acia.uaf.edu/pages/scientific.html>

[133] Stroeve, J., M.M. Holland, W. Meier, T. Scambos, and M. Serreze, 2007: Arctic sea ice decline: faster than forecast. *Geophysical Research Letters*, **34**, L09501, doi:10.1029/2007GL029703.

[134] L'Heureux, M.L., A. Kumar, G.D. Bell, M.S. Halpert, and R.W. Higgins, 2008: Role of the Pacific-North American (PNA) pattern in the 2007 Arctic sea ice decline. *Geophysical Research Letters*, **35**, L20701, doi:10.1029/2008GL035205.

[135] Johannessen, O.M., 2008: Decreasing Arctic sea ice mirrors increasing CO_2 on decadal time scale. *Atmospheric and Oceanic Science Letters*, **1(1)**, 51-56.

[136] National Snow and Ice Data Center, 2008: *Arctic Sea Ice Down to Second-Lowest Extent; Likely Record-Low Volume*. Press release October 2, 2008. <http://nsidc.org/news/press/20081002_seaice_pressrelease.html>

[137] Polyak, L., J. Andrews, J. Brigham-Grette, D. Darby, A. Dyke, S. Funder, M. Holland, A. Jennings, J. Savelle, M. Serreze, and E. Wolff, 2009: History of sea ice in the Arctic. In: *Past Climate Variability and Change in the Arctic and at High Latitude*. Synthesis and Assessment Product 1.2. U.S. Geological Survey, Reston, VA, pp. 358-420.

[138] Images from *Sea Ice Yearly Minimum 1979-2007*. [Web site] NASA/Goddard Space Flight Center Scientific Visualization Studio. Thanks to Rob Gerston (GSFC) for providing the data. <http://svs.gsfc.nasa.gov/goto?3464>

[139] Fetterer, F., K. Knowles, W. Meier, and M. Savoie, 2002: updated 2008: *Sea Ice Index*. [Web site] National Snow and Ice Data Center, Boulder, CO. <http://nsidc.org/data/seaice_index/>

[140] Pacala, S., R. Birdsey, S. Bridgham, R.T. Conant, K. Davis, B. Hales, R. Houghton, J.C. Jenkins, M. Johnston, G. Marland, and K. Paustian, 2007: The North American carbon budget past and present. In: *The First State of the Carbon Cycle Report (SOCCR): The North American Carbon Budget and Implications for the Global Carbon Cycle* [King, A.W., L. Dilling, G.P. Zimmerman, D.F. Fairman, R.A. Houghton, G. Marland, A.Z. Rose, and T.J. Wilbanks (eds.)]. Synthesis and Assessment Product 2.2. National Oceanic and Atmospheric Administration, National Climatic Data Center, Asheville, NC, pp. 29-36.

[141] Marland, G., R.J. Andres, T.J. Blasing, T.A. Boden, C.T. Broniak, J.S. Gregg, L.M. Losey, and K. Treanton, 2007: Energy, industry, and waste management activities: an introduction to CO_2 emissions from fossil fuels. In: *The First State of the Carbon Cycle Report (SOCCR): The North American Carbon Budget and Implications for the Global Carbon Cycle* [King, A.W., L. Dilling, G.P. Zimmerman, D.M. Fairman, R.A. Houghton, G. Marland, A.Z. Rose, and T.J. Wilbanks (eds.)]. Synthesis and Assessment Product 2.2. National Oceanic and Atmospheric Administration, National Climatic Data Center, Asheville, NC, pp. 57-64.

[142] Bates, B.C., Z.W. Kundzewicz, S. Wu, and J.P. Palutikof (eds.), 2008: *Climate Change and Water*. Technical paper of the Intergovernmental Panel on Climate Change. IPCC Secretariat, Geneva, Switzerland, 210 pp.

[143] Feng, S. and Q. Hu, 2007: Changes in winter snowfall/precipitation ratio in the contiguous United States. *Journal of Geophysical Research*, **112**, D15109, doi:10.1029/2007JD008397.

[144] Guttman, N.B. and R.G. Quayle, 1996: A historical perspective of U.S. climate divisions. *Bulletin of the American Meteorological Society*, **77(2)**, 293-303. Operational practices described in this paper continue.

[145] Groisman, P.Ya., R.W. Knight, D.R. Easterling, T.R. Karl, G.C. Hegerl, and V.N. Razuvaev, 2005: Trends in intense precipitation in the climate record. *Journal of Climate*, **18(9)**, 1326-1350.

[146] Kundzewicz, Z.W., L.J. Mata, N.W. Arnell, P. Döll, P. Kabat, B. Jiménez, K.A. Miller, T. Oki, Z. Sen, and I.A. Shiklomanov, 2007: Freshwater resources and their management. In: *Climate Change 2007: Impacts, Adaptation and Vulnerability*. Contribution of Working Group II to the Fourth Assessment Report of the Intergovernmental Panel on Climate Change [Parry, M.L., O.F. Canziani, J.P. Palutikof, P.J. van der Linden, and C.E. Hanson (eds.)]. Cambridge University Press, Cambridge, UK, and New York, pp. 173-210.

[147] Groisman, P.Ya. and R.W. Knight, 2008: Prolonged dry episodes over the conterminous United States: new tendencies emerging during the last 40 years. *Journal of Climate*, **21(9)**, 1850-1862.

[148] Tebaldi, C., K. Hayhoe, J.M. Arblaster, and G.A. Meehl, 2006: Going to the extremes: an intercomparison of model-simulated historical and future changes in extreme events. *Climatic Change*, **79(3-4)**, 185-211.

[149] Lettenmaier, D., D. Major, L. Poff, and S. Running, 2008: Water resources. In: *The Effects of Climate Change on Agriculture, Land Resources, Water Resources, and Biodiversity in the United States* [Backlund, P., A. Janetos, D. Schimel, J. Hatfield, K. Boote, P. Fay, L. Hahn, C. Izaurralde, B.A. Kimball, T. Mader, J. Morgan, D. Ort, W. Polley, A. Thomson, D. Wolfe, M.G. Ryan, S.R. Archer, R. Birdsey, C. Dahm, L. Heath, J. Hicke, D. Hollinger, T. Huxman, G. Okin, R. Oren, J. Randerson, W. Schlesinger, D. Lettenmaier, D. Major, L. Poff, S. Running, L. Hansen, D. Inouye, B.P. Kelly, L. Meyerson, B. Peterson, and R. Shaw (eds.)]. Synthesis and Assessment Product 4.3. U.S. Department of Agriculture, Washington, DC, pp. 121-150.

[150] Hayhoe, K., C.P. Wake, T.G. Huntington, L. Luo, M.D. Schwartz, J. Sheffield, E. Wood, B. Anderson, J. Bradbury, A. DeGaetano, T.J. Troy, and D. Wolfe, 2007: Past and future changes in climate and hydrological indicators in the U.S. Northeast. *Climate Dynamics*, **28(4)**, 381-407.

[151] Milly, P.C.D., J. Betancourt, M. Falkenmark, R.M. Hirsch, Z.W. Kundzewicz, D.P. Lettenmaier, and R.J. Stouffer, 2008: Stationarity is dead: Whither water management? *Science*, **319(5863)**, 573-574.

[152] Christensen, N.S., A.W. Wood, N. Voisin, D.P. Lettenmaier, and R.N. Palmer, 2004: The effects of climate change on the hydrology

and water resources of the Colorado River basin. *Climatic Change*, **62(1-3)**, 337-363.

[153] Mote, P., A. Hamlet, and E. Salathé, 2008: Has spring snowpack declined in the Washington Cascades? *Hydrology and Earth System Sciences*, **12(1)**, 193-206.

[154] Knowles, N., M.D. Dettinger, and D.R. Cayan, 2006: Trends in snowfall versus rainfall in the western United States. *Journal of Climate*, **19(18)**, 4545-4559.

[155] Huntington T.G., G.A. Hodgkins, B.D. Keim, and R.W. Dudley, 2004: Changes in the proportion of precipitation occurring as snow in New England (1949 to 2000). *Journal of Climate*, **17(13)**, 2626-2636.

[156] Burakowski, E.A., C.P. Wake, B. Braswell, and D.P. Brown, 2008: Trends in wintertime climate in the northeastern United States: 1965–2005. *Journal of Geophysical Research*, **113**, D20114, doi:10.1029/2008JD009870/

[157] Stewart, I.T., D.R. Cayan, and M.D. Dettinger, 2004: Changes in snowmelt runoff timing in western North America under a 'business as usual' climate change scenario. *Climatic Change*, **62(1-3)**, 217-232.

[158] Stewart, I.T., D.R. Cayan, and M.D. Dettinger, 2005: Changes toward earlier streamflow timing across western North America. *Journal of Climate*, **18(8)**, 1136-1155.

[159] Rauscher, S.A., J.S. Pal, N.S. Diffenbaugh, and M.M. Benedetti, 2008: Future changes in snowmelt-driven runoff timing over the western United States. *Geophysical Research Letters*, **35**, L16703, doi:10.1029/2008GL034424.

[160] Pierce, D.W., T.P. Barnett, H.G. Hidalgo, T. Das, C. Bonfils, B.D. Santer, G. Bala, M.D. Dettinger, D.R. Cayan, A. Mirin, A.W. Wood, and T. Nozawa, 2008: Attribution of declining western U.S. snowpack to human effects. *Journal of Climate*, **21(23)**, 6425-6444.

[161] Bonfils, C., B.D. Santer, D.W. Pierce, H.G. Hidalgo, G. Bala, T. Das, T.P. Barnett, D.R. Cayan, C. Doutriaux, A.W. Wood, A. Mirin, and T. Nozawa, 2008: Detection and attribution of temperature changes in the mountainous western United States. *Journal of Climate*, **21(23)**, 6404-6424.

[162] U.S. Environmental Protection Agency, 2008: *National Water Program Strategy: Response to Climate Change*. U.S. Environmental Protection Agency, Washington, DC, 97 pp. <http://www.epa.gov/water/climatechange/>

[163] Ebi, K.L., J. Balbus, P.L. Kinney, E. Lipp, D. Mills, M.S. O'Neill, and M. Wilson, 2008: Effects of global change on human health. In: *Analyses of the Effects of Global Change on Human Health and Welfare and Human Systems* [Gamble, J.L. (ed.), K.L. Ebi, F.G. Sussman, and T.J. Wilbanks (authors)]. Synthesis and Assessment Product 4.6. U.S. Environmental Protection Agency, Washington, DC, pp. 39-87.

[164] Field, C.B., L.D. Mortsch, M. Brklacich, D.L. Forbes, P. Kovacs, J.A. Patz, S.W. Running, and M.J. Scott, 2007: North America. In: *Climate Change 2007: Impacts, Adaptation and Vulnerability*. Contribution of Working Group II to the Fourth Assessment Report of the Intergovernmental Panel on Climate Change [Parry, M.L., O.F. Canziani, J.P. Palutikof, P.J. van der Linden, and C.E. Hanson (eds.)]. Cambridge University Press, Cambridge, UK, and New York, pp. 617-652.

[165] Winter, J., J.W. Harvey, O.L. Franke, and W.M. Alley, 1998: *Ground Water and Surface Water: A Single Resource*. USGS circular 1139. U.S. Geological Survey, Denver, CO, 79 pp. <http://pubs.usgs.gov/circ/circ1139/>

[166] Austin, J.A. and S.M. Colman, 2007: Lake Superior summer water temperatures are increasing more rapidly than regional air temperatures: a positive ice-albedo feedback. *Geophysical Research Letters*, **34**, L06604, doi:10.1029/2006GL029021.

[167] U.S. General Accounting Office, 2003: *Freshwater Supply: States' Views of How Federal Agencies Could Help Them Meet the Challenges of Expected Shortages*. GAO-03-514. General Accounting Office, Washington, DC, 110 pp. <http://www.gao.gov/new.items/d03514.pdf>

[168] U.S. Environmental Protection Agency, 2002: *The Clean Water and Drinking Water Infrastructure Gap Analysis*. EPA-816-R-02-020. U.S. Environmental Protection Agency, Washington, DC, 50 pp. <http://www.epa.gov/safewater/gapreport.pdf>

[169] National Research Council, 2004: *Confronting the Nation's Water Problems: The Role of Research*. National Academies Press, Washington, DC, 310 pp.

[170] Yohe, G.W., R.D. Lasco, Q.K. Ahmad, N.W. Arnell, S.J. Cohen, C. Hope, A.C. Janetos, and R.T. Perez, 2007: Perspectives on climate change and sustainability. In: *Climate Change 2007: Impacts, Adaptation and Vulnerability*. Contribution of Working Group II to the Fourth Assessment Report of the Intergovernmental Panel on Climate Change [Parry, M.L., O.F. Canziani, J.P. Palutikof, P.J. van der Linden, and C.E. Hanson (eds.)]. Cambridge University Press, Cambridge, UK, and New York, pp. 811-841.

[171] U.S. Bureau of Reclamation, 2005: *Water 2025: Preventing Crises and Conflict in the West*. U.S. Bureau of Reclamation, Washington, DC, 32 pp. Updated from USBR <http://www.usbr.gov/uc/crsp/GetSiteInfo>

[172] Wilbanks, T.J., P. Romero Lankao, M. Bao, F. Berkhout, S. Cairncross, J.-P. Ceron, M. Kapshe, R. Muir-Wood, and R. Zapata-Marti, 2007: Industry, settlement and society. In: *Climate Change 2007: Impacts, Adaptation and Vulnerability*. Contribution of Working Group II to the Fourth Assessment Report of the Intergovernmental Panel on Climate Change [Parry, M.L., O.F. Canziani, J.P. Palutikof, P.J. van der Linden, and C.E. Hanson (eds.)]. Cambridge University Press, Cambridge, UK, and New York, pp. 357-390.

[173] Adger, W.N., S. Agrawala, M.M.Q. Mirza, C. Conde, K. O'Brien, J. Pulhin, R. Pulwarty, B. Smit, and K. Takahashi, 2007: Assessment of adaptation practices, options, constraints and capacity. In: *Climate Change 2007: Impacts, Adaptation and Vulnerability*. Contribution of Working Group II to the Fourth Assessment Report of the Intergovernmental Panel on Climate Change [Parry, M.L., O.F. Canziani, J.P. Palutikof, P.J. van der Linden, and C.E. Hanson (eds.)]. Cambridge University Press, Cambridge, UK, and New York, pp. 717-743.

[174] Hartmann, H.C., 2008: Decision support for water resources management. In: *Uses and Limitations of Observations, Data, Forecasts and Other Projections in Decision Support for Selected Sectors and Regions*. Synthesis and Assessment Product 5.1. U.S. Climate Change Science Program, Washington, DC, pp. 45-55.

[175] Ruhl, J.B., 2005: Water wars, eastern style: divvying up the Apalachicola-Chattahoochee-Flint River Basin. *Journal of Contemporary Water Research & Education*, **131**, 47-54.

[176] Leitman, S., 2008: Lessons learned from transboundary management efforts in the Apalachicola-Chattahoochee-Flint Basin, USA. In: *Transboundary Water Resources: A Foundation for Regional Stability in Central Asia* [Moerlins, J.E., M.K. Khankhasayev, S.F. Leitman, and E.J. Makhmudov (eds.)]. Springer, Dordrecht and London, pp. 195-208.

[177] Gobalet, K., P. Schulz, T. Wake, and N. Siefkin, 2004: Archaeological perspectives on Native American fisheries of California, with emphasis on steelhead and salmon. *Transactions of the American Fisheries Society*, **133(4)**, 801-833.

[178] Hammersmark, C., W. Fleenor, and S. Schladow, 2005: Simulation of flood impact and habitat extent for a tidal freshwater marsh restoration. *Ecological Engineering*, **25(2)**, 137-152.

[179] Kondolf, G., P. Angermeier, K. Cummins, T. Dunne, M. Healey, W. Kimmerer, P. Moyle, D. Murphy, D. Patten, S. Railsback, D. Reed, R. Spies, and R. Twiss, 2008: Projecting cumulative benefits of multiple river restoration projects: An example from the Sacramento-San Joaquin River system in California. *Environmental Management*, **42(6)**, 933-945.

[180] McKee, L., N. Ganju, and D. Schoellhamer, 2006: Estimates of suspended sediment entering San Francisco Bay from the Sacramento and San Joaquin Delta, San Francisco Bay, California. *Journal of Hydrology*, **323(1-4)**, 335-352.

[181] Trenham, P., H. Shaffer, and P. Moyle, 1998: Biochemical identification and assessment of population subdivision in morphologically similar native and invading smelt species (*Hypomesus*) in the Sacramento San Joaquin estuary, California. *Transactions of the American Fisheries Society*, **127(3)**, 417-424.

[182] Vengosh, A., J. Gill, M. Davisson, and G. Hudson, 2002: A multi-isotope (B, Sr, O, H, and C) and age dating (^{3}H-^{3}He and ^{14}C) study of groundwater from Salinas Valley, California: hydrochemistry, dynamics, and contamination processes. *Water Resources Research*, **38(1)**, 1008, doi:10.1029/2001WR000517.

[183] Haggerty, G.M., D. Tave, R. Schmidt-Petersen, and J. Stomp, 2008: Raising endangered fish in New Mexico. *Southwest Hydrology*, **7(4)**, 20-21.

[184] National Research Council, 2004: *Endangered and Threatened Fishes in the Klamath River Basin: Causes of Decline and Strategies for Recovery*. National Academies Press, Washington, DC, 398 pp.

[185] National Research Council, 2008: *Hydrology, Ecology, and Fishes of the Klamath River Basin*. National Academies Press, Washington, DC, 272 pp.

[186] National Research Council, 2007: *Colorado River Basin Water Management: Evaluating and Adjusting to Hydroclimatic Variability*. National Academies Press, Washington, DC, 218 pp.

[187] U.S. Bureau of Reclamation, 2007: *Colorado River Interim Guidelines for Lower Basin Shortages and Coordinated Operations for Lake Powell and Lake Mead: Final Environmental Impact Statement*. [U.S. Bureau of Reclamation, Boulder City, NV], 4 volumes. <http://www.usbr.gov/lc/region/programs/strategies/FEIS/index.html>

[188] University of Arizona, Undated: *Native American Water Rights in Arizona*. [Web site] <http://www.library.arizona.edu/about/libraries/govdocs/waterdoc.html>

[189] Cook, E.R., C.A. Woodhouse, C.M. Eakin, D.M. Meko, and D.W. Stahle, 2004: Long-term aridity changes in the western United States. *Science*, **306(5698)**, 1015-1018.

[190] Ingram, H., D. Feldman, N. Mantua, K.L. Jacobs, D. Fort, N. Beller-Simms, and A.M. Waple, 2008: The changing context. In: *Decision-Support Experiments and Evaluations Using Seasonal-to-Interannual Forecasts and Observational Data: A Focus on Water Resources* [Beller-Simms, N., H. Ingram, D. Feldman, N. Mantua, K.L. Jacobs, and A.M. Waple (eds.)]. Synthesis and Assessment Product 5.3. NOAA's National Climatic Data Center, Asheville, NC, pp. 7-28.

[191] Bull, S.R., D.E. Bilello, J. Ekmann, M.J. Sale, and D.K. Schmalzer, 2007: Effects of climate change on energy production and distribution in the United States. In: *Effects of Climate Change on Energy Production and Use in the United States* [Wilbanks, T.J., V. Bhatt, D.E. Bilello, S.R. Bull, J. Ekmann, W.C. Horak, Y.J. Huang, M.D. Levine, M.J. Sale, D.K. Schmalzer, and M.J. Scott (eds.)]. Synthesis and Assessment Product 4.5. U.S. Climate Change Science Program, Washington, DC, pp. 45-80.

[192] Hyman, R.C., J.R. Potter, M.J. Savonis, V.R. Burkett, and J.E. Tump, 2008: Why study climate change impacts on transportation? In: *Impacts of Climate Change and Variability on Transportation Systems and Infrastructure: Gulf Coast Study, Phase I* [Savonis, M.J., V.R. Burkett, and J.R. Potter (eds.)]. Synthesis and Assessment Product 4.7. U.S. Department of Transportation, Washington, DC, pp. 1-1 to 1F-2 [48 pp.]

[193] Hatfield, J., K. Boote, P. Fay, L. Hahn, C. Izaurralde, B.A. Kimball, T. Mader, J. Morgan, D. Ort, W. Polley, A. Thomson, and D. Wolfe, 2008: Agriculture. In: *The Effects of Climate Change on Agriculture, Land Resources, Water Resources, and Biodiversity in the United States* [Backlund, P., A. Janetos, D. Schimel, J. Hatfield, K. Boote, P. Fay, L. Hahn, C. Izaurralde, B.A. Kimball, T. Mader, J. Morgan, D. Ort, W. Polley, A. Thomson, D. Wolfe, M.G. Ryan, S.R. Archer, R. Birdsey, C. Dahm, L. Heath, J. Hicke, D. Hollinger, T. Huxman, G. Okin, R. Oren, J. Randerson, W. Schlesinger, D. Lettenmaier, D. Major, L. Poff, S. Running, L. Hansen, D. Inouye, B.P. Kelly, L. Meyerson, B. Peterson, and R. Shaw (eds.)]. Synthesis and Assessment Product 4.3. U.S. Department of Agriculture, Washington, DC, pp. 21-74.

[194] Feldman, D.L., K.L. Jacobs, G. Garfin, A. Georgakakos, B. Morehouse, P. Restrepo, R. Webb, B. Yarnal, D. Basketfield, H.C. Hartmann, J. Kochendorfer, C. Rosenzweig, M. Sale, B. Udall, and C. Woodhouse, 2008: Making decision-support information useful, useable, and responsive to decision-maker needs In: *Decision-Support Experiments and Evaluations Using Seasonal-to-Interannual Forecasts and Observational Data: A Focus on Water Resources* [Beller-Simms, N., H. Ingram, D. Feldman, N. Mantua, K.L. Jacobs, and A.M. Waple (eds.)]. Synthesis and Assessment Product 5.3. NOAA's National Climatic Data Center, Asheville, NC, pp. 101-140.

[195] McCabe, G.J. and D.M. Wolock, 2007: Warming may create substantial water supply shortages in the Colorado River basin. *Geophysical Research Letters*, **34**, L22708, doi:10.1029/2007GL031764.

[196] Brekke, L., B. Harding, T. Piechota, B. Udall, C. Woodhouse, and D. Yates (eds.), 2007: Appendix U: Climate Technical Work Group Report: Review of science and methods for incorporating climate change information into Bureau of Reclamation's Colorado River Basin planning studies. In: *Colorado River Interim Guidelines for Lower Basin Shortages and Coordinated Operations for Lake Powell and Lake Mead: Final Environmental Impact Statement*. U.S. Bureau of Reclamation, Boulder City, NV, 110 pp. <http://www.usbr.gov/lc/region/programs/strategies/FEIS/index.html>

[197] U.S. Department of Energy, 2006: *Energy Demands on Water Resources. Report to Congress on the Interdependency of Energy and Water*. Sandia National Laboratories, Albuquerque, NM, 80 pp. <http://www.sandia.gov/energy-water/docs/121-RptToCongress-EWwEIAcomments-FINAL.pdf>

[198] California Energy Commission, 2005: *California's Water -- Energy Relationship*. CEC-700-2005-011-SF. California Energy Commission, [Sacramento], 174 pp. <http://www.energy.ca.gov/2005publications/CEC-700-2005-011/CEC-700-2005-011-SF.PDF>

[199] California Energy Commission, 2006: *Refining Estimates of Water-related Energy Use in California*. CEC-500-2006-118. California Energy Commission, [Sacramento]. <http://www.energy.ca.gov/2006publications/CEC-500-2006-118/CEC-500-2006-118.PDF>

[200] U.S. Energy Information Administration, 2008: *Energy in Brief: What are Greenhouse Gases and How Much are Emitted by the United States?* [Web site] Energy Information Administration, Washington, DC. <http://tonto.eia.doe.gov/energy_in_brief/greenhouse_gas.cfm>

[201] Wilbanks, T.J., *et al.*, 2007: Executive summary. In: *Effects of Climate Change on Energy Production and Use in the United States* [Wilbanks, T.J., V. Bhatt, D.E. Bilello, S.R. Bull, J. Ekmann, W.C. Horak, Y.J. Huang, M.D. Levine, M.J. Sale, D.K. Schmalzer,

and M.J. Scott (eds.)]. Synthesis and Assessment Product 4.5. U.S. Climate Change Science Program, Washington, DC, pp. x-xii.

202 From *Inventory of U.S. Greenhouse Gas Emissions and Sinks: 1990-2003*, U.S. EPA. Allocations from "Electricity & Heat" and "Industry" to end uses are WRI estimates based on energy use data from the International Energy Agency (IEA, 2005). All data is for 2003. All calculations are based on CO_2 equivalents, using 100-year global warming potentials from the IPCC (1996).

203 U.S. Energy Information Administration, 2008: *Annual Energy Review 2007*. U.S. Department of Energy, Washington, DC, 400 pp. <http://www.eia.doe.gov/emeu/aer/pdf/aer.pdf>

204 Bhatt, V., J. Eckmann, W.C. Horak, and T.J. Wilbanks, 2007: Possible indirect effects on energy production and distribution in the United States. In: *Effects of Climate Change on Energy Production and Use in the United States* [Wilbanks, T.J., V. Bhatt, D.E. Bilello, S.R. Bull, J. Ekmann, W.C. Horak, Y.J. Huang, M.D. Levine, M.J. Sale, D.K. Schmalzer, and M.J. Scott (eds.)]. Synthesis and Assessment Product 4.5. U.S. Climate Change Science Program, Washington, DC, pp. 81-97.

205 Scott, M.J. and Y.J. Huang, 2007: Effects of climate change on energy use in the United States. In: *Effects of Climate Change on Energy Production and Use in the United States* [Wilbanks, T.J., V. Bhatt, D.E. Bilello, S.R. Bull, J. Ekmann, W.C. Horak, Y.J. Huang, M.D. Levine, M.J. Sale, D.K. Schmalzer, and M.J. Scott (eds.)]. Synthesis and Assessment Product 4.5. U.S. Climate Change Science Program, Washington, DC, pp. 8-44.

206 U.S. Department of Energy, 2008: *Buildings Energy Data Book*. U.S. Department of Energy, Office of Energy Efficiency and Renewable Energy, [Washington, DC]. <http://buildingsdatabook.eere.energy.gov/>

207 U.S. Census Bureau, 2002: *Population of States and Counties of the United States: 1790-2000*. U.S. Census Bureau, Washington, DC, 226 pp. <http://www.census.gov/population/www/censusdata/hiscendata.html> See <http://www.census.gov/popest/counties/> for 2008 estimates.

208 Feldman, D.L., K.L. Jacobs, G. Garfin, A. Georgakakos, B. Morehouse, R. Webb, B. Yarnal, J. Kochendorfer, C. Rosenzweig, M. Sale, B. Udall, and C. Woodhouse, 2008: Decision-support experiments within the water resource management sector. In: *Decision-Support Experiments and Evaluations Using Seasonal-to-Interannual Forecasts and Observational Data: A Focus on Water Resources* [Beller-Simms, N., H. Ingram, D. Feldman, N. Mantua, K.L. Jacobs, and A.M. Waple (eds.)]. Synthesis and Assessment Product 5.3. NOAA's National Climatic Data Center, Asheville, NC, pp. 65-100.

209 Hightower, M. and S.A. Pierce, 2008: The energy challenge. *Nature*, **452(7185)**, 285-286.

210 Wilbanks, T.J., *et al.*, 2007: Conclusions and research priorities. In: *Effects of Climate Change on Energy Production and Use in the United States* [Wilbanks, T.J., V. Bhatt, D.E. Bilello, S.R. Bull, J. Ekmann, W.C. Horak, Y.J. Huang, M.D. Levine, M.J. Sale, D.K. Schmalzer, and M.J. Scott (eds.)]. Synthesis and Assessment Product 4.5. U.S. Climate Change Science Program, Washington, DC, pp. 98-108.

211 Maulbetsch, J.S. and M.N. DiFilippo, 2006: *Cost and Value of Water Use at Combined Cycle Power Plants*. CEC-500-2006-034. California Energy Commission, PIER Energy-Related Environmental Research. <http://www.energy.ca.gov/2006publications/CEC-500-2006-034/CEC-500-2006-034.PDF>

212 U.S. Energy Information Administration, 2002: *National Trends in Coal Transportation*. [Web site] Energy Information Administration, Washington, DC. <http://www.eia.doe.gov/cneaf/coal/ctrdb/natltrends.html>

213 NOAA's National Climatic Data Center, 2008: *Climate of 2008: Midwestern U.S. Flood Overview*. [Web site] <http://www.ncdc.noaa.gov/oa/climate/research/2008/flood08.html>

214 CBO Testimony, 2005: *Macroeconomic and Budgetary Effects of Hurricanes Katrina and Rita*. Statement of Douglas Holtz-Easkin, Director, before the Committee on the Budget, U.S. House of Representatives. Congressional Budget Office, Washington, DC, 21 pp. <http://www.cbo.gov/doc.cfm?index=6684>

215 Potter, J.R., V.R. Burkett, and M.J. Savonis, 2008: Executive summary. In: *Impacts of Climate Change and Variability on Transportation Systems and Infrastructure: Gulf Coast Study, Phase I* [Savonis, M. J., V.R. Burkett, and J.R. Potter (eds.)]. Synthesis and Assessment Product 4.7. U.S. Department of Transportation, Washington, DC, pp. ES-1 to ES-10.

216 U.S. Energy Information Administration, U.S. Department of Energy [data]; assembled by Evan Mills, Lawrence Berkeley National Laboratory. Data available at <http://www.eia.doe.gov/cneaf/electricity/page/disturb_events.html>

217 Kafalenos, R.S., K.J. Leonard, D.M. Beagan, V.R. Burkett, B.D. Keim, A. Meyers, D.T. Hunt, R.C. Hyman, M.K. Maynard, B. Fritsche, R.H. Henk, E.J. Seymour, L.E. Olson, J.R. Potter, and M.J. Savonis, 2008: What are the implications of climate change and variability for Gulf coast transportation? In: *Impacts of Climate Change and Variability on Transportation Systems and Infrastructure: Gulf Coast Study, Phase I* [Savonis, M.J., V.R. Burkett, and J.R. Potter (eds.)]. Synthesis and Assessment Product 4.7. U.S. Department of Transportation, Washington, DC, pp. 4-1 to 4F-27 [104 pp.]

218 U.S. Energy Information Administration, 2008: *Monthly Energy Review*. U.S. Department of Energy, Energy Information Administration, Washington, DC. <http://www.eia.doe.gov/emeu/mer/contents.html>

219 National Assessment Synthesis Team (NAST), 2001: *Climate Change Impacts on the United States: The Potential Consequences of Climate Variability and Change*. Cambridge University Press, Cambridge, UK, and New York, 612 pp. <http://www.usgcrp.gov/usgcrp/Library/nationalassessment/>

220 ACIA, 2004: *Impacts of a Warming Arctic: Arctic Climate Impact Assessment*. Cambridge University Press, Cambridge, UK, and New York, 139 pp. <http://www.acia.uaf.edu>

221 U.S. Environmental Protection Agency, 2008: *Inventory of U.S. Greenhouse Gas Emissions and Sinks: 1990-2006*. USEPA 430-R-08-005. U.S. Environmental Protection Agency, Washington, DC, 473 pp. <http://www.epa.gov/climatechange/emissions/usinventoryreport.html>

222 National Research Council, 2008: *Potential Impacts of Climate Change on U.S. Transportation*. Transportation Research Board special report 290. Transportation Research Board, Washington, DC, 280 pp. <http://onlinepubs.trb.org/onlinepubs/sr/sr290.pdf>

223 Schimel, D., A. Janetos, P. Backlund, J. Hatfield, M.G. Ryan, S.R. Archer, and D. Lettenmaier, 2008: Synthesis. In: *The Effects of Climate Change on Agriculture, Land Resources, Water Resources, and Biodiversity in the United States* [Backlund, P., A. Janetos, D. Schimel, J. Hatfield, K. Boote, P. Fay, L. Hahn, C. Izaurralde, B.A. Kimball, T. Mader, J. Morgan, D. Ort, W. Polley, A. Thomson, D. Wolfe, M.G. Ryan, S.R. Archer, R. Birdsey, C. Dahm, L. Heath, J. Hicke, D. Hollinger, T. Huxman, G. Okin, R. Oren, J. Randerson, W. Schlesinger, D. Lettenmaier, D. Major, L. Poff, S. Running, L. Hansen, D. Inouye, B.P. Kelly, L. Meyerson, B. Peterson, and R. Shaw (eds.)]. Synthesis and Assessment Product 4.3. U.S. Department of Agriculture, Washington, DC, pp. 183-193.

224 Burkett, V.R., R.C. Hyman, R. Hagelman, S.B. Hartley, M. Sheppard, T.W. Doyle, D.M. Beagan, A. Meyers, D.T. Hunt, M.K. Maynard, R.H. Henk, E.J. Seymour, L.E. Olson, J.R. Potter, and

N.N. Srinivasan, 2008: Why study the Gulf Coast? In: *Impacts of Climate Change and Variability on Transportation Systems and Infrastructure: Gulf Coast Study, Phase I* [Savonis, M.J., V.R. Burkett, and J.R. Potter (eds.)]. Synthesis and Assessment Product 4.7. U.S. Department of Transportation, Washington, DC, pp. 2-1 to 2F-26. [66 pp.]

225 Peterson, T.C., M. McGuirk, T.G. Houston, A.H. Horvitz, and M.F. Wehner, 2008: *Climate Variability and Change with Implications for Transportation.* National Research Council, Washington, DC, 90 pp. <http://onlinepubs.trb.org/onlinepubs/sr/sr290Many.pdf>

226 Hay, J.E., R. Warrick, C. Cheatham, T. Manarangi-Trott, J. Konno, and P. Hartley, 2005: *Climate Proofing: A Risk-based Approach to Adaptation.* Asian Development Bank, Manila, The Philippines, 191 pp. <http://www.adb.org/Documents/Reports/Climate-Proofing/default.asp>

227 OSHA, 2008: Heat stress. In: *OSHA Technical Manual, Section III: Chapter 4.* Occupational Safety & Health Administration, Washington, DC. <http://www.osha.gov/dts/osta/otm/otm_iii/otm_iii_4.html>

228 Landsea, C.W., 1993: A climatology of intense (or major) Atlantic hurricanes. *Monthly Weather Review,* 121(6), 1710-1713.

229 NOAA's National Climatic Data Center, 2008: *Billion Dollar U.S. Weather Disasters* [Web site] <http://www.ncdc.noaa.gov/oa/reports/billionz.html>

230 Larsen, P.H., S. Goldsmith, O. Smith, M.L. Wilson, K. Strzepek, P. Chinowsky, and B. Saylor, 2008: Estimating future costs for Alaska public infrastructure at risk from climate change. *Global Environmental Change,* 18(3), 442-457.

231 U.S. Environmental Protection Agency, 2008: *Inventory of U.S. Greenhouse Gas Emissions and Sinks: 1990-2006.* USEPA 430-R-08-005. U.S. Environmental Protection Agency, Washington, DC, 473 pp. <http://www.epa.gov/climatechange/emissions/usinventoryreport.html>

232 National Agricultural Statistics Service, 2002: *2002 Census of Agriculture.* USDA National Agricultural Statistics Service, Washington, DC. <http://www.agcensus.usda.gov/Publications/2002/index.asp>

233 Wolfe, W., L. Ziska, C. Petzoldt, A. Seaman, L. Chase, and K. Hayhoe, 2007: Projected change in climate thresholds in the northeastern U.S.: implications for crops, pests, livestock, and farmers. *Mitigation and Adaptation Strategies for Global Change,* 13(5-6), 555-575.

234 Frumhoff, P.C., J.J. McCarthy, J.M. Melillo, S.C. Moser, and D.J. Wuebbles, 2007: *Confronting Climate Change in the U.S. Northeast: Science, Impacts and Solutions.* Synthesis report of the Northeast Climate Impacts Assessment. Union of Concerned Scientists, Cambridge, MA, 146 pp.

235 Gu, L., P.J. Hanson, W.M. Post, D.P. Kaiser, B. Yang, R. Nemani, S.G. Pallardy, and T. Meyers, 2008: The 2007 eastern U.S. spring freeze: increased cold damage in a warming world? *BioScience,* 58(3), 253-262.

236 Inouye, D.W., 2008: Effects of climate change on phenology, frost damage, and floral abundance of montane wildflowers. *Ecology,* 89(2), 353-362.

237 Either hydrogen dioxide or oxygenated hydrocarbons.

238 Peet, M.M. and D.W. Wolf, 2000: Crop ecosystem responses to climate change: vegetable crops. In: *Climate Change and Global Crop Productivity* [Reddy, K.R. and H.F. Hodges (eds.)]. CABI Publishing, New York, and Wallingford, UK, 472 pp.

239 Bridges, D.C. (ed.), 1992: *Crop Losses Due to Weeds in the United States.* Weed Science Society of America, Champaign, IL, 403 pp.

240 Joyce, L.A., G.M. Blate, J.S. Littell, S.G. McNulty, C.I. Millar, S.C. Moser, R.P. Neilson, K. O'Halloran, and D.L. Peterson, 2008: National forests. In: *Preliminary Review of Adaptation Options for Climate-sensitive Ecosystems and Resources.* [Julius, S.H., J.M. West (eds.), J.S. Baron, B. Griffith, L.A. Joyce, P. Kareiva, B.D. Keller, M.A. Palmer, C.H. Peterson, and J.M. Scott (authors)]. Synthesis and Assessment Product 4.4. U.S. Environmental Protection Agency, Washington, DC, pp. 3-1 to 3-127.

241 Kiely, T., D. Donaldson, and A. Grube, 2004: *Pesticides Industry Sales and Usage: 2000 and 2001 Market Estimates.* U.S. Environmental Protection Agency, Washington, DC, 33 pp. <http://www.epa.gov/oppbead1/pestsales/>

242 Natural Resources Conservation Service, 1997: *1997 Five-Year Natural Resources Inventory.* USDA Natural Resources Conservation Service, Washington, DC. <http://www.nrcs.usda.gov/technical/NRI/1997/index.html>

243 Ryan, M.G., S.R. Archer, R. Birdsey, C. Dahm, L. Heath, J. Hicke, D. Hollinger, T. Huxman, G. Okin, R. Oren, J. Randerson, and W. Schlesinger, 2008: Land resources. In: *The Effects of Climate Change on Agriculture, Land Resources, Water Resources, and Biodiversity in the United States* [Backlund, P., A. Janetos, D. Schimel, J. Hatfield, K. Boote, P. Fay, L. Hahn, C. Izaurralde, B.A. Kimball, T. Mader, J. Morgan, D. Ort, W. Polley, A. Thomson, D. Wolfe, M.G. Ryan, S.R. Archer, R. Birdsey, C. Dahm, L. Heath, J. Hicke, D. Hollinger, T. Huxman, G. Okin, R. Oren, J. Randerson, W. Schlesinger, D. Lettenmaier, D. Major, L. Poff, S. Running, L. Hansen, D. Inouye, B.P. Kelly, L. Meyerson, B. Peterson, and R. Shaw (eds.)]. Synthesis and Assessment Product 4.3. U.S. Department of Agriculture, Washington, DC, pp. 75-120.

244 Parmesan, C., 1996: Climate and species range. *Nature,* 382(6594), 765-766.

245 Hinzman, L.D., N.D. Bettez, W.R. Bolton, F.S. Chapin, M.B. Dyurgerov, C.L. Fastie, B. Griffith, R.D. Hollister, A. Hope, H.P. Huntington, A.M. Jensen, G.J. Jia, T. Jorgenson, D.L. Kane, D.R. Klein, G. Kofinas, A.H. Lynch, A.H. Lloyd, A.D. McGuire, F.E. Nelson, M. Nolan, W.C. Oechel, T.E. Osterkamp, C.H. Racine, V.E. Romanovsky, R.S. Stone, D.A. Stow, M. Sturm, C.E. Tweedie, G.L. Vourlitis, M.D. Walker, D.A. Walker, P.J. Webber, J.M. Welker, K.S. Winker, and K. Yoshikawa, 2005: Evidence and implications of recent climate change in northern Alaska and other Arctic regions. *Climatic Change,* 72(3), 251-298.

246 Mimura, N., L. Nurse, R.F. McLean, J. Agard, L. Briguglio, P. Lefale, R. Payet, and G. Sem, 2007: Small islands. In: *Climate Change 2007: Impacts, Adaptation and Vulnerability.* Contribution of Working Group II to the Fourth Assessment Report of the Intergovernmental Panel on Climate Change [Parry, M.L., O.F. Canziani, J.P. Palutikof, P.J. van der Linden, and C.E. Hanson (eds.)]. Cambridge University Press, Cambridge, UK, and New York, pp. 687-716.

247 Millennium Ecosystem Assessment, 2005: *Ecosystems and Human Well-being: Biodiversity Synthesis.* World Resources Institute, Washington, DC, 86 pp. <http://www.millenniumassessment.org>

248 Fischlin, A., G.F. Midgley, J.T. Price, R. Leemans, B. Gopal, C. Turley, M.D.A. Rounsevell, O.P. Dube, J. Tarazona, and A.A. Velichko, 2007: Ecosystems, their properties, goods, and services. In: *Climate Change 2007: Impacts, Adaptation and Vulnerability.* Contribution of Working Group II to the Fourth Assessment Report of the Intergovernmental Panel on Climate Change [Parry, M.L., O.F. Canziani, J.P. Palutikof, P.J. van der Linden and C.E. Hanson (eds.)]. Cambridge University Press, Cambridge, UK, and New York, pp. 211-272.

249 National Interagency Fire Center, [2008]: *Total Wildland Fires and Acres (1960-2007).* National Interagency Coordination Center, Boise, ID. Data at <http://www.nifc.gov/fire_info/fire_stats.htm>

250 CCSP, 2009: Case studies. In: *Thresholds of Climate Change in Ecosystems* [Fagre, D.B., C.W. Charles, C.D. Allen, C. Birkeland, F.S. Chapin III, P.M. Groffman, G.R. Guntenspergen, A.K. Knapp,

A.D. McGuire, P.J. Mulholland, D.P.C. Peters, D.D. Roby, and G. Sugihara] Synthesis and Assessment Product 4.2. U.S. Geological Survey, Reston, VA, pp. 32-73.

[251] Pounds, J.A., M.R. Bustamante, L.A. Coloma, J.A. Consuegra, M.P. L. Fogden, P.N. Foster, E. La Marca, K.L. Masters, A. Merino-Viteri, R. Puschendorf, S.R. Ron, G.A. Sánchez-Azofeifa, C.J. Still, and B.E. Young, 2006: Widespread amphibian extinctions from epidemic disease driven by global warming. *Nature*, **439(7073)**, 161-167.

[252] Peterson, T.C., D.M. Anderson, S.J. Cohen, M. Cortez-Vázquez, R.J. Murnane, C. Parmesan, D. Phillips, R.S. Pulwarty, and J.M.R. Stone, 2008: Why weather and climate extremes matter. In: *Weather and Climate Extremes in a Changing Climate. Regions of Focus: North America, Hawaii, Caribbean, and U.S. Pacific Islands* [Karl, T.R., G.A. Meehl, C.D. Miller, S.J. Hassol, A.M. Waple, and W.L. Murray (eds.)]. Synthesis and Assessment Product 3.3. U.S. Climate Change Science Program, Washington, DC, pp. 11-34.

[253] Backlund, P., D. Schimel, A. Janetos, J. Hatfield, M.G. Ryan, S.R. Archer, and D. Lettenmaier, 2008: Introduction. In: *The Effects of Climate Change on Agriculture, Land Resources, Water Resources, and Biodiversity in the United States* [Backlund, P., A. Janetos, D. Schimel, J. Hatfield, K. Boote, P. Fay, L. Hahn, C. Izaurralde, B.A. Kimball, T. Mader, J. Morgan, D. Ort, W. Polley, A. Thomson, D. Wolfe, M.G. Ryan, S.R. Archer, R. Birdsey, C. Dahm, L. Heath, J. Hicke, D. Hollinger, T. Huxman, G. Okin, R. Oren, J. Randerson, W. Schlesinger, D. Lettenmaier, D. Major, L. Poff, S. Running, L. Hansen, D. Inouye, B.P. Kelly, L. Meyerson, B. Peterson, and R. Shaw (eds.)]. Synthesis and Assessment Product 4.3. U.S. Department of Agriculture, Washington, DC, pp. 11-20.

[254] Burkett, V.R., D.A. Wilcox, R. Stottlemeyer, W. Barrow, D. Fagre, J. Baron, J. Price, J.L. Nielsen, C.D. Allen, D.L. Peterson, G. Ruggerone, and T. Doyle, 2005: Nonlinear dynamics in ecosystem response to climatic change: case studies and policy implications. *Ecological Complexity*, **2(4)**, 357-394.

[255] Anderson, K.E., D.R. Cahoon, S.K. Gill, B.T. Gutierrez, E.R. Thieler, J.G. Titus, and S.J. Williams, 2009: Executive summary. In: *Coastal Elevations and Sensitivity to Sea-level Rise: A Focus on the Mid-Atlantic Region* [J.G. Titus (coordinating lead author), K.E. Anderson, D.R. Cahoon, D.B. Gesch, S.K. Gill, B.T. Gutierrez, E.R. Thieler, and S.J. Williams (lead authors)]. Synthesis and Assessment Product 4.1. U.S. Environmental Protection Agency, Washington, DC, pp. 1-8.

[256] Park, R.A., M.S. Trehan, P.W. Mausel, and R.C. Howe, 1989: *The Effects of Sea Level Rise on U.S. Coastal Wetlands*. U.S. Environmental Protection Agency, Washington, DC, 55 pp.

[257] Gutierrez, B.T., S.J. Williams, and E.R Thieler, 2009: Ocean coasts. In: *Coastal Elevations and Sensitivity to Sea-level Rise: A Focus on the Mid-Atlantic Region* [J.G. Titus (coordinating lead author), K.E. Anderson, D.R. Cahoon, D.B. Gesch, S.K. Gill, B.T. Gutierrez, E.R. Thieler, and S.J. Williams (lead authors)]. Synthesis and Assessment Product 4.1. U.S. Environmental Protection Agency, Washington, DC, pp. 43-56.

[258] *Monaco Declaration*, 2009: Developed at the *Second International Symposium on the Ocean in a High-CO₂ World*, Monaco, 6-9 October 2008. <http://ioc3.unesco.org/oanet/Symposium2008/MonacoDeclaration.pdf>

[259] Feely, R.A., C.L. Sabine, J.M. Hernandez-Ayon, D. Ianson, and B. Hales, 2008: Evidence for upwelling of corrosive "acidified" water onto the continental shelf. *Science*, **320(5882)**, 1490-1492.

[260] Beckage, B., B. Osborne, D.G. Gavin, C. Pucko, T. Siccama, and T. Perkins, 2008: A rapid upward shift of a forest ecotone during 40 years of warming in the Green Mountains of Vermont. *Proceedings of the National Academy of Sciences*, **105(11)**, 4197-4202.

[261] Beever, E.A., P.F. Brussard, and J. Berger, 2003: Patterns of apparent extirpation among isolated populations of pikas (*Ochotona princeps*) in the Great Basin. *Journal of Mammology*, **84(1)**, 37-54.

[262] Krajick, K., 2004: All downhill from here? *Science*, **303(5664)**, 1600-1602.

[263] Battin, J., M.W. Wiley, M.H. Ruckelshaus, R.N. Palmer, E. Korb, K.K. Bartz, and H. Imaki, 2007: Projected impacts of climate change on salmon habitat restoration. *Proceedings of the National Academy of Sciences*, **104(16)**, 6720-6725.

[264] Williams, J.E., A.L. Haak, N.G. Gillespie, H.M. Neville, and W.T. Colyer, 2007: *Healing Troubled Waters: Preparing Trout and Salmon Habitat for a Changing Climate*. Trout Unlimited, Arlington, VA, 12 pp. <http://www.tu.org/climatechange>

[265] Daily, G.C., T. Soderqvist, S. Aniyar, K. Arrow, P. Dasgupta, P.R. Ehrlich, C. Folke, A. Jansson, B.O. Jansson, N. Kautsky, S. Levin, J. Lubchenco, K.G. Maler, D. Simpson, D. Starrett, D. Tilman, and B. Walker, 2000: Ecology - The value of nature and the nature of value. *Science*, **289(5478)**, 395-396.

[266] Hamilton, J.M. and R.S.J. Tol, 2004: The impact of climate change on tourism and recreation. In: *Human-Induced Climate Change - An Interdisciplinary Assessment* [Schlesinger, M., H.S. Kheshgi, J. Smith, F.C. de la Chesnaye, J.M. Reilly, T. Wilson, and C. Kolstad (eds.)]. Cambridge University Press, Cambridge (UK), pp. 147-155.

[267] Cordell, H.K., B. McDonald, R.J. Teasley, J.C. Bergstrom, J. Martin, J. Bason, and V.R. Leeworthy, 1999: Outdoor recreation participation trends. In: *Outdoor Recreation in American Life: A National Assessment of Demand and Supply Trends* [Cordell, H.K. and S.M. McKinney (eds.)]. Sagamore Publishing, Champaign, IL, pp. 219-321.

[268] Wall, G., 1998: Implications of global climate change for tourism and recreation in wetland areas. *Climatic Change*, **40(2)**, 371-389.

[269] Titus, J.G. and M. Craghan, 2009: Shore protection and retreat. In: *Coastal Elevations and Sensitivity to Sea-level Rise: A Focus on the Mid-Atlantic Region* [J.G. Titus (coordinating lead author), K.E. Anderson, D.R. Cahoon, D.B. Gesch, S.K. Gill, B.T. Gutierrez, E.R. Thieler, and S.J. Williams (lead authors)]. Synthesis and Assessment Product 4.1. U.S. Environmental Protection Agency, Washington, DC, pp. 87-104.

[270] Peterson, C.H., R.T. Barber, K.L. Cottingham, H.K. Lotze, C.A. Simenstad, R.R. Christian, M.F. Piehler, and J. Wilson, 2008: National estuaries. In: *Preliminary Review of Adaptation Options for Climate-sensitive Ecosystems and Resources*. [Julius, S.H., J.M. West (eds.), J.S. Baron, B. Griffith, L.A. Joyce, P. Kareiva, B.D. Keller, M.A. Palmer, C.H. Peterson, and J.M. Scott (authors)]. Synthesis and Assessment Product 4.4. U.S. Environmental Protection Agency, Washington, DC, pp. 7-1 to 7-108.

[271] Titus, J.G. and J.E. Neumann, 2009: Implications for decisions. In: *Coastal Elevations and Sensitivity to Sea-level Rise: A Focus on the Mid-Atlantic Region* [J.G. Titus (coordinating lead author), K.E. Anderson, D.R. Cahoon, D.B. Gesch, S.K. Gill, B.T. Gutierrez, E.R. Thieler, and S.J. Williams (lead authors)]. Synthesis and Assessment Product 4.1. U.S. Environmental Protection Agency, Washington, DC, pp. 141-156.

[272] Confalonieri, U., B. Menne, R. Akhtar, K.L. Ebi, M. Hauengue, R.S. Kovats, B. Revich, and A. Woodward, 2007: Human health. In: *Climate Change 2007: Impacts, Adaptation and Vulnerability*. Contribution of Working Group II to the Fourth Assessment Report of the Intergovernmental Panel on Climate Change [Parry, M.L., O.F. Canziani, J.P. Palutikof, P.J. van der Linden, and C.E. Hanson (eds.)]. Cambridge University Press, Cambridge, UK, and New York, pp. 391-431.

273 Borden, K.A. and S.L. Cutter, 2008: Spatial patterns of natural hazards mortality in the United States. *International Journal of Health Geographics*, **7:64**, doi:10.1186/1476-072X-7-64.

274 Gamble, J.L., K.L. Ebi, A. Grambsch, F.G. Sussman, T.J. Wilbanks, C.E. Reid, K. Hayhoe, J.V. Thomas, and C.P. Weaver, 2008: Introduction. In: *Analyses of the Effects of Global Change on Human Health and Welfare and Human Systems* [Gamble, J.L. (ed.), K.L. Ebi, F.G. Sussman, and T.J. Wilbanks (authors)]. Synthesis and Assessment Product 4.6. U.S. Environmental Protection Agency, Washington, DC, pp. 13-37.

275 Brikowski, T.H., Y. Lotan, and M.S. Pearle, 2008: Climate-related increase in the prevalence of urolithiasis in the United States. *Proceedings of the National Academy of Sciences*, **105(28)**, 9841-9846.

276 Semenza, J.C., 1999: Acute renal failure during heat waves. *American Journal of Preventive Medicine*, **17(1)**, 97.

277 Luber, G.E. and L.M Conklin, 2006: Heat-related deaths: United States, 1999–2003. *Morbidity and Mortality Weekly Report*, **55(29)**, 796-798.

278 Zanobetti, A. and J. Schwartz, 2008: Temperature and mortality in nine US cities. *Epidemiology*, **19(4)**, 563-570.

279 Davis, R.E., P.C. Knappenberger, P.J. Michaels, and W.M. Novicoff, 2003: Changing heat-related mortality in the United States. *Environmental Health Perspectives*, **111(14)**, 1712-1718.

280 U.S. Energy Information Administration, 2005: Table HC10.6: Air conditioning characteristics by U.S. census region, 2005. In: *2005 Residential Energy Consumption Survey*. Data available at <http://www.eia.doe.gov/emeu/recs/recs2005/hc2005_tables/detailed_tables2005.html>

281 U.S. Energy Information Administration, 2005: Table 2: Type of air-conditioning equipment by census region and survey year. In: *2005 Residential Energy Consumption Survey*. Data available at <http://www.eia.doe.gov/emeu/consumptionbriefs/recs/actrends/recs_ac_trends_table2.html>

282 Sheridan, S.C., A.J. Kalkstein, and L.S. Kalkstein, 2008: Trends in heat-related mortality in the United States, 1975–2004. *Natural Hazards*, **50(1)**, 145-160.

283 Hayhoe, K., K. Cherkauer, N. Schlegal, J. VanDorn, S. Vavrus, and D. Wuebbles, 2009: Regional climate change projections for Chicago and the Great Lakes. *Journal of Great Lakes Research*, in press.

284 Hayhoe, K., D. Cayan, C.B. Field, P.C. Frumhoff, E.P. Maurer, N.L. Miller, S.C. Moser, S.H. Schneider, K.N. Cahill, E.E. Cleland, L. Dale, R. Drapek, R.M. Hanemann, L.S. Kalkstein, J. Lenihan, C.K. Lunch, R.P. Neilson, S.C. Sheridan, and J.H. Verville, 2004: Emissions pathways, climate change, and impacts on California. *Proceedings of the National Academy of Sciences*, **101(34)**, 12422-12427.

285 Vavrus, S. and J. van Dorn, 2008: Projected future temrperature and precipitation extremes in Chicago. *Journal of Great Lakes Research*, in press.

286 Lemmen, D.S. and F.J. Warren (eds.), 2004: *Climate Change Impacts and Adaptation: A Canadian Perspective*. Climate Change Impacts and Adaptation Program, Natural Resources Canada, Ottawa, ON, 174 pp. <http://adaptation.nrcan.gc.ca/perspective/pdf/report_e.pdf>

287 Ebi, K.L., J. Smith, I. Burton, and J. Scheraga, 2006: Some lessons learned from public health on the process of adaptation. *Mitigation and Adaptation Strategies for Global Change*, **11(3)**, 607-620.

288 Ebi, K.L., T.J. Teisberg, L.S. Kalkstein, L. Robinson, and R.F. Weiher, 2004: Heat watch/warning systems save lives: estimated costs and benefits for Philadelphia 1995-98. *Bulletin of the American Meteorological Society*, **85(8)**, 1067-1073.

289 Medina-Ramon, M. and J. Schwartz, 2007: Temperature, temperature extremes, and mortality: a study of acclimatisation and effect modification in 50 U.S. cities. *Occupational and Environmental Medicine*, **64(12)**, 827-833.

290 U.S. Environmental Protection Agency, 2008: *National Air Quality: Status and Trends through 2007*. U.S. EPA Air Quality Assessment Division, Research Triangle Park, NC, 48 pp. <http://www.epa.gov/airtrends/2008/index.html>

291 Tao, Z., A. Williams, H.-C. Huang, M. Caughey, and X.-Z. Liang, 2007: Sensitivity of U.S. surface ozone to future emissions and climate changes. *Geophysical Research Letters*, **34**, L08811, doi:10.1029/2007GL029455.

292 California Air Resources Board, 2007: *Recent Research Findings: Health Effects of Particulate Matter and Ozone Air Pollution*. [Online fact sheet] California Air Resources Board, [Sacramento], 7 pp. <http://www.arb.ca.gov/research/health/fs/pm_ozone-fs.pdf>

293 California Climate Action Team, 2006: *Climate Action Team Report to Governor Schwarzenegger and the Legislature*. California Environmental Protection Agency, Sacramento, 107 pp. <http://www.climatechange.ca.gov/climate_action_team/reports/index.html>

294 Westerling A.L., H.G. Hidalgo, D.R. Cayan, and T.W. Swetnam, 2006: Warming and earlier spring increase western U.S. forest wildfire activity. *Science*, **313(5789)**, 940-943.

295 Rhode Island Public Transit Authority, Undated: *Air Quality Alert Days Program*. [Web site] <http://www.ripta.com/content259.html>

296 Friedman, M.S., K.E. Powell, L. Hutwagner, L.M. Graham, and W.G. Teague, 2001: Impact of changes in transportation and commuting behaviors during the 1996 Summer Olympic Games in Atlanta on air quality and childhood asthma. *Journal of the American Medical Association*, **285(7)**, 897-905.

297 Wuebbles, D.J., H. Lei, and J.-T. Lin, 2007: Intercontinental transport of aerosols and photochemical oxidants from Asia and its consequences. *Environmental Pollution*, **150**, 65-84.

298 Bell, M.L., R. Goldberg, C. Hogrefe, P. Kinney, K. Knowlton, B. Lynn, J. Rosenthal, C. Rosenzweig, and J.A. Patz, 2007: Climate change, ambient ozone, and health in 50 U.S. cities. *Climatic Change*, **82(1-2)**, 61-76.

299 U.S. Environmental Protection Agency, 2008: *National Pollutant Discharge Elimination System (NPDES) Combined Sewer Overflows Demographics*. [Web site] <http://cfpub.epa.gov/npdes/cso/demo.cfm?program_id=5>

300 Tibbetts, J., 2005: Combined sewer systems: down, dirty, and out of date. *Environmental Health Perspectives*, **113(7)**, A464-A467.

301 U.S. Environmental Protection Agency, 2008: *Clean Watersheds Needs Survey 2004: Report to Congress*. U.S. Environmental Protection Agency, Washington, DC. <http://www.epa.gov/cwns/2004rtc/cwns2004rtc.pdf>

302 Patz, J.A., S.J. Vavrus, C.K. Uejio, and S.L. McClellan, 2008: Climate change and waterborne disease risk in the Great Lakes region of the U.S. *American Journal of Preventive Medicine*, **35(5)**, 451-458.

303 Centers for Disease Control and Prevention, 2009: *West Nile Virus: Statistics, Surveillance, and Control*. [Web site] <http://www.cdc.gov/ncidod/dvbid/westnile/surv&control.htm>

304 Kilpatrick, A.M., M.A. Meola, R.M. Moudy, and L.D. Kramer, 2008: Temperature, viral genetics, and the transmission of West Nile virus by *Culex pipiens* mosquitoes. *PLoS Pathogens*, **4(6)**, e1000092. doi:10.1371/journal.ppat.1000092

305 Sussman, F.G., M.L. Cropper, H. Galbraith, D. Godschalk, J. Loomis, G. Luber, M. McGeehin, J.E. Neumann, W.D. Shaw, A. Vedlitz, and S. Zahran, 2008: Effects of global change on human welfare. In: *Analyses of the Effects of Global Change on Human*

Health and Welfare and Human Systems [Gamble, J.L. (ed.), K.L. Ebi, F.G. Sussman, and T.J. Wilbanks (authors)]. Synthesis and Assessment Product 4.6. U.S. Environmental Protection Agency, Washington, DC, pp. 111-168.

306 Smith, J.B. and D. Tirpak (eds.), 1989: *The Potential Effects of Global Climate Change on the United States* U.S. Environmental Protection Agency, Washington, DC, 413 pp.

307 Ziska, L.H. and F.A. Caulfield, 2000: Rising CO_2 and pollen production of common ragweed (*Ambrosia artemisiifolia L.*), a known allergy-inducing species: implications for public health. *Australian Journal of Plant Physiology*, **27(10)**, 893-898.

308 Mohan J.E., L.H. Ziska, W.H. Schlesinger, R.B. Thomas, R.C. Sicher, K. George, and J.S. Clark, 2006: Biomass and toxicity responses of poison ivy (*Toxicodendron radicans*) to elevated atmospheric CO_2. *Proceedings of the National Academy of Sciences*, **103(24)**, 9086-9089.

309 Hunt, R., D.W. Hand, M.A. Hannah, and A.M. Neal, 1991: Response to CO_2 enrichment in 27 herbaceous species. *Functional Ecology*, **5(3)**, 410-421.

310 Ziska, L.H., 2003: Evaluation of the growth response of six invasive species to past, present and future atmospheric carbon dioxide. *Journal of Experimental Botany*, **54(381)**, 395-404.

311 CDC, 2007: *2007 National Diabetes Fact Sheet*. Centers for Disease Control and Prevention, Atlanta, GA, 14 pp. <http://www.cdc.gov/diabetes/pubs/factsheet07.htm>

312 Patz, J.A., H.K. Gibbs, J.A. Foley, J.V. Rogers, and K.R. Smith, 2007: Climate change and global health: quantifying a growing ethical crisis. *EcoHealth*, **4(4)**, 397-405.

313 Wilbanks, T.J., P. Kirshen, D. Quattrochi, P. Romero-Lankao, C. Rosenzweig, M. Ruth, W. Solecki, J. Tarr, P. Larsen, and B. Stone, 2008: Effects of global change on human settlements. In: *Analyses of the Effects of Global Change on Human Health and Welfare and Human Systems* [Gamble, J.L. (ed.), K.L. Ebi, F.G. Sussman, and T.J. Wilbanks (authors)]. Synthesis and Assessment Product 4.6. U.S. Environmental Protection Agency, Washington, DC, pp. 89-109.

314 U.S. Census Bureau, 2005: *Domestic Net Migration in the United States: 2000 to 2004: Population Estimates and Projections*. <http://www.census.gov/prod/2006pubs/p25-1135.pdf>

315 U.S. General Accounting Office, 2003: *Alaska Native Villages: Most Are Affected by Flooding and Erosion, but Few Qualify for Federal Assistance*. GAO-04-142. U.S. General Accounting Office, Washington, DC, 82 pp. <http://purl.access.gpo.gov/GPO/LPS42077>

316 Gamble, J.L., K.L. Ebi, F.G. Sussman, T.J. Wilbanks, C. Reid, J.V. Thomas, and C.P. Weaver, 2008: Common themes and research recommendations. In: *Analyses of the Effects of Global Change on Human Health and Welfare and Human Systems* [Gamble, J.L. (ed.), K.L. Ebi, F.G. Sussman, and T.J. Wilbanks (authors)]. Synthesis and Assessment Product 4.6. U.S. Environmental Protection Agency, Washington, DC, pp. 169-176.

317 United States Conference of Mayors, 2005: *U.S. Conference of Mayors Climate Protection Agreement*, as endorsed by the 73rd Annual U.S. Conference of Mayors meeting, Chicago, 2005. <http://usmayors.org/climateprotection/agreement.htm>

318 Semenza, J.C., J.E. McCullough, W.D. Flanders, M.A. McGeehin, and J.R. Lumpkin, 1999: Excess hospital admissions during the July 1995 heat wave in Chicago. *American Journal of Preventive Medicine*, **16(4)**, 269-277.

319 Borden, K.A., M.C. Schmidtlein, C.T. Emrich, W.W. Piegorsch, and S.L. Cutter, 2007: Vulnerability of U.S. cities to environmental hazards. *Journal of Homeland Security and Emergency Management*, **4(2)**, article 5. <http://www.bepress.com/jhsem/vol4/iss2/5>

320 van Kamp, I., K. Leidelmeijer, G. Marsman, and A. de Hollander, 2003: Urban environmental quality and human well-being: towards a conceptual framework and demarcation of concepts; a literature study. *Landscape and Urban Planning*, **65(1-2)**, 5-18.

321 Grimmond, S., 2007: Urbanization and global environmental change: local effects of urban warming. *Geographical Journal*, **173(1)**, 83-88.

322 Anderson, C.A., 2001: Heat and violence. *Current Directions in Psychological Science*, **10(1)**, 33-38.

323 Anderson, C.A., B.J. Bushman, and R.W. Groom, 1997: Hot years and serious and deadly assault: empirical test of the heat hypothesis. *Journal of Personality and Social Psychology*, **73(6)**, 1213-1223.

324 Milly, P.C.D., R.T. Wetherald, K.A. Dunne, and T.L. Delworth, 2002: Increasing risk of great floods in a changing climate. *Nature*, **415(6871)**, 514-517.

325 Miller, N.L., K. Hayhoe, J. Jin, and M. Auffhammer, 2008: Climate, extreme heat, and electricity demand in California. *Journal of Applied Meteorology and Climatology*, **47(6)**, 1834-1844.

326 Wang, J.X.L. and J.K. Angell, 1999: *Air Stagnation Climatology for the United States (1948-1998)*. NOAA/Air Resources Laboratory atlas no.1. NOAA Air Resources Laboratory, Silver Spring, MD, 74 pp.

327 Riebsame, W.E., S.A. Changnon Jr., and T.R. Karl, 1991: *Drought and Natural Resources Management in the United States: Impacts and Implications of the 1987–89 Drought*. Westview Press, Boulder, CO, 174 pp.

328 NOAA's National Climatic Data Center, 2009: *NCDC Climate Monitoring: U.S. Records* [Web site] <http://www.ncdc.noaa.gov/oa/climate/research/records/>

329 NOAA's National Climatic Data Center, 2009: *U.S. Percent Area Wet or Dry*. [Web site] <http://www.ncdc.noaa.gov/img/climate/research/2008/ann/Reg110_wet-dry_bar01001208-mod_pg.gif>

330 Leung L.R. and W.I. Gustafson Jr., 2005: Potential regional climate change and implications to U.S. air quality. *Geophysical Research Letters*, **32**, L16711, doi:10.1029/2005GL022911.

331 Several U.S. Department of Energy reports document the increase in electricity demand to provide air-conditioning in hotter summers in many regions of the country: Chapter 2: Carbon dioxide emissions. In: *Emissions of Greenhouse Gases in the United States 2002*. Report released 2004. <http://www.eia.doe.gov/oiaf/1605/archive/gg03rpt/index.html>; *South Atlantic Household Electricity Report*. Release date: 2006. <http://www.eia.doe.gov/emeu/reps/enduse/er01_so-atl.html>; *South Central Appliance Report 2001*. Updated 2005. <http://www.eia.doe.gov/emeu/reps/appli/w_s_c.html>

332 CCSP, 2007: *Effects of Climate Change on Energy Production and Use in the United States*. [Wilbanks, T.J., V. Bhatt, D.E. Bilello, S.R. Bull, J. Ekmann, W.C. Horak, Y.J. Huang, M.D. Levine, M.J. Sale, D.K. Schmalzer, and M.J. Scott (eds.)]. Synthesis and Assessment Product 4.5. U.S. Department of Energy, Office of Biological & Environmental Research, Washington, DC, 160 pp. See chapters 2 and 3.

333 Daily data were used for both air stagnation and heat waves:
1. Heat waves:
   • The GHCN-Daily dataset from NCDC was used <http://www.ncdc.noaa.gov/oa/climate/ghcn-daily/>
   • Data from 979 U.S. stations having long periods of record and high quality.
   • At each station, a day was considered hot if the maximum temperature for that day was at or above the 90% of daily maximum temperatures at that station.
2. Air stagnation:

• For each day in summer and at each air-stagnation grid point, it was determined if that location had stagnant air:

  ○ The stagnation index was formulated by Wang, J.X.L. and J.K. Angell, 1999: *Air Stagnation Climatology for the United States (1948-1998)*. NOAA/Air Resources Laboratory atlas no.1 NOAA Air Resources Laboratory, Silver Spring, MD, 74 pp. <http://www.arl.noaa.gov/documents/reports/atlas.pdf>

  ○ Operational implementation of this index is described at <http://www.ncdc.noaa.gov/oa/climate/research/stagnation/index.php>

Note: Although Wang and Angell used a criteria of four day stagnation periods, single stagnation days were used for this analysis.

3. For each location in the air stagnation grid, the nearest station (of the aforementioned 979 U.S. stations) was used to determine the coincidence of summer days having stagnant air and excessive heat as a percentage of the number of days having excessive heat.

334 Solecki, W.D. and C. Rosenzweig, 2006: Climate change and the city: observations from metropolitan New York. In: *Cities and Environmental Change* [Bai, X. (ed.)]. Yale University Press, New York.

335 Rosenzweig, C. and W. Solecki (eds.), 2001: *Climate Change and a Global City: The Potential Consequences of Climate Variability and Change – Metro East Coast*. Columbia Earth Institute, New York. <http://metroeast_climate.ciesin.columbia.edu/>

336 Kirshen, P., M. Ruth, W. Anderson, T.R. Lakshmanan, S. Chapra, W. Chudyk, L. Edgers, D. Gute, M. Sanayei, and R. Vogel, 2004: *Climate's Long-term Impacts on Metro Boston (CLIMB) Final Report*. Civil and Environmental Engineering Department, Tufts University, 165 pp. <http://www.clf.org/uploadedFiles/CLIMB_Final_Report.pdf>

337 Grimm, N.B., S.H. Faeth, N.E. Golubiewski, C.L. Redman, J. Wu, X. Bai, and J.M. Briggs, 2008: Global change and the ecology of cities. *Science*, **319(5864)**, 756-760.

338 Kuo, F.E. and W.C. Sullivan, 2001: Environment and crime in the inner city: Does vegetation reduce crime? *Environment and Behavior*, **33(3)**, 343-367.

339 Julius, S.H., J.M. West, G. Blate, J.S. Baron, B. Griffith, L.A. Joyce, P. Kareiva, B.D. Keller, M. Palmer, C. Peterson, and J.M. Scott, 2008: Executive summary. In: *Preliminary Review of Adaptation Options for Climate-sensitive Ecosystems and Resources* [Julius, S.H. and J.M. West (eds.), J.S. Baron, B. Griffith, L.A. Joyce, P. Kareiva, B.D. Keller, M.A. Palmer, C.H. Peterson, and J.M. Scott (authors)]. Synthesis and Assessment Product 4.4. U.S. Environmental Protection Agency, Washington, DC, pp. 1-1 to 1-6.

340 Baker, L.A., A.J. Brazel, N. Selover, C. Martin, N. McIntyre, F.R. Steiner, A. Nelson, and L. Musacchio, 2002: Urbanization and warming of Phoenix (Arizona, USA): impacts, feedbacks, and mitigation. *Urban Ecosystems*, **6(3)**, 183-203.

341 LoVecchio, F., J.S. Stapczynski, J. Hill, A.F. Haffer, J.A. Skindlov, D. Engelthaler, C. Mrela, G.E. Luber, M. Straetemans, and Z. Duprey, 2005: Heat-related mortality – Arizona, 1993-2002, and United States, 1979-2002. *Morbidity and Mortality Weekly Report*, **54(25)**, 628-630.

342 Scott, D., J. Dawson, and B. Jones, 2008: Climate change vulnerability of the US Northeast winter recreation–tourism sector. *Mitigation and Adaptation Strategies for Global Change*, **13(5-6)**, 577-596.

343 Bin, O., C. Dumas, B. Poulter, and J. Whitehead, 2007: *Measuring the Impacts of Climate Change on North Carolina Coastal Resources*. National Commission on Energy Policy, Washington, DC, 91 pp. <http://econ.appstate.edu/climate/>

344 Mills, E., 2005: Insurance in a climate of change. *Science*, **309(5737)**, 1040-1044.

345 Adapted from U.S. Government Accountability Office, 2007: *Climate Change: Financial Risks to Federal and Private Insurers in Coming Decades are Potentially Significant*. U.S. Government Accountability Office, Washington, DC, 68 pp. <http://purl.access.gpo.gov/GPO/LPS89701> Data shown are not adjusted for inflation.

346 Pielke Jr., R.A., J. Gratz, C.W. Landsea, D. Collins, M. Saunders, and R. Musulin, 2008: Normalized hurricane damages in the United States: 1900-2005. *Natural Hazards Review*, **9(1)**, 29-42.

347 Rosenzweig, C., G. Casassa, D.J. Karoly, A. Imeson, C. Liu, A. Menzel, S. Rawlins, T.L. Root, B. Seguin, and P. Tryjanowski, 2007: Assessment of observed changes and responses in natural and managed systems. In: *Climate Change 2007: Impacts, Adaptation and Vulnerability*. Contribution of Working Group II to the Fourth Assessment Report of the Intergovernmental Panel on Climate Change [Parry, M.L., O.F. Canziani, J.P. Palutikof, P.J. van der Linden, and C.E. Hanson, (eds.)]. Cambridge University Press, Cambridge, UK, and New York, pp. 79-131.

348 Pielke Jr., R.A., 2005: Attribution of disaster losses. *Science*, **310(5754)**, 1615; and Mills, E., 2005: Response. *Science*, **310(5754)**, 1616.

349 Mills, E., 2009: A global review of insurance industry responses to climate change. *The Geneva Papers on Risk and Insurance–Issues and Practice*, in press.

350 Meehl, G.A. and C. Tebaldi, 2004: More intense, more frequent, and longer lasting heat waves in the 21st century. *Science*, **305(5686)**, 994-997.

351 Mills, E., 2006: Synergisms between climate change mitigation and adaptation: an insurance perspective. *Mitigation and Adaptation Strategies for Global Change*, **12(5)**, 809-842.

352 U.S. Government Accountability Office, 2007 [data]; assembled by Evan Mills, Lawrence Berkeley National Laboratory.

353 Reeve, N. and R. Toumi, 1999: Lightning activity as an indicator of climate change. *Quarterly Journal of the Royal Meteorological Society*, **125(555)**, 893-903.

354 Price, C. and D. Rind, 1994: Possible implications of global climate change on global lightning distributions and frequencies. *Journal of Geophysical Research*, **99(D5)**, 10823-10831.

355 Ross, C., E. Mills, and S. Hecht, 2007: Limiting liability in the greenhouse: insurance risk-management in the context of global climate change. *Stanford Environmental Law Journal and Stanford Journal of International Law, Symposium on Climate Change Risk*, **26A/43A**, 251-334.

356 Nutter, F.W., 1996: Insurance and the natural sciences: partners in the public interest. *Research Review: Journal of the Society of Insurance Research*, **Fall**, 15-19.

357 Federal Emergency Management Agency, 2008: *Emergency Management Institute*. [Web site] <http://training.fema.gov/EMIWeb/CRS/>

358 Bernstein, L., J. Roy, K.C. Delhotal, J. Harnisch, R. Matsuhashi, L. Price, K. Tanaka, E. Worrell, F. Yamba, and Z. Fengqi, 2007: Industry. In: *Climate Change 2007: Mitigation of Climate Change*. Contribution of Working Group III to the Fourth Assessment Report of the Intergovernmental Panel on Climate Change [Metz, B., O.R. Davidson, P.R. Bosch, R. Dave, and L.A. Meyer (eds.)]. Cambridge University Press, Cambridge, UK, and New York, pp. 447-496.

359 Hayhoe, K., C.P. Wake, B. Anderson, X.-Z. Liang, E. Maurer, J. Zhu, J. Bradbury, A. DeGaetano, A. Hertel, and D. Wuebbles, 2008: Regional climate change projections for the northeast U.S. *Mitigation and Adaptation Strategies for Global Change*, **13(5-6)**, 425-436.

360 New York City Department of Health and Mental Hygiene, 2006: Deaths associated with heat waves in 2006. In: *NYC Vital Signs:*

Investigation Report, Special Report. Department of Health and Mental Hygiene, New York, 4 pp. <http://www.nyc.gov/html/doh/downloads/pdf/survey/survey-2006heatdeaths.pdf>

[361] Kunkel, K.E., H.-C. Huang, X.-Z. Liang, J.-T. Lin, D. Wuebbles, Z. Tao, A. Williams, M. Caughey, J. Zhu, and K. Hayhoe, 2008: Sensitivity of future ozone concentrations in the northeast U.S. to regional climate change. *Mitigation and Adaptation Strategies for Global Change*, **13(5-6)**, 597-606.

[362] Hauagge, R. and J.N. Cummins, 1991: Phenotypic variation of length of bud dormancy in apple cultivars and related malus species. *Journal of the American Society for Horticultural Science*, **116(1)**, 100-106.

[363] DeMoranville, C., 2007: Personal communication from May 29, 2008. Experts at the University of Massachusetts Cranberry Station estimate cranberry chilling requirements to be around 1,200-1,400 hours, but they advise growers to seek 1,500 hours to avoid crop failure. There are 4-5 commonly grown cultivars but no low-chill varieties. Dr. Carolyn DeMoranville is the director of the UMass Cranberry Station, a research and extension center of UMass-Amherst.

[364] Iverson, L., A. Prasad, and S. Matthews, 2008: Potential changes in suitable habitat for 134 tree species in the northeastern United States. *Mitigation and Adaptation Strategies for Global Change*, **13(5-6)**, 487-516.

[365] U.S. Department of Agriculture (USDA) National Agriculture Statistics Service (NASS), 2002: *Statistics by State.* [Web site] <http://www.nass.usda.gov/Statistics_by_State/>

[366] St. Pierre, N.R., B. Cobanov, and G. Schnitkey, 2003: Economic losses from heat stress by U.S. livestock industries. *Journal of Dairy Science*, **86(E Sup)**, E52- E77.

[367] Gornitz, V., S. Couch, and E.K. Hartig, 2001: Impacts of sea level rise in the New York City metropolitan area. *Global and Planetary Change*, **32(1)**, 61-88.

[368] AIR Worldwide Corporation, 2008: *The Coastline at Risk: 2008 Update to the Estimated Insured Value of U.S. Coastal Properties.* AIR Worldwide Corporation, Boston, MA, 3 pp. <http://www.air-worldwide.com/download.aspx?c=388&id=15836>

[369] Kirshen, P., C. Watson, E. Douglas, A. Gontz, J. Lee, and Y. Tian, 2008: Coastal flooding in the northeastern United States due to climate change. *Mitigation and Adaptation Strategies for Global Change*, **13(5-6)**, 437-451.

[370] Bowman, M., D. Hill, F. Buonaiuto, B. Colle, R. Flood, R. Wilson, R. Hunter, and J. Wang, 2008: Threats and responses associated with rapid climate change in metropolitan New York. In: *Sudden and Disruptive Climate Change: Exploring the Real Risks and How We Can Avoid Them.* [MacCracken, M.C., F. Moore, and J.C. Topping Jr. (eds.)]. Earthscan, London and Sterling, VA, pp. 119-142.

[371] Titus, J.G., 2009: Ongoing adaptation. In: *Coastal Elevations and Sensitivity to Sea-level Rise: A Focus on the Mid-Atlantic Region* [J.G. Titus (coordinating lead author), K.E. Anderson, D.R. Cahoon, D.B. Gesch, S.K. Gill, B.T. Gutierrez, E.R. Thieler, and S.J. Williams (lead authors)]. Synthesis and Assessment Product 4.1. U.S. Environmental Protection Agency, Washington, DC, pp. 157-162.

[372] International Snowmobile Manufacturers Association, 2006: *International Snowmobile Industry Facts and Figures.* [Web site] <http://www.snowmobile.org/pr_snowfacts.asp>

[373] Northeast Climate Impact Assessment (NECIA), 2006: *Climate Change in the U.S. Northeast: A Report of the Northeast Climate Impacts Assessment.* Union of Concerned Scientists, Cambridge, MA, 35 pp.

[374] Atlantic States Marine Fisheries Commission, 2005: *American Lobster.* [Web site] <http://www.asmfc.org/americanLobster.htm>

[375] Fogarty, M.J., 1995: Populations, fisheries, and management. In: *The Biology of the American Lobster* Homarus americanus. [Factor, J.R. (ed.)]. Academic Press, San Diego, CA, pp. 111-137.

[376] Glenn, R.P. and T.L. Pugh, 2006: Epizootic shell disease in American lobster (*Homarus americanus*) in Massachusetts coastal waters: interactions of temperature, maturity, and intermolt duration. *Journal of Crustacean Biology*, **26(4)**, 639-645.

[377] Fogarty, M., L. Incze, K. Hayhoe, D. Mountain, and J. Manning, 2008: Potential climate change impacts on Atlantic cod (*Gadus morhua*) off the northeastern United States. *Mitigation and Adaptation Strategies for Global Change*, **13(5-6)**, 453-466.

[378] Dutil, J.-D. and K. Brander, 2003: Comparing productivity of North Atlantic cod (*Gadus morhua*) stocks and limits to growth production. *Fisheries Oceanography*, **12(4-5)**, 502-512.

[379] Drinkwater, K.F., 2005: The response of Atlantic cod (*Gadus morhua*) to future climate change. *ICES Journal of Marine Science*, **62(7)**, 1327-1337.

[380] Karl, T.R. and R.W. Knight, 1998: Secular trends of precipitation amount, frequency, and intensity in the United States. *Bulletin of the American Metrological Society*, **79(2)**, 231-241.

[381] Keim, B.D., 1997: Preliminary analysis of the temporal patterns of heavy rainfall across the southeastern United States. *Professional Geographer*, **49(1)**, 94-104.

[382] Observed changes in precipitation for the Southeast were calculated from the US Historical Climatology Network Version 2. See Menne, M.J., C.N. Williams, and R.S. Vose, 2009: The United States Historical Climatology Network Monthly Temperature Data - Version 2. *Bulletin of the American Meteorological Society*, Early online release, 25 February 2009, doi:10.1175/2008BAMS2613.1

[383] Temperature:
Menne, M.J., C.N. Williams, and R.S. Vose, 2009: The United States Historical Climatology Network Monthly Temperature Data - Version 2. *Bulletin of the American Meteorological Society*, Early online release, 25 February 2009, doi:10.1175/2008BAMS2613.1
Precipitation:
NOAA's National Climatic Data Center, 2008: *The USHCN Version 2 Serial Monthly Dataset.* [Web site] <http://www.ncdc.noaa.gov/oa/climate/research/ushcn/>

[384] Hoyos, C.D., P.A. Agudelo, P.J. Webster, and J.A. Curry, 2006: Deconvolution of the factors contributing to the increase in global hurricane intensity. *Science*, **312(577)**, 94-97.

[385] Mann, M.E. and K.A. Emanuel, 2006: Atlantic hurricane trends linked to climate change. *EOS Transactions of the American Geophysical Union*, **87(24)**, 233, 244.

[386] Trenberth, K.E. and D.J. Shea, 2006: Atlantic hurricanes and natural variability in 2005. *Geophysical Research Letters*, **33**, L12704, doi:10.1029/2006GL026894.

[387] Webster, P.J., G.J. Holland, J.A. Curry, and H.-R. Chang, 2005: Changes in tropical cyclone number, duration, and intensity in a warming environment. *Science*, **309(5742)**, 1844-1846.

[388] Komar, P.D. and J.C. Allan, 2007: Higher waves along U.S. East Coast linked to hurricanes. *EOS Transactions of the American Geophysical Union*, **88(30)**, 301.

[389] Change calculated from daily minimum temperatures from NCDC's Global Historical Climatology Network - Daily data set. <http://www.ncdc.noaa.gov/oa/climate/ghcn-daily/>

[390] Nicholls, R.J., P.P. Wong, V.R. Burkett, J.O. Codignotto, J.E. Hay, R.F. McLean, S. Ragoonaden, and C.D. Woodroffe, 2007: Coastal systems and low-lying areas. In: *Climate Change 2007: Impacts, Adaptations and Vulnerability.* Contribution of Working Group II to the Fourth Assessment Report of the Intergovernmental Panel on Climate Change [Parry, M.L., O.F. Canziani, J.P. Palutikof, P.J. van der Linden, and C.E. Hanson (eds.)]. Cambridge University Press, Cambridge, UK, and New York, pp. 316-356.

[391] Boyles, S., 2008: *Heat Stress and Beef Cattle*. Ohio State University Extension Service. <http://beef.osu.edu/library/heat.html>

[392] Convention on Biological Diversity, 2006: *Guidance for Promoting Synergy Among Activities Addressing Biological Diversity, Desertification, Land Degradation and Climate Change*. CBD technical series 25. Secretariat of the Convention on Biological Diversity, Montreal, Canada, 43 pp. <http://www.biodiv.org/doc/publications/cbd-ts-25.pdf>

[393] Burkett, V., 2008: The northern Gulf of Mexico coast: human development patterns, declining ecosystems, and escalating vulnerability to storms and sea level rise. In: *Sudden and Disruptive Climate Change: Exploring the Real Risks and How We Can Avoid Them*. [MacCracken, M.C., F. Moore, and J.C. Topping (eds.)]. Earthscan Publications, London [UK], and Sterling, VA, pp. 101-118.

[394] Twilley, R.R., E. Barron, H.L. Gholz, M.A. Harwell, R.L. Miller, D.J. Reed, J.B. Rose, E. Siemann, R.G. Welzel, and R.J. Zimmerman, 2001: *Confronting Climate Change in the Gulf Coast Region: Prospects for Sustaining Our Ecological Heritage*. Union of Concerned Scientists, Cambridge, MA, and Ecological Society of America, Washington, DC, 82 pp.

[395] USGS, photos and images:
Photos on the left:
Tihansky, A.B., 2005: Before-and-after aerial photographs show coastal impacts of Hurricane Katrina. *Sound Waves*, September. <http://soundwaves.usgs.gov/2005/09/fieldwork2.html>
Images on the right:
Adapted from Fauver, L., 2007: Predicting flooding and coastal hazards: USGS hydrologists and geologists team up at the National Hurricane Conference to highlight data collection. *Sounds Waves*, June. <http://soundwaves.usgs.gov/2007/06/meetings.html>

[396] Barras, J.A., 2006: *Land Area Change in Coastal Louisiana After the 2005 Hurricanes: A Series of Three Maps*. U.S. Geological Survey open-file report 2006-1274. <http://pubs.usgs.gov/of/2006/1274>

[397] Main Development Region of the Atlantic Ocean is defined in: Bell, G.D., E. Blake, C.W. Landsea, S.B. Goldenberg, R. Pasch, and T. Kimberlain, 2008: Tropical cyclones: Atlantic basin. In: Chapter 4: The Tropics, of *State of the Climate in 2007* [Levinson, D.H. and J.H. Lawrimore (eds.)]. *Bulletin of the American Meteorological Society*, **89(Supplement)**, S68-S71.

[398] Williams, K.L., K.C. Ewel, R.P. Stumpf, F.E. Putz, and T.W. Workman, 1999: Sea-level rise and coastal forest retreat on the west coast of Florida. *Ecology*, **80(6)**, 2045-2063.

[399] McNulty, S.G., J.M. Vose, and W.T. Swank, 1996: Potential climate change affects on loblolly pine productivity and hydrology across the southern United States. *Ambio*, **25(7)**, 449-453.

[400] Zimmerman, R.J., T.J. Minello, and L.P. Rozas, 2002: Salt marsh linkages to productivity of penaeid shrimps and blue crabs in the northern Gulf of Mexico. In: *Concepts and Controversies in Tidal Marsh Ecology* [Weinstein, M.P. and D.A. Kreeger (eds.)]. Kluwer, Dordrecht and Boston, pp. 293-314.

[401] U.S. Army Corps of Engineers, 2009: *Risk Reduction Plan: Levees/Floodwalls/Armoring*. [Web site] U.S. Army Corps of Engineers New Orleans District. <http://www.mvn.usace.army.mil/hps2/hps_levees_flood_armor.asp>

[402] Kling, G.W., K. Hayhoe, L.B. Johnson, J.J. Magnuson, S. Polasky, S.K. Robinson, B.J. Shuter, M.M. Wander, D.J. Wuebbles, and D.R. Zak, 2003: *Confronting Climate Change in the Great Lakes Region: Impacts on Our Communities and Ecosystems*. Union of Concerned Scientists, Cambridge, MA, and Ecological Society of America, Washington, DC, 92 pp. <http://www.ucsusa.org/greatlakes/>

[403] Hayhoe, K., D. Wuebbles, and the Climate Science Team, 2008: *Climate Change and Chicago: Projections and Potential Impacts*.
City of Chicago, [175 pp.] <http://www.chicagoclimateaction.org/>

[404] Wuebbles, D.J. and K. Hayhoe, 2004: Climate change projections for the United States Midwest. *Mitigation and Adaptation Strategies for Global Change*, **9(4)**, 335-363.

[405] Ebi, K.L. and G.A. Meehl, 2007: The heat is on: climate change and heat waves in the Midwest. In: *Regional Impacts of Climate Change: Four Case Studies in the United States*. Pew Center on Global Climate Change, Arlington, VA, pp. 8-21. <http://www.pewclimate.org/regional_impacts>

[406] Kalkstein, L.S., J.S. Green, D.M. Mills, A.D. Perrin, J.P. Samenow, and J.-C. Cohen, 2008: Analog European heat waves for U.S. cities to analyze impacts on heat-related mortality. *Bulletin of the American Meteorological Society*, **89(1)**, 75-85.

[407] Hayhoe, K., S. Sheridan, J.S. Greene and L. Kalkstein, 2009: Climate change, heat waves, and mortality projections for Chicago. *Journal of Great Lakes Research*, in press.

[408] Lin, J.-T., K.O. Patten, X.-Z. Liang, and D.J. Wuebbles, 2008: Climate change effects on ozone air quality in the United States and China with constant precursor emissions. *Journal of Applied Meteorology and Climatology*, **47(7)**, 1888-1909.

[409] Holloway, T., S.N. Spak, D. Barker, M. Bretl, K. Hayhoe, J. Van Dorn, and D. Wuebbles, 2008: Change in ozone air pollution over Chicago associated with global climate change. *Journal of Geophysical Research*, **113**, D22306, doi:10.1029/2007JD009775.

[410] Hedegaard, G.B., J. Brandt, J.H. Christensen, L.M. Frohn, C. Geels, K.M. Hansen, and M. Stendel, 2008: Impacts of climate change on air pollution levels in the Northern Hemisphere with special focus on Europe and the Arctic. *Atmospheric Chemistry and Physics*, **8(12)**, 3337-3367.

[411] Several different city of Chicago analyses have substantiated the finding of up to 77 degree difference – for example, see City of Chicago, Department of Environment, Undated: *City Hall Rooftop Garden*. [Web site] <http://egov.cityofchicago.org/city/webportal/portalDeptCategoryAction.do?deptCategoryOID=-536889314&contentType=COC_EDITORIAL&topChannelName=Dept&entityName=Environment&deptMainCategoryOID=-536887205>

[412] Angel, J. and K. Kunkel, 2009: The response of Great Lakes water levels to future climate scenarios with an emphasis of Lake Michigan. *Journal of Great Lakes Research*, in press.

[413] Annin, P., 2006: *The Great Lakes Water Wars*. Island Press, Washington, DC, 303 pp.

[414] Assel, R.A., 2003: *An Electronic Atlas of Great Lakes Ice Cover, Winters: 1973-2002*. NOAA Great Lakes Environmental Research Laboratory, Ann Arbor, MI, 2 CD-ROM set or DVD. <http://www.glerl.noaa.gov/data/ice/atlas>

[415] Changnon, S.A. (ed.), 1996: *The Great Flood of 1993: Causes, Impacts and Responses*. Westview Press, Boulder, CO, 321 pp.

[416] Changnon, S.A. and K.E. Kunkel, 2006: *Severe Storms in the Midwest*. Illinois State Water Survey, Champaign, IL, 74 pp. <http://www.sws.uiuc.edu/pubdoc/IEM/ISWSIEM2006-06.pdf>

[417] Kunkel, K.E., K. Andsager, G. Conner, W.L. Decker, H.J. Hilaker Jr., P.N. Knox, F.V. Nurnberger, J.C. Rogers, K. Scheeringa, W.M. Wendland, J. Zandlo, and J.R. Angel, 1998: An expanded digital daily database for climatic resources applications in the midwestern United States. *Bulletin of the American Meteorological Society*, **79(7)**, 1357-1366.

[418] Agricultural Research Service, 1990: *USDA Plant Hardiness Zone Map*. Agricultural Research Service, Washington, DC, 1 map.
Arbor Day Foundation, 2006: *Hardiness Zones*. [Web site] The Arbor Day Foundation, Nebraska City, NE. <http://www.arborday.org/media/zones.cfm>

419 Woodhouse, C.A. and J.T. Overpeck, 1998: 2000 years of drought variability in the central United States. *Bulletin of the American Meteorological Society,* **79(12),** 2693-2714.

420 DeGaetano, A.T. and R.J. Allen, 2002: Trends in twentieth-century temperature extremes across the United States. *Journal of Climate,* **15(22),** 3188-3205.

421 Garbrecht, J., M. Van Liew, and G.O. Brown, 2004: Trends in precipitation, streamflow, and evapotranspiration in the Great Plains of the United States. *Journal of Hydrologic Engineering,* **9(5),** 360-367.

422 McGuire, V., 2007: *Water-level Changes in the High Plains Aquifer, Predevelopment to 2005 and 2003 to 2005.* U.S. Geological Survey scientific investigations report 2006–5324. U.S. Geological Survey, Reston, VA, 7 pp. <http://pubs.usgs.gov/sir/2006/5324/>

423 Dennehy, K. 2000: *High Plains Regional Ground-Water Study.* U.S. Geological Survey fact sheet FS-091-00, 6 pp. <http://pubs.er.usgs.gov/usgspubs/fs/fs09100>

424 Gurdak, J.J., R.T. Hanson, P.B. McMahon, B.W. Bruce, J.E. McCray, G.D. Thyne, and R.C. Reedy, 2007: Climate variability controls on unsaturated water and chemical movement, High Plains aquifer, USA. *Vadose Zone Journal,* **6(3),** 533-547.

425 Green, T.R., M. Taniguchi, and H. Kooi, 2007: Potential impacts of climate change and human activity on subsurface water resources. *Vadose Zone Journal,* **6(3),** 531-532.

426 Data from the PRISM Group, Oregon State University <http://www.prismclimate.org>; Map created by National Climatic Data Center, March 2009.

427 Mahmood, R., S.A. Foster, T. Keeling, K.G. Hubbard, C. Carlson, and R. Leeper, 2006: Impacts of irrigation on 20th century temperature in the northern Great Plains. *Global and Planetary Change,* **54(1-2),** 1-18.

428 Moore, N. and S. Rojstaczer, 2002: Irrigation's influence on precipitation: Texas High Plains, USA. *Geophysical Research Letters,* **29(16),** 1755, doi:10.1029/2002GL014940.

429 Schubert, S.D., M.J. Suarez, P.J. Pegion, R.D. Koster, and J.T. Bacmeister, 2004: On the cause of the 1930s Dust Bowl. *Science,* **303(5665),** 1855-1859.

430 Motha, R.P. and W. Baier, 2005: Impacts of present and future climate change and climate variability on agriculture in the temperate regions: North America. *Climatic Change,* **70(1-2),** 137-164.

431 Izaurralde, R.C., N.J. Rosenberg, R.A. Brown, and A.M. Thomson, 2003: Integrated assessment of Hadley Center (HadCM2) climate-change impacts on agricultural productivity and irrigation water supply in the conterminous United States: Part II. Regional agricultural production in 2030 and 2095. *Agricultural and Forest Meteorology,* **117(1-2),** 97-122.

432 Ziska, L. and K. George, 2004: Rising carbon dioxide and invasive, noxious plants: potential threats and consequences. *World Resource Review,* **16,** 427-447.

433 Parton, W., M. Gutmann, and D. Ojima, 2007: Long-term trends in population, farm income, and crop production in the Great Plains. *Bioscience,* **57(9),** 737-747.

434 Reilly, J., F. Tubiello, B. McCarl, D. Abler, R. Darwin, K. Fuglie, S. Hollinger, C. Izaurralde, S. Jagtap, J. Jones, L. Mearns, D. Ojima, E. Paul, K. Paustian, S. Riha, N. Rosenberg, and C. Rosenzweig, 2003: U.S. agriculture and climate change: new results. *Climatic Change,* **57(1-2),** 43-69.

435 Allen, V.G., C.P. Brown, R. Kellison, E. Segarra, T. Wheeler, P.A. Dotray, J.C. Conkwright, C.J. Green, and V. Acosta-Martinez, 2005: Integrating cotton and beef production to reduce water withdrawal from the Ogallala aquifer in the southern high plains. *Agronomy Journal,* **97(2),** 556-567.

436 Hanson, J.D., M.A. Liebig, S.D. Merrill, D.L. Tanaka, J.M. Krupinsky, and D.E. Stott, 2007: Dynamic cropping systems:
increasing adaptability amid an uncertain future. *Agronomy Journal,* **99(4),** 939-943.

437 Cameron, G.N. and D. Scheel, 2001: Getting warmer: effect of global climate change on distribution of rodents in Texas. *Journal of Mammalogy,* **82(3),** 652-680.

438 Levia, D.F. and E.E. Frost, 2004: Assessment of climatic suitability for the expansion of *Solenopsis invicta* Buren in Oklahoma using three general circulation models. *Theoretical and Applied Climatology,* **79(1-2),** 23-30.

439 Peterson, A.T., 2003: Projected climate change effects on Rocky Mountain and Great Plains birds: generalities of biodiversity consequences. *Global Change Biology,* **9(5),** 647-655.

440 Niemuth, N.D. and J.W. Solberg, 2003: Response of waterbirds to number of wetlands in the prairie pothole region of North Dakota, USA. *Waterbirds,* **26(2),** 233-238.

441 Conway, W.C., L.M. Smith, and J.D. Ray, 2005: Shorebird breeding biology in wetlands of the Playa Lakes, Texas, USA. *Waterbirds,* **28(2),** 129-138.

442 Scanlon, B., R. Reedy, and J. Tachovsky, 2007: Semiarid unsaturated zone chloride profiles: archives of past land use change impacts on water resources in the southern High Plains, United States. *Water Resources Research,* **43,** W06423, doi:10.1029/2006WR005769.

443 Haukos, D.A. and L.M. Smith, 2003: Past and future impacts of wetland regulations on playa ecology in the southern Great Plains. *Wetlands,* **23(3),** 577-589.

444 Matthews, J., 2008: *Anthropogenic Climate Change in the Playa Lakes Joint Venture Region: Understanding Impacts, Discerning Trends, and Developing Responses.* Playa Lakes Joint Venture, Lafayette, CO, 40 pp. <http://www.pljv.org/cms/climate-change>

445 Playa Lakes Joint Venture, Undated: *Playas and the Ogallala Aquifer – What's the Connection?* Playa Lakes Joint Venture, Lafayette, CO, 2 pp. <http://www.pljv.org/assets/Media/Recharge.pdf>

446 U.S. Census Bureau, 2007: *Census Bureau Announces Most Populous Cities.* Press release June 28, 2007. <http://www.census.gov/Press-Release/www/releases/archives/population/010315.html>

447 Ebi, K.L., D.M. Mills, J.B. Smith, and A. Grambsch, 2006: Climate change and human health impacts in the United States: an update on the results of the US national assessment. *Environmental Health Perspectives,* **114(9),** 1318-1324.

448 Nielsen, D.C., P.W. Unger, and P.R. Miller, 2005: Efficient water use in dryland cropping systems in the Great Plains. *Agronomy Journal,* **97(2),** 364-372.

449 NOAA's National Climatic Data Center, 2008: Southwest region Palmer Hydrological Drought index (PHDI). In: *Climate of 2008 – October: Southwest Region Moisture Status.* [Web site] NOAA National Climatic Data Center, Asheville, NC. <http://www.ncdc.noaa.gov/img/climate/research/prelim/drought/Reg107Dv00_palm06_pg.gif>

450 Guhathakurta, S. and P. Gober, 2007: The impact of the Phoenix urban heat island on residential water use. *Journal of the American Planning Association,* **73(3),** 317-329.

451 Gleick, P.H., 1988: Regional hydrologic consequences of increases in atmospheric CO_2 and other trace gases. *Climatic Change,* **10(2),** 137-160.

452 Woodhouse, C.A., S.T. Gray, and D.M. Meko, 2006: Updated streamflow reconstructions for the upper Colorado River basin. *Water Resources Research,* **42,** W05415, doi:10.1029/2005WR004455.

453 Meko, D.M., C.A. Woodhouse, C.A. Basisan, T. Knight, J.J. Lukas, M.K. Hughes, and M.W. Salzer, 2007: Medieval drought in the upper Colorado River basin. *Geophysical Research Letters,* **34,** L10705, doi:10.1029/2007GL029988.

454 Stine, S., 1994: Extreme and persistent drought in California and Patagonia during mediaeval time. *Nature,* **369(6481),** 546-549.

455 Breshears, D.D., N.S. Cobb, P.M. Rich, K.P. Price, C.D. Allen, R.G. Balice, W.H. Romme, J.H. Hastens, M.L. Floyd, J. Belnap, J.J. Anderson, O.B. Myers, and C.W. Meyer, 2005: Regional vegetation die-off in response to global-change drought. *Proceedings of the National Academy of Sciences*, **102(42)**, 15144-15148.

456 Archer, C.L. and K. Caldiera, 2008: Historical trends in the jet streams. *Geophysical Research Letters*, **35**, L08803, doi:10.1029/2008GL033614.

457 McAfee, S.A. and J.L. Russell, 2008: Northern annular mode impact on spring climate in the western United States. *Geophysical Research Letters*, **35**, L17701, doi:10.1029/2008GL034828.

458 Westerling, A.L., H.G. Hidalgo, D.R. Cayan, and T.W. Swetnam, 2006: Warming and earlier spring increase western U.S. forest wildfire activity. *Science*, **313(5789)**, 940-943.

459 Lenihan, J.M., D. Bachelet, R.P. Neilson, and R. Drapek, 2008: Response of vegetation distribution, ecosystem productivity, and fire to climate change scenarios for California. *Climatic Change*, **87(Supplement 1)**, S215-S230.

460 Westerling, A.L. and B.P. Bryant, 2008: Climate change and wildfire in California. *Climatic Change*, **87(Supplement 1)**, S231-S249.

461 Fried, J.S., J.K. Gilless, W.J. Riley, T.J. Moody, C.S. de Blas, K. Hayhoe, M. Moritz, S. Stephens and M. Torn, 2008: Predicting the effect of climate change on wildfire behavior and initial attack success. *Climatic Change*, **87(Supplement 1)**, S251-S264.

462 Moritz, M.A. and S.L. Stephens, 2008: Fire and sustainability: considerations for California's altered future climate. *Climatic Change*, **87(Supplement 1)**, S265-S271.

463 Rehfeldt, G.E., N.L. Crookston, M.V. Warwell, and J.S. Evans, 2006: Empirical analyses of plant-climate relationships for the western United States. *International Journal of Plant Sciences*, **167(6)**, 1123-1150.

464 Weiss, J. and J.T. Overpeck, 2005: Is the Sonoran Desert losing its cool? *Global Change Biology*, **11(12)**, 2065-2077.

465 Dole, K.P., M.E. Loik, and L.C. Sloan, 2003: The relative importance of climate change and the physiological effects CO_2 on freezing tolerance for the future distribution of *Yucca brevifolia*. *Global and Planetary Change*, **36(1-2)**, 137-146.

466 Loarie, S.R., B.E. Carter, K. Hayhoe, S. McMahon, R. Moe, C.A. Knight, and D.D. Ackerley, 2008: Climate change and the future of California's endemic flora. *PLoS ONE*, **3(6)**, e2502, doi:10.1371/journal.pone.0002502.

467 Myers, N., R.A. Mittermeier, C.G. Mittermeier, G.A.B. daFonseca, and J. Kent, 2000: Biodiversity hotspots for conservation priorities. *Nature*, **403(6772)**, 853-858.

468 Mittermeier R.A., P. Robles Gil, M. Hoffman, J. Pilgrim, T. Brooks, C. Goettsch Mittermeier, J. Lamoreux, and G.A.B. da Fonseca, 2005: *Hotspots Revisited: Earth's Biologically Richest and Most Endangered Terrestrial Ecoregions*. Conservation International, Washington DC, 392 pp.

469 Farjon, A. and C.N. Page (eds.), 1999: *Conifers: Status Survey and Conservation Action Plan*. IUCN/SSC Conifer Special Group. International Union for Conservation of Nature and Natural Resources, Gland, Switzerland, and Cambridge, UK, 121 pp.

470 Nixon, K.C., 1993: The genus *Quercus* in Mexico. In: *Biological Diversity of Mexico: Origins and Distribution* [Ramamoorthy, T.P., R. Bye, A. Lot, and J. Fa (eds.)]. Oxford University Press, New York, 812 pp.

471 Brower, L.P., G. Castilleja, A. Peralta, J. López-García, L. Bojórquez-Tapia, S. Díaz, D. Melgarejo, and M. Missrie, 2002: Quantitative changes in forest quality in a principal overwintering area of the monarch butterfly in Mexico, 1971–1999. *Conservation Biology*, **16(2)**, 346-359.

472 Goodrich, G. and A. Ellis, 2008: Climatic controls and hydrologic impacts of a recent extreme seasonal precipitation reversal in Arizona. *Journal of Applied Meteorology and Climatology*, **47(2)**, 498-508.

473 Allan, R.P. and B.J. Soden, 2008: Atmospheric warming and the amplifications of precipitation extremes. *Science*, **321(5895)**, 1481-1484.

474 Bales, R.C., N.P. Molotch, T.H. Painter, M.D. Dettinger, R. Rice, and J. Dozier, 2006: Mountain hydrology of the western United States. *Water Resources Research*, **42**, W08432, doi:10.1029/2005WR004387.

475 Delta Risk Management Strategy, 2008: Section 2: Sacramento/San Joaquin Delta and Suisun Marsh. In: *Phase I Report: Risk Analysis*. California Department of Water Resources, [13 pp.] <http://www.drms.water.ca.gov>

476 Delta Risk Management Strategy, 2008: Summary report. In: *Phase I Report: Risk Analsysis*. California Department of Water Resources, [42 pp.]. <http://www.drms.water.ca.gov>

477 Zimmerman, G., C. O'Brady, and B. Hurlbutt, 2006: Climate change: modeling a warmer Rockies and assessing the implications. In: *The 2006 State of the Rockies Report Card*. Colorado College, Colorado Springs, pp. 89-102. <http://www.coloradocollege.edu/stateoftherockies/06ReportCard.html>

478 Lazar, B. and M. Williams, 2008: Climate change in western ski areas: potential changes in the timing of wet avalanches and snow quality for the Aspen ski areas in the years 2030 and 2100. *Cold Regions Science and Technology*, **51(2-3)**, 219-228.

479 Kleeman, M.J., 2008: A preliminary assessment of the sensitivity of air quality in California to global change. *Climatic Change*, **87(Supplement 1)**, S273-S292.

480 Vicuna, S., R. Leonardson, M.W. Hanemann, L.L. Dale, and J.A. Dracup, 2008: Climate change impacts on high elevation hydropower generation in California's Sierra Nevada: a case study in the upper American River. *Climatic Change*, **87(Supplement 1)**, S123-S137.

481 Medellín-Azuara, J., J.J. Harou, M.A. Olivares, K. Madani, J.R. Lund, R.E. Howitt, S.K. Tanaka, M.W. Jenkins, and T. Zhu, 2008: Adaptability and adaptations of California's water supply system to dry climate warming. *Climatic Change*, **87(Supplement 1)**, S75-S90.

482 Baldocchi, D. and S. Wong, 2008: Accumulated winter chill is decreasing in the fruit growing regions of California. *Climatic Change*, **87(Supplement 1)**, S153-S166.

483 Lobell, D., C. Field, K. Nicholas Cahill, and C. Bonfils, 2006: Impacts of future climate change on California perennial crop yields: model projections with climate and crop uncertainties. *Agricultural and Forest Meteorology*, **141(2-4)**, 208-218.

484 Purkey, D.R., B. Joyce, S. Vicuna, M.W. Hanemann, L.L. Dale, D. Yates, and J.A. Dracup, 2008: Robust analysis of future climate change impacts on water for agriculture and other sectors: a case study in the Sacramento Valley. *Climatic Change*, **87(Supplement 1)**, S109-S122.

485 Mote, P.W., 2003: Trends in temperature and precipitation in the Pacific Northwest during the twentieth century. *Northwest Science*, **77(4)**, 271-282.

486 Mote, P., E. Salathé, V. Dulière, and E. Jump, 2008: *Scenarios of Future Climate for the Pacific Northwest*. Climate Impacts Group, University of Washington, Seattle, 12 pp. <http://cses.washington.edu/db/pubs/abstract628.shtml>

487 Hamlet, A.F., P.W. Mote, M. Clark, and D.P. Lettenmaier, 2005: Effects of temperature and precipitation variability on snowpack trends in the western United States. *Journal of Climate*, **18(21)**, 4545-4561.

488 Mote, P.W., 2006: Climate-driven variability and trends in mountain snowpack in western North America. *Journal of Climate*, **19(23)**, 6209-6220.

489 Payne, J.T., A.W. Wood, A.F. Hamlet, R.N. Palmer, and D.P. Lettenmaier, 2004: Mitigating the effects of climate change on the water resources of the Columbia River basin. *Climatic Change*, **62(1-3)**, 233-256.

490 Figure provided by the Climate Impacts Group, University of Washington, Seattle. <http://cses.washington.edu/cig/>

491 Mote, P.W., A.F. Hamlet, M.P. Clark, and D.P. Lettenmaier, 2005: Declining mountain snowpack in western North America. *Bulletin of the American Meteorological Society*, **86(1)**, 39-49.

492 Hamlet, A.F. and D.P. Lettenmaier, 2007: Effects of 20th century warming and climate variability on flood risk in the western U.S. *Water Resources Research*, **43**, W06427, doi:10.1029/2006WR005099.

493 Bonneville Power Administration, 2001: *The Columbia River System Inside Story*. Internal report DOE/BP-3372. Bonneville Power Administration, Portland OR, 2nd edition, 78 pp. <http://www.bpa.gov/corporate/Power_of_Learning/docs/columbia_river_inside_story.pdf>

494 Casola, J.H., J.E. Kay, A.K. Snover, R.A. Norheim, L.C. Whitely Binder, and Climate Impacts Group, 2005: *Climate Impacts on Washington's Hydropower, Water Supply, Forests, Fish, and Agriculture*. Center for Science in the Earth System, Joint Institute for the Study of the Atmosphere and Ocean, University of Washington, Seattle, 43 pp. <http://cses.washington.edu/db/pdf/kc05whitepaper459.pdf>

495 Ministry of the Environment, British Columbia, Canada, 2007: *Environmental Trends 2007: The Mountain Pine Beetle in British Columbia*. <http://www.env.gov.bc.ca/soe/et07/pinebeetle.html>

496 Francis, R.C. and N.J. Mantua, 2003: Climatic influences on salmon populations in the northeast Pacific. In: *Assessing Extinction Risk for West Coast Salmon* [MacCall, A.D. and T.C. Wainwright (eds.)]. NOAA technical memo NMFS-NWFSC-56. National Marine Fisheries Service, [Washington, DC], pp. 37-67. <http://www.nwfsc.noaa.gov/assets/25/3946_06162004_130044_tm56.pdf>

497 Salathé, E.P., 2005: Downscaling simulations of future global climate with application to hydrologic modelling. *International Journal of Climatology*, **25(4)**, 419-436.

498 Mote, P.W., A. Petersen, S. Reeder, H. Shipman, and L. Whitely Binder, 2008: *Sea Level Rise Scenarios for Washington State*. Center for Science in the Earth System, Joint Institute for the Study of the Atmosphere and Oceans, University of Washington, Seattle, and Washington Department of Ecology, Lacey, 11 pp. <http://www.cses.washington.edu/db/pdf/moteetalslr579.pdf>

499 Petersen, A.W., 2007: *Anticipating Sea Level Rise Response in Puget Sound*. M.M.A. thesis, School of Marine Affairs. University of Washington, Seattle, 73 pp.

500 Snover, A.K., L. Whitely Binder, J. Lopez, E. Willmott, J. Kay, D. Howell, and J. Simmonds, 2007: *Preparing for Climate Change: A Guidebook for Local, Regional, and State Governments*. The Climate Impacts Group, University of Washington, and King County, Washington, in association with and published by ICLEI – Local Governments for Sustainability, Oakland, CA, 172 pp.

501 Fitzpatrick, J., R.B. Alley, J. Brigham-Grette, G.H. Miller, L. Polyak, and M. Serreze, 2008: Preface: Why and how to use this synthesis and assessment report. In: *Past Climate Variability and Change in the Arctic and at High Latitude*. Synthesis and Assessment Product 1.2. U.S. Geological Survey, Reston, VA, pp. 8-21.

502 Data provided by Dr. Glenn Juday, School of Natural Resources and Agricultural Science, Agricultural and Forestry Experiment Station, University of Alaska, Fairbanks.

503 Euskirchen, E.S., A.D. McGuire, D.W. Kicklighter, Q. Zhuang, J.S. Clein, R.J. Dargaville, D.G. Dye, J.S. Kimball, K.C. McDonald, V.E. Melillo, V.E. Romanovsky, and N.V. Smith, 2006: Importance of recent shifts in soil thermal dynamics on growing season length, productivity, and carbon sequestration in terrestrial high-latitude ecosystems. *Global Change Biology*, **12(4)**, 731-750.

504 Euskirchen, E.S., A.D. McGuire, and F.S. Chapin III, 2007: Energy feedbacks of northern high-latitude ecosystems to the climate system due to reduced snow cover during 20th century warming. *Global Change Biology*, **13(11)**, 2425-2438.

505 Barber, V.A., G.P. Juday, and B.P. Finney, 2000: Reduced growth of Alaskan white spruce in the twentieth century from temperature-induced drought stress. *Nature*, **405 (6787)**, 668-673.

506 Juday, G.P., V. Barber, P. Duffy, H. Linderholm, T.S. Rupp, S. Sparrow, E. Vaganov, and J. Yarie, 2005: Forests, land management, and agriculture. In: *Arctic Climate Impact Assessment*. Cambridge University Press, Cambridge, UK, and New York, pp. 781-862. <http://www.acia.uaf.edu/pages/scientific.html>

507 Balshi, M.S., A.D. McGuire, P. Duffy, M. Flannigan, J. Walsh, and J.M. Melillo, 2008: Assessing the response of area burned to changing climate in western boreal North America using a Multivariate Adaptive Regression Splines (MARS) approach. *Global Change Biology*, **15(3)**, 578-600.

508 Berman, M., G.P. Juday, and R. Burnside, 1999: Climate change and Alaska's forests: people, problems, and policies. In: *Assessing the Consequences of Climate Change in Alaska and the Bering Sea Region*. Proceedings of a workshop at the University of Alaska Fairbanks, 29-30 October 1998. Center for Global Change and Arctic System Research, University of Alaska, pp. 21-42.

509 Fleming, R.A. and W.J.A. Volney, 1995: Effects of climate change on insect defoliator population processes in Canada's boreal forest: some plausible scenarios. *Water, Soil, and Air Pollution*, **82(1-2)**, 445-454.

510 Kasischke, E.S. and M.R. Turetsky, 2006: Recent changes in the fire regime across the North American boreal region - spatial and temporal patterns of burning across Canada and Alaska. *Geophysical Research Letters*, **33**, L09703, doi:10.1029/2006GL025677.

511 Chapin, F.S., III, S.F. Trainor, O. Huntington, A.L. Lovecraft, E. Zavaleta, D.C. Natcher, A.D. McGuire, J.L. Nelson, L. Ray, M. Calef, N. Fresco, H. Huntington, T.S. Rupp, L. DeWilde, and R.A. Naylor, 2008: Increasing wildfire in Alaska's boreal forest: pathways to potential solutions of a wicked problem. *BioScience*, **58(6)**, 531-540.

512 Berner, J. and C. Furgal, 2005: Human health. In: *Arctic Climate Impact Assessment*. Cambridge University Press, Cambridge, UK, and New York, pp. 863-906. <http://www.acia.uaf.edu/pages/scientific.html>

513 Klein, E., E.E. Berg, and R. Dial, 2005: Wetland drying and succession across the Kenai Peninsula Lowlands, south-central Alaska. *Canadian Journal of Forest Research*, **35(8)**, 1931-1941.

514 Riordan, B., D. Verbyla, and A.D. McGuire, 2006: Shrinking ponds in subarctic Alaska based on 1950–2002 remotely sensed images. *Journal of Geophysical Research*, **111**, G04002, doi:10.1029/2005JG000150.

515 Osterkamp, T., 2007: Characteristics of the recent warming of permafrost in Alaska. *Journal of Geophysical Research*, **112**, F02S02, doi:10.1029/2006JF000578.

516 Instanes, A., O. Anisimov, L. Brigham, D. Goering, L.N. Khrustalev, B. Ladanyi, and J.O. Larsen, 2005: Infrastructure: buildings, support systems, and industrial facilities. In: *Arctic Climate Impact Assessment*. Cambridge University Press, Cambridge, UK, and New York, pp. 907-944. <http://www.acia.uaf.edu/pages/scientific.html>

517 Brown, J. and V.E. Romanovsky, 2008: Report from the International Permafrost Association: State of permafrost in the first decade of the 21st century. *Permafrost and Periglacial Processes*, **19(2)**, 255-260.

518 Busey, R.C., L.D. Hinzman, J.J. Cassano, and E. Cassano, 2008: Permafrost distributions on the Seward Peninsula: past, present, and future. In: *Proceedings of the Ninth International Conference on Permafrost*, University of Alaska Fairbanks, June 29-July 3, 2008 [Kane, D.L. and K.M. Hinkel (eds.)]. Institute of Northern Engineering, University of Alaska, Fairbanks, pp. 215-220.

519 Shumaker, V.O., 1979: *The Alaska Pipeline*. Julian Messner, New York, 64 pp.

520 Bengtsson, L., K.I. Hodges, and E. Roeckner, 2006: Storm tracks and climate change. *Journal of Climate*, **19(15)**, 3518-3543.

521 U.S. Army Corps of Engineers, 2008: *Newtok Evacuation Center, Mertarvik, Nelson Island, Alaska*. Revised environmental assessment. U.S. Army Corps of Engineers, Alaska District, [Elmendorf AFB, 64 pp.]

522 Jones, B.M., C.D. Arp, M.T. Jorgenson, K.M. Hinkel, J.A. Schmutz, and P.L. Flint, 2009: Increase in the rate and uniformity of coastline erosion in Arctic Alaska. *Geophysical Research Letters*, **36**, L03503, doi:10.1029/2008GL036205.

523 Yin, J.H., 2005: A consistent poleward shift of the storm tracks in simulations of 21st century climate. *Geophysical Research Letters*, **32**, L18701, doi:10.1029/2005GL023684.

524 Salathé, E.P., Jr., 2006: Influences of a shift in North Pacific storm tracks on western North American precipitation under global warming. *Geophysical Research Letters*, **33**, L19820, doi:10.1029/2006GL026882.

525 Data provided by Dr. David Atkinson, International Arctic Research Center, University of Alaska, Fairbanks.

526 Feely, R.A., C.L. Sabine, K. Lee, W. Berelson, J. Kleypas, V.J. Fabry and F.J. Millero, 2004: Impact of anthropogenic CO_2 on $CaCO_3$ system in the oceans. *Science*, **305(5682)**, 362-366.

527 Grebmeier, J.M., J.E. Overland, S.E. Moore, E.V. Farley, E.C. Carmack, L.W. Cooper, K.E. Frey, J.H. Helle, F.A. McLaughlin, and S.L. McNutt, 2006: A major ecosystem shift in the northern Bering Sea. *Science*, **311(5766)**, 1461-1464.

528 Ray, G.C., J. McCormick-Ray, P. Berg, and H.E. Epstein, 2006: Pacific walrus: benthic bioturbator of Beringia. *Journal of Experimental Marine Biology and Ecology*, **330(1)**, 403-419.

529 Mueter, F.J. and M.A. Litzow, 2007: Sea ice retreat alters the biogeography of the Bering Sea continental shelf. *Ecological Applications*, **18(2)**, 309-320.

530 Shea, E.L., G. Dolcemascolo, C.L. Anderson, A. Barnston, C.P. Guard, M.P. Hamnett, S.T. Kubota, N. Lewis, J. Loschnigg, and G. Meehl, 2001: *Preparing for a Changing Climate: The Potential Consequences of Climate Variability and Change: Pacific Islands*. East-West Center, Honolulu, HI, 102 pp. <http://www2.eastwestcenter.org/climate/assessment/>

531 Guidry, M.W. and F.T. Mackenzie, 2006: *Climate Change, Water Resources, and Sustainability in the Pacific Basin: Emphasis on O'ahu, Hawai'I and Majuro Atoll, Republic of the Marshall Islands*. University of Hawaii Sea Grant Program, Honolulu, 100 pp. <http://nsgl.gso.uri.edu/hawau/hawaut06001.pdf>

532 CEO, 2004: *Caribbean Environmental Outlook*. [Heileman, S., L.J. Walling, C. Douglas, M. Mason, and M. Chevannes-Creary (eds.)]. United Nations Environmental Programme, Kingston, Jamaica, 114 pp. <http://www.unep.org/geo/pdfs/caribbean_eo.pdf>

533 Church, J.A., N.J. White, and J.R. Hunter, 2006: Sea-level rise at tropical Pacific and Indian Ocean islands. *Global and Planetary Change*, **53(3)**, 155-168.

534 Burns, W.C.G., 2002: Pacific island developing country water resources and climate change. In: *The World's Water* [Gleick, P. (ed.)]. Island Press, Washington, DC, 3rd edition, pp. 113-132.

535 Baker, J.D., C.L. Littnan, and D.W. Johnston, 2006: Potential effects of sea level rise on the terrestrial habitats of endangered and endemic megafauna in the northwestern Hawaiian Islands. *Endangered Species Research*, **2**, 21-30.

536 Firing, Y. and M.A. Merrifield, 2004: Extreme sea level events at Hawaii: influence of mesoscale eddies. *Geophysical Research Letters*, **31**, L24306, doi:10.1029/2004GL021539.

537 Scott, D., M. Overmars, T. Falkland, and C. Carpenter, 2003: *Pacific Dialogue on Water and Climate, Synthesis Report*. South Pacific Applied Geoscience Commission, Fiji Islands, 28 pp. <http://www.oas.org/CDWC/Documents/SIDS%20Paper/Pacific%20Report%20-%20Final.pdf>

538 Hay, J., N. Mimura, J. Campbell, S. Fifita, K. Koshy, R.F. McLean, T. Nakalevu, P. Nunn, and N. deWet, 2003: *Climate Variability and Change and Sea Level Rise in the Pacific Islands Regions: A Resource Book for Policy and Decision Makers, Educators and Other Stakeholders*. South Pacific Regional Environmental Programme (SPREP), Apia, Samoa, 94 pp. <http://www.sprep.org/publication/pub_detail.asp?id=181>

539 Gillespie, R.G., E.M. Claridge, and G.K. Roderick, 2008: Biodiversity dynamics in isolated island communities: interaction between natural and human-mediated processes. *Molecular Ecology*, **17(1)**, 45-57.

540 Becken, S. and J.E. Hay, 2007: *Tourism and Climate Change: Risks and Opportunities*. Channel View Publications, Clevedon, UK, and Buffalo, NY, 329 pp.

541 State of Hawaii, Division of Business, Economic Development, and Tourism, 2008: *Facts and Figures: State of Hawaii*. [Web site] <http://hawaii.gov/dbedt/info/economic/library/facts/state>

542 Cesar, H.S.F. and F.H. van Beukering, 2004: Economic valuation of the coral reefs of Hawaii. *Pacific Science*, **58(2)**, 231-242.

543 Hoegh-Guldberg, O., P.J. Mumby, A.J. Hooten, R.S. Steneck, P. Greenfield, E. Gomez, C.D. Harvell, P.F. Sale, A.J. Edwards, K. Caldeira, N. Knowlton, C.M. Eakin, R. Iglesias-Prieto, N. Muthiga, R.H. Bradbury, A. Dubi, and M.E. Hatziolos, 2007: Coral reefs under rapid climate change and ocean acidification. *Science*, **318(5857)**, 1737-1742.

544 Donner, S.D., W.J. Skirving, C.M. Little, M. Oppenheimer, and O. Hoegh-Guldberg, 2005: Global assessment of coral bleaching and required rates of adaptation under climate change. *Global Change Biology*, **11(12)**, 2251-2265.

545 Paulay, G., L. Kirkendale, G. Lambert, and C. Meyer, 2002: Anthropogenic biotic interchange in a coral reef ecosystem: a case study from Guam. *Pacific Science*, **56(4)**, 403-422.

546 Hotta, M., 2000: The sustainable contribution of fisheries to food security in the Asia and Pacific region: regional synthesis. In: *Sustainable Contribution of Fisheries to Food Security*. Food and Agriculture Organization of the United Nations, Bangkok, Thailand, pp. 1-28. <http://www.fao.org/DOCREP/003/X6956E/x6956e02.htm>

547 Graham, N.A.J., S.K. Wilson, S. Jennings, N.V.C. Polunin, J.P. Bijoux, and J. Robinson, 2006: Dynamic fragility of oceanic coral reef ecosystems. *Proceedings of the National Academy of Sciences*, **103(22)**, 8425-8429.

548 Lehodey, O., M. Bertignac, J. Hampton, A. Lewis, and J. Picaut, 1997: El Niño Southern Oscillation and tuna in the western Pacific. *Nature*, **389(6652)**, 715-718.

549 Crowell, M., S. Edelman, K. Coulton, and S. McAfee, 2007: How many people live in coastal areas? *Journal of Coastal Research*, **23(5)**, iii-vi.

550 Gill, S.K., R. Wright, J.G. Titus, R. Kafalenos, and K. Wright, 2009: Population, land use, and infrastructure. In: *Coastal Elevations and Sensitivity to Sea-level Rise: A Focus on the Mid-Atlantic Region* [J.G. Titus (coordinating lead author), K.E. Anderson, D.R. Cahoon, D.B. Gesch, S.K. Gill, B.T. Gutierrez, E.R. Thieler, and

S.J. Williams (lead authors)]. Synthesis and Assessment Product 4.1. U.S. Environmental Protection Agency, Washington, DC, pp. 105-116.

[551] U.S. Commission on Ocean Policy, 2004: *An Ocean Blueprint for the 21st Century.* U.S. Commission on Ocean Policy, Washington, DC. <http://www.oceancommission.gov/documents/full_color_rpt/welcome.html>

[552] Day, J.W., Jr., D.F. Boesch, E.J. Clairain, G.P. Kemp, S.B. Laska, W.J. Mitsch, K. Orth, H. Mashriqui, D.J. Reed, L. Shabman, C.A. Simenstad, B.J. Streever, R.R. Twilley, C.C. Watson, J.T. Wells, and D.F. Whigham, 2007: Restoration of the Mississippi Delta: essons from hurricanes Katrina and Rita. *Science,* **315(5819),** 1679-1684.

[553] Boesch, D.F. (ed.), 2008: *Global Warming and the Free State: Comprehensive Assessment of Climate Change Impacts in Maryland.* University of Maryland Center for Environmental Science, Cambridge, MA, 85 pp. <http://www.umces.edu/climateimpacts/>

[554] Cazenave, A., K. Dominh, S. Guinehut, E. Berthier, W. Llovel, G. Ramillien, M. Ablain, and G. Larnicol, 2009: Sea level budget over 2003–2008: a reevaluation from GRACE space gravimetry, satellite altimetry and Argo. *Global and Planetary Change,* **65(1-2),** 83-88.

[555] Vaughan, D.G. and R. Arthern, 2007: Why is it hard to predict the future of ice sheets? *Science,* **315(5818),** 1503-1504.

[556] Justić, D., N.N. Rabalais, and R.E. Turner, 2003: Simulated responses of the Gulf of Mexico hypoxia to variations in climate and anthropogenic nutrient loading. *Journal of Marine Systems,* **42(2-3),** 115-126.

[557] Boesch, D.F., V.J. Coles, D.G. Kimmel, and W.D. Miller, 2007: Coastal dead zones and global climate change: ramifications of climate change for Chesapeake Bay hypoxia. In: *Regional Impacts of Climate Change: Four Case Studies in the United States.* Pew Center for Global Climate Change, Arlington, VA, pp. 57-70. <http://www.pewclimate.org/regional_impacts>

[558] Wicks, C., D. Jasinski, and B. Longstaff, 2007: *Breath of Life: Dissolved Oxygen in Chesapeake Bay.* EcoCheck, Oxford, MD, 4 pp. <http://www.eco-check.org/pdfs/do_letter.pdf>

[559] Kimmerer, W.J., E. Garside, and J.J. Orsi, 1994: Predation by an introduced clam as the likely cause of substantial declines in zooplankton of San Francisco Bay. *Marine Ecology Progress Series,* **113,** 81-93.

[560] Carpenter, K.E., M. Abrar, G. Aeby, R.B. Aronson, S. Banks, A. Bruckner, A. Chiriboga, J. Cortés, J.C. Delbeek, L. DeVantier, G.J. Edgar, A.J. Edwards, D. Fenner, H.M. Guzmán, B.W. Hoeksema, G. Hodgson, O. Johan, W.Y. Licuanan, S.R. Livingstone, E.R. Lovell, J.A. Moore, D.O. Obura, D. Ochavillo, B.A. Polidoro, W.F. Precht, M.C. Quibilan, C. Reboton, Z.T. Richards, A.D. Rogers, J. Sanciangco, A. Sheppard, C. Sheppard, J. Smith, S. Stuart, E. Turak, J.E.N. Veron, C. Wallace, E. Weil, and E. Wood, 2008: One-third of reef-building corals face elevated extinction risk from climate change and local impacts. *Science,* **321(5888),** 560-563.

[561] Chan, F., J.A. Barth, J. Lubchenco, A. Kirincich, H. Weeks, W.T. Peterson, and B.A. Menge, 2008: Emergence of anoxia in the California current large marine ecosystem. *Science,* **319(5865),** 920.

[562] Data for hypoxia distribution maps made available by Dr. Bill Peterson, (NOAA), the Bonneville Power Administration, Drs. Francis Chan and Jane Lubchenco, (PISCO, Partnership for Interdisciplinary Studies of Coastal Oceans), Drs. Jack Barth and Steve Pierce (Oregon State University/College of Oceanic and Atmospheric Sciences) and NOAA Northwest Fisheries Science Center.

[563] Crozier, L.G., A.P. Hendry, P.W. Lawson, T.P. Quinn, N.J. Mantua, J. Battin, R.G. Shaw, and R.B. Huey, 2008: Potential responses to climate change in organisms with complex life histories: evolution and plasticity in Pacific salmon. *Evolutionary Applications,* **1(2),** 252-270.

[564] Hondzo, M. and H.G. Stefan, 1991: Three case studies of lake temperature and stratification response to warmer climate. *Water Resoures Research,* **27(8),** 1837-1846.

[565] Morel, F.M.M., A.M.L. Kraepiel, and M. Amyot, 1998: The chemical cycle and bioaccumulation of mercury. *Annual Review of Ecology and Systematics,* **29,** 543-566.

[566] Williamson, C.E., J.E. Saros, and D.W. Schindler, 2009: Sentinels of change. *Science,* **323(5916),** 887-888.

[567] Stachowicz, J.J., J.R. Terwin, R.B. Whitlatch, and R.W. Osman, 2002: Linking climate change and biological invasions: ocean warming facilitates nonindigenous species invasions. *Proceedings of the National Academy of Sciences,* **99(24),** 15497-15500.

[568] Keleher, C.J. and F.J. Rahel, 1996: Thermal limits to salmonid distributions in the Rocky Mountain region and potential habitat loss due to global warming: A Geographic Information System (GIS) approach. *Transactions of the American Fisheries Society,* **125(1),** 1-13.

[569] McCullough, D., S. Spalding, D. Sturdevant, and M. Hicks, 2001: *Issue Paper 5: Summary of Technical Literature Examining the Physiological Effects of Temperature on Salmonids.* EPA-910-D-01-005, prepared as part of U.S. EPA Region 10 Temperature Water Quality Criteria Guidance Development Project. [EPA Pacific Northwest Regional Office (Region 10), Seattle, WA], 114 pp.

PHOTO AND FIGURE CREDITS

Contact Information

Global Change Research Information Office
c/o U.S. Global Change Research Program Office
1717 Pennsylvania Avenue, NW
Suite 250
Washington, DC 20006
202-223-6262 (voice)
202-223-3065 (fax)

To obtain a copy of this document, place
an order at the Global Change Research
Information Office (GCRIO) web site:
http://www.gcrio.org/orders

U.S. Global Change Research Program and the Subcommittee on Global Change Research

Jack Kaye, Vice Chair,
National Aeronautics and Space Administration,
Acting Director, U.S. Global Change Research
Program

Allen Dearry
Department of Health and Human Services

Anna Palmisano
Department of Energy

Mary Glackin
National Oceanic and Atmospheric Administration

Charles Vincent
Department of Defense

William Hohenstein
Department of Agriculture

Linda Lawson
Department of Transportation

Thomas Armstrong
U.S. Geological Survey

Tim Killeen
National Science Foundation

Patrick Neale
Smithsonian Institution

William Breed
U.S. Agency for International Development

Joel Scheraga
Environmental Protection Agency

Jonathan Pershing
Department of State

EXECUTIVE OFFICE AND OTHER LIAISONS

Robert Marlay
Climate Change Technology Program

Katharine Gebbie
National Institute of Standards & Technology

Jason Bordoff
Council on Environmental Quality

Philip DeCola
Office of Science and Technology Policy

Stuart Levenbach
Office of Management and Budget

Margaret McCalla
Office of the Federal Coordinator for Meteorology

Howard Frumkin
Center for Disease Control

www.ingramcontent.com/pod-product-compliance
Lightning Source LLC
Chambersburg PA
CBHW080638180526
45168CB00008B/3213